Soils of the Laurentian Great Lakes, USA and Canada

James G. Bockheim

Soils of the Laurentian Great Lakes, USA and Canada

Springer

James G. Bockheim
Department of Soil Science
University of Wisconsin-Madison
Madison, WI, USA

ISBN 978-3-030-52424-1 ISBN 978-3-030-52425-8 (eBook)
https://doi.org/10.1007/978-3-030-52425-8

@NOAA-GLERL

This Springer imprint is published by the registered company Springer Nature Switzerland AG
The registered company address is: Gewerbestrasse 11, 6330 Cham, Switzerland

This book is dedicated to the professional soil mappers and data managers of the US Department of Agriculture, Natural Resources Conservation Service. This organization began in 1899 as the Division of Soils, became of Bureau of Soils in 1901, the Soil Conservation Service in 1935, and the Natural Resources Conservation Service in 1994. The report that follows draws heavily on information gathered from 82 counties surrounding the Great Lakes that were mapped by NRCS personnel, as well as other data.

Preface

Despite the availability of considerable data, there has not been a comprehensive treatment of the soils of the Great Lakes. This book is intended to introduce individuals with some understanding of science to the soils of the Great Lakes, including their history, soil-forming factors, soil taxonomic structure, soil geography, pedodiversity, and importance of soils for protection of the Great Lakes Coastal Zone. A glossary is provided. Hopefully, the book will create an interest in soils that rivals that of birders, lighthouse seekers, waterfall enthusiasts, kayakers, and geo-cachers in the Great Lakes region.

Madison, WI, USA James G. Bockheim

Acknowledgments

The following individuals graciously contributed digital images: Larissa Hindman, Soil Scientist, NRCS, Duluth, MN; Dwight Jerome, Resource Soil Scientist, NRCS, Marquette, MI; Gerald Smith, Soil Scientist, NRCS, Paul Smiths, NY; Matt Bromley, Soil Scientist, NRCS, Grand Rapids, MI; Mike Walczynski, Resource Soil Scientist, NRCS, Duluth, MN.

Author's Note

The Laurentian Great Lakes are an important part of the US and Canada. Home to 33 million people, including 90% of all Canadians, the Great Lakes account for 20% of the world and 90% of the US surface freshwater. Key industries in the Great Lakes region include shipping, steel and automobile production, energy generation, fishing, pulp and paper making, agriculture, and recreation. To date, there has been no comprehensive inventory of the soils of the region, which is undergoing dramatic climate change and environmental degradation.

This book was prepared from US Department of Agriculture, Natural Resources Conservation Service databases, including the Web Soil Survey, Soil Series Extent Explorer, soil classification and characterization databases, and county soil surveys, supplemented by shoreline viewer software, a comprehensive review of the scientific literature, the author's research, consultation with colleagues, and trips around the Great Lakes.

Methodology

Identification of Soil Series

Soil series within 5 km of the coastline of the US portion of the Great Lakes were identified by following the coastline with Soil Extent Explorer (Natural Resources Conservation Service, 2018a) set at the broadest viewing scale of 1:50,000. For the Canadian GLCZ, the coastlines of 24 counties or districts (Table 1.4) were followed manually and soil series were identified. The scales of these maps were generally 1:63,360 but occasionally were 1:126,720.

Soil Classification

Spreadsheets listing soil series for US and Canada were prepared. For the US Great Lakes coastline, the NRCS Soil Series Classification Database (SC) was used to create a report for the list of series in the US with the download option (NRCS, 2018b). For the Canadian GLCZ, soil series were examined in the Agriculture and Agri-Food Canada, Soils of Ontario database (http://sis.agr.gc.ca/cansis/soils/on/soils.html). Soil classification data were entered onto the spreadsheet manually.

Official Soil Series Descriptions

For the US GLCZ, Official Soil Series Descriptions (NRCS, 2018c) of 664 soil series were examined, and the following data were added manually to the spreadsheet: parent materials, landform, vegetation, type and thickness of diagnostic horizons, whether or not there were competing soil series, and geographically associated soils. For the Canadian GLCZ, soil descriptions of 144 soil series were examined, and the following data were added manually to the spreadsheet: parent materials, landform, and texture.

Soil Characterization

For the US GLCZ, primary soil characterization was obtained for representative pedons of key soil series from the NRCS Soil Characterization Database (NRCS, 2018c). Only basic data were included in Tables 5.1 to 5.5, including percent clay, silt, sand, organic C, base saturation, Fe_d or Fe_o, Al_d or Al_o, and cation-exchange capacity at pH 7 and pH in water. Preference was given to pedons in the GLCZ and with complete data.

Pedodiversity

Pedodiversity of the US GLCZ was determined from mean density of soil series (number of soil series, divided by area) (Bockheim, 2018). Endemic soils are defined as the only soil in a family (Bockheim, 2005). Rare soils are those with an area less than 10,000 ha (Ditzler, 2003). Endangered soils are defined here as those that are endemic and rare.

Soil-Forming Processes

Dominant soil-forming processes were identified based on taxonomic category (Bockheim and Gennadiyev, 2000). Primary processes were identified at the order (e.g., Spodosol = podsolization) and suborder (e.g., Aquept = gleization) levels; and secondary processes were assigned from the great-group (e.g., Argiudoll = argilluviation) and subgroup (Spodic = podsolization) levels.

Soil Evolution

Soil series were assigned to proglacial lakes in the Great Lakes basin by comparing soil extent maps with maps of prominent shorelines (Fig. 2.15). Ages were assigned using the post-glacial lake chronology given in Table 2.1. The evolution of soils derived from stabilized sand dunes, beach ridges, and sandy lake terraces was determined from an integration of seven published soil chronosequences-chronofunctions conducted along the Lake Michigan and Lake Huron shores.

Soil Photographs

County soil survey reports were an excellent source for color images of key soil series. Additionally, I used personal photographs and those from colleagues.

Additional Information

Other materials that were consulted in preparing this report included research publication identified electronically from the Web of Science, county soil survey reports, the Web Soil Survey (https://websoilsurvey.sc.egov.usda.gov/App/HomePage.htm), the US Geological Survey national topographic map (https://viewer.nationalmap.gov/launch/), the Great Lakes Shoreviewer (http://superiorwatersheds.org/shorelineviewer2011/), and the *Quaternary Geologic Atlas of the USA* (US Geological Survey, 2012).

Contents

1 **Introduction to the Soils of the Great Lakes Coastal Zone** . . . 1
1.1 Concept of Soil . . . 1
1.2 History of Soils Investigations in the Great Lakes Region . . . 1
1.3 Great Lakes Coastal Zone . . . 10
1.4 Conclusions . . . 12
References . . . 14

2 **Soil-Forming Factors of the Great Lakes Coastal Zone** . . . 17
2.1 General . . . 17
2.2 Climate . . . 17
2.3 Vegetation . . . 17
2.4 Relief and Elevation . . . 19
2.5 Bedrock Geology . . . 22
2.6 Surficial Geology . . . 22
2.7 Land Use and Population . . . 29
2.8 Conclusions . . . 30
References . . . 33

3 **Soil Taxonomic Systems Used in the Great Lakes Coastal Zone** . . . 35
3.1 Soil Horizons . . . 35
3.2 Approaches to Soil Classification . . . 35
3.3 Soil Taxonomy . . . 35
3.3.1 Hierarchical Levels . . . 35
3.3.2 Diagnostic Horizons . . . 35
3.3.3 Constructing a Soil Name . . . 36
3.4 The Canadian System of Soil Classification . . . 39
3.5 Linking the US and Canadian Soil Taxonomic Systems . . . 39
3.6 Conclusions . . . 40
References . . . 40

4 **Soil Taxonomic Structure and Factors Affecting Soil Distribution in the Great Lakes Coastal Zone** . . . 43
4.1 Soil Taxonomic Structure in the US Great Lakes Coastal Zone . . . 43
4.2 Soil Taxonomic Structure in the Canadian Great Lakes Coastal Zone . . . 43
4.3 Factors Affecting Soil Distribution in the Great Lakes Coastal Zone . . . 46
4.3.1 Climate and Soil Distribution . . . 46
4.3.2 Vegetation and Soil Distribution . . . 46
4.3.3 Landforms, Parent Material and Soil Distribution . . . 46
4.3.4 The Time Factor and Soil Distribution . . . 49
4.4 Conclusions . . . 49
References . . . 58

5 Soils of the Great Lakes Coastal Zone 59
5.1 Introduction 59
5.2 Alfisols 59
5.3 Spodosols 61
5.4 Inceptisols 65
5.5 Entisols 68
5.6 Mollisols 70
5.7 Histosols 75
5.8 Conclusions 81
References 88

6 Soils of the Lake Superior Coastal Zone 89
6.1 Introduction 89
6.2 Soils by Natural Resource Sement 89
6.3 Conclusions 95
References 98

7 Soils of the Lake Michigan Coastal Zone 99
7.1 Introduction 99
7.2 Soils by Natural Resource Segment 99
7.3 Conclusions 104
References 109

8 Soils of the Lake Huron Coastal Zone 111
8.1 Introduction 111
8.2 Soils by Natural Resource Segment 111
8.3 Conclusions 115
References 119

9 Soils of the Lake Erie Coastal Zone 121
9.1 Introduction 121
9.2 Soils by Natural Resource Segment 121
9.3 Conclusions 127
References 127

10 Soils of the Lake Ontario Coastal Zone 129
10.1 Introduction 129
10.2 Soils by Natural Resource Segment 129
10.3 Conclusions 135
References 135

11 Soil-Forming Processes in the Great Lakes Coastal Zone 137
11.1 Introduction 137
11.2 Gleization 137
11.3 Argilluviation 137
11.4 Podzolization 137
11.5 Cambisolization 137
11.6 Melanization 137
11.7 Glossification 138
11.8 Paludization 138
11.9 Vertization 138
11.10 Anthrosolization 138
11.11 Linkage Soil-Forming Processes and Soil Taxa 138
11.12 Conclusions 139
References 139

12 Pedodiversity of the Great Lakes Coastal Zone 141
12.1 Soil Series Density 141
12.2 Factors Influencing Pedodiversity in the Great Lakes Region 141
12.2.1 Parent Materials 141
12.2.2 Climatic Gradient 141
12.2.3 Vegetation Gradient 142
12.3 Benchmark, Rare, Endemic, and Endangered Soils of the Great Lakes Coastal Zone 143
12.4 Conclusions 143
References 144

13 Protection of Great Lakes Soils 145
13.1 Introduction 145
13.2 Key Environmental Problems in the Great Lakes Coastal Zone 145
13.3 Conclusions 146
References 148

14 Some Unusual Relict Features in the Great Lakes Coastal Zone 149
14.1 Introduction 149
14.2 Glacial and Periglacial 149
14.3 Glaciolacustrine and Lacustrine 149
14.4 Early People 153
14.5 Conclusions 159
References 159

15 Conclusions 161

Appendix A 163

Appendix B 185

Appendix C 203

Appendix D 211

Index 225

About the Author

Dr. James G. Bockheim was Professor of Soil Science, Forest and Wildlife Ecology, and Environmental Studies at the University of Wisconsin from 1975 until his retirement in 2015. He has conducted research in forestry, soils, and surficial geology throughout the upper Great Lakes region. His previous books include *Pedodiversity* (2013; with J. J. Ibáñez); *Soil Geography of the USA: a Diagnostic-Horizon Approach* (2014); *Cryopedology* (2015); *The Soils of Antarctica* (2015); and *The Soils of Wisconsin* (2017; with A. E. Hartemink).

1 Introduction to the Soils of the Great Lakes Coastal Zone

1.1 Concept of Soil

There are many definitions for soil ranging from the utilitarian to a description that focuses on material. Soil has been recognized as (i) a natural body, (ii) a medium for plant growth, (iii) an ecosystem component, (iv) a vegetated water-transmitting mantle and (iv) an archive of past climate and processes. In this book, the definition given in the *Keys to Soil Taxonomy* (Soil Survey Staff 2014, p. 1) is followed, i.e., that the soil "is a natural body comprised of solids (minerals and organic matter), liquid, and gases that occurs on the land surface, occupies space, and is characterized by one or both of the following: horizons, or layers, that are distinguishable from the initial material as a result of additions, losses, transfers, and transformations of energy and matter or the ability to support rooted plants in a natural environment".

1.2 History of Soils Investigations in the Great Lakes Region

Few of the early geologists included soils in their analysis of Great Lakes landforms (Table 1.1). An exception is Frank Leverett, who compiled massive tomes elucidating the surficial geology and ancient shorelines of Lakes Erie and Ontario (Leverett 1902), the southern (Leverett 1912) and northern (Leverett 1910) Peninsulas of Michigan, Lake Superior (Leverett 1929), and the Great Lakes as a whole (Leverett and Taylor 1915). Leverett estimated that he had walked the equivalent of four times the circumference of the Earth while doing his fieldwork. He published more than 75 papers that totaled more than 5,000 printed pages! He cooperated with the newly established (1899) US Department of Agriculture's Bureau of Soils. Leverett (1902) included a brief section on soils in his discussion of the surficial geology of Lakes Erie and Ontario, following the Bureau of Soils in delineating soils on the basis of their origin and texture.

County soil surveys published between 1901 and 1920 for the US Great Lakes region are listed in Table 1.2. The early surveys recognized soil series and soil types (textural subdivisions of soil series), along with one or more land types, such as beach sand, marsh, peat, muck, and meadow. For the first two decades of the Bureau of Soils surveys, 14 or fewer soil series per county were recognized, with most having widespread extents. Some of the earliest soil series included the Volusia (1901), Saugatuck (1901), Dunkirk (1901), Alton (1902), Warners (1903), Superior (1904), Vergennes (1904), Newton (1905), and Allis (1906). Soil mapping in the US Great Lakes initially focused on Lake Ontario; counties along Lake Superior generally were mapped last (Table 1.3).

Examples of early soil surveys in the Great Lakes region with their more recent counterparts are given in Figs. 1.1 through 1.5. The earliest soil survey on a Great Lakes coast was of Allegan County (Pippen and Rice 1901) along the southeastern shore of Lake Michigan (Fig. 1.1). Fifteen map units were recognized, including six soil series. The Allegan soil series, which is no longer recognized, was divided into seven soil types, with textures ranging from sand to clay. Dune sand occurred along the northeast shore (orange); the Saugatuck sand (yellow) and Allegan sand (brownish tinge) along the central coast, and finer-textured versions of the Allegan series in the south. The 1984 general soil map of the county shows the Oakville soil series along the north and central coast and the Capac-Rimer-Pipestone Association in the south (Knapp 1987). The 2019 Web Soil Survey lists the Plainfield soil series as being dominant, with the Capac, Chelsea, and Rimer being common.

The first soil map along Lake Superior was of the Superior, WI area in 1905 (Maynadier et al. 1912) (Fig. 1.2). The map has eight units and two soil series. Dune sand (gold color) is shown on the bay-mouth bar linking Wisconsin Point and Minnesota Point; the remaining coastline contains the Superior clay (light green). The prominent moraine to the south (pink color) features the Miami sandy loam, which is now recognized as containing loess over loamy till and only

J. G. Bockheim, *Soils of the Laurentian Great Lakes, USA and Canada*,
https://doi.org/10.1007/978-3-030-52425-8_1

Table 1.1 History of soil investigations in the Great Lakes Coastal Zone

Year	Author	Topic
1901	Bureau of Soils	Soil survey of Allegan County, MI, the first along the Great Lakes shoreline
1912	Leverett	used soils to differentiate among drift sheets; worked closely with newly established Bureau of Soils
1922	Downing	recognized soils as key component of Great Lakes ecosystems
1939	Bretz	viewed soil as weathered "mantle rock"; used soils to distinguish among lake terraces near Chicago
1958	Olson	classic study of dune succession S Lake Michigan; developed chronosequence-chronofunction concepts
1970	Lee and Hole	characterized Two Creeks buried soil along E Lake Michigan
1979	Cary et al.	related soil properties to surficial geology of features in Apostle Islands, Lake Superior
1980	Berg	studied soil development on an age-series of beach ridges S Lake Michigan
1995	Anderton and Loope	characterized buried soils in coastal dune sequences
1995	Albert	included soils in regional landscape ecology classification
2018		79% of Great Lakes coastline in US and Canada mapped; more than 800 soil series recognized

Note early geomorphologists, such as A. P. Coleman, G. K. Gilbert, A. C. Lawson, J. W. Goldthwait, W. H. Hobbs, J. W. Spencer, G.R. Stanley, J. L. Hough, rarely mention soils except in context of buried organic matter and red clay

Table 1.2 Soil surveys in the US Great Lakes region from 1901 to 1920

County	Lake	State	Date	No. soil series	No. land types	Color soil map
Allegan	Michigan	MI	1901	7	2	x
Saginaw Area	Huron	MI	1904	3	3	x
Superior Area	Superior	WI	1905	3	3	x
Racine	Michigan	WI	1907	6	2	x
Niagara	Ontario	NY	1908	5	3	x
Erie	Erie	PA	1910	5	1	x
Monroe	Ontario	NY	1910	9	4	x
Marinette	Michigan	WI	1911	5	3	x
Bayfield	Superior	WI	1912	3	1	x
Kewaunee	Michigan	WI	1913	9	1	x
Lake	Michigan	IL	1915	nd	nd	x
Chautauqua	Erie	NY	1916	9	3	x
Door	Michigan	WI	1918	7	1	x
Milwaukee	Michigan	WI	1918	9	1	x
Kenosha & Racine	Michigan	WI	1919	13	3	x
Oswego	Ontario	NY	1919	8	3	x
Sandusky	Erie	OH	1920	14	3	x

occurring in the southern Great Lakes region. The most recent soil map (Web Soil Survey 2019) of the Superior area shows more than 30 soil series. Udipsamments are shown on the bay-mouth bar; the Miskoaki, Bergland, Cuttre, and Amnicon soil series (Glossudalfs and Glossaqualfs) on the clayey till and lake plain depicted in green; and the Keweenaw and associated soil series on the disintegration moraine shown in pink to the south.

A soil map of the Saginaw Bay area was published by McLendon and Carter in 1904 and represents the first map of

Table 1.3 County soil surveys for the US Great Lakes Coastal Zone

County	State	Great Lake	Date soil survey	Scale	Landforms/soil associations
Macomb	MI	Erie	1971, current		Lakeplains (Toledo, Paulding, Lenawee, Corunna, Lamson, Selfridge, Au Gres, Metamora); 1–4 mi inland (avg. = 2)
Monroe	MI	Erie	1981, current		Lakeplains (Lenawee, Del Rey); 2–6 mi inland (avg. = 2)
St. Clair	MI	Erie	1974, current		Glacial lake beaches (Eastport, Wainola, Tobico, Deford); lakeplains (Bach, Latty, Paulding, Wasepi); till plains (Hoytville, Allendale, Nappanee); 1–10 mi inland
Wayne	MI	Erie	1977, current		Lakeplains (Belleville, Selfridge); till plains (Pewamo, Blount, Hoytville, Nappanee); 0–2 mi; wave-washed
Chautauqua	NY	Erie	1994, current		Lakeplains (Niagara, Canandaigua, Minoa, Barcelona, Rhinebeck); 1–3 mi i; glaciolacustrine-outwash (Valois, Chenango, Pompton); backed by till plain
Erie	NY	Erie	1986, current		Lakeplains (Hudson, Varysburg, Valois, Niagara, Canandaigua, Cosad, Odessa, Schoharie, Rhinebeck); 0–5 mi inland; till plains (Darien, Remsen, Angola, Orpark, Manlius, Derb)
Ashtabula	OH	Erie	2007, current		Lakeplains (Conneaut, Painsesville, Elnora); 3–5 mi inland
Cuyahoga	OH	Erie	1980, current		Lakeplains and beach ridges (Oshtemo, Chili, Geeburg, Mentor, Elnora, Jimtown, Mitiwanga); 4–12 + mi inland; till plains (Mahoning)
Erie	OH	Erie	2006, current		Lakeplains (Toledo, Fulton, Colwood, Del Rey, Milford); 0–5 mi inland; relict near-shore lakeplains (Pewamo, Bennington, Allis); beach ridges (Jimtown, Oshtemo, Millgrove); till on bedrock benches (Hornell, Fries)
Lake	OH	Erie	1979, current		Lakeplains and offshore bars (Conneaut, Painesville, Red Hook); beach ridges, terraces and offshore bars (Elnora, Stafford, Tyner, Otisville, Conotton, Oshtemo); 5 mi inland; till plains (Platea, Pierpont, Darien, Mahoning, Ellsworth)
Lorain	OH	Erie	1976, current		Lakeplains (Fitchville, Luray, Sebring); till on lakeplains (Allis, Mitiwanga, Miner); beach ridges and outwash plains (Haskins, Jimtown, Oshtemo); 4–10 mi inland
Lucas	OH	Erie	1980, current		Lakeplains (Latty, Toledo, Fulton, Del Rey, Lenawee); 5 + mi inland
Ottawa	OH	Erie	1985, current		Glacial lakebed sediments (Toledo, Nappanee, Lenawee, Kibbie, Colwood); thin till over ls bedrock (Castalia, Milton, Hoytville); 0–11 + mi inland
Sandusky	OH	Erie	1987, current		Lakeplains (Lenawee, Del Rey, Toledo, Fulton); beach ridges, lakeplains, deltas, former offshore bars (Kibbie, Dixboro, Rimer); 10 + mi inland
Erie	PA	Erie	1960, current		Lakeplains (Wallington, Birdsall, Williamson, Collamer, Rimer, Wauseon, Berrien); beach ridges (Conoton, Ottawa, Fredon); 2–4 mi inland; till upland
Niagara	NY	Erie, Ontario	1972, current		Clayey and silty lakeplains Erie and Ontario (Canandaigua, Raynham, Rhinebeck, Odessa, Lakemont, Ovid); 4–6 mi inland Erie; 3–6 mi inland Ontario; beach and bar deposits Ontario (Howard, Arkport, Phelps)
Alcona	MI	Huron	1998, current	1:20000	Lakeplains (Au Gres, Wakeley, Tawas, Lupton, Leafriver, Algonquin, Negwegon, Springport); 2–6 mi inland to till plain
Alpena	MI	Huron	2004, current		Lakeplains (Deford, Au Gres, Croswell, Tacoda, Tawas); beach ridges (Proper, Deford, Rousseau); 0–4 mi inland; glacial lake benches (Namur, Chippeny, Alpena)
Arenac	MI	Huron	1967, current		Lakeplains (Roscommon, Au Gres, Deford, Croswell, Epoufette, Charity, Pickford); beach ridges (Eastport); 20 + mi inland
Bay	MI	Huron	1980, current		Lakeplains (Pipestone, Tobico, Rousseau); wave-washed till plain (Tappan, Belleville, Londo, Poseyville); 0–2 mi inland lacustrine, 22 + wave-washed till

(continued)

Table 1.3 (continued)

County	State	Great Lake	Date soil survey	Scale	Landforms/soil associations
Cheboygan	MI	Huron	1991, current		Lakeplains (Roscommon, Charity, Au Gres, Rubicon, Rudyard, Bruce, Ontonagon); beach ridges (East Lake); glacial lake benches (Fairport, Onaway); 0–7 mi inland
Huron	MI	Huron	1980, current		Lakeplain islands (Bach, Sanilac); wave-washed till plain (Aubarque, Filion, Tappan); glacial lake benches (Tobico); mainly 0 mi inland
Iosco	MI	Huron	2002, current		Beach ridges and dunes (Deer Park, Meehan, Wurtsmith, Finch, Deford, Proper); lakeplains (Manary, Whittemore, Iargo, Algonquin, Allendale, Springport, McIvor, Wakeley, Udorthents); 0.35–7 mi inland to till
Presque Isle	MI	Huron	1993, current		Lakeplains (Moltke, Grace, Glawe, Au Gres, Roscommon, Pinconning, Hettinger); beach ridges and dunes (East Lake, Deer Park, Summerville, Croswell); glacial lake benches (Alpena, Hessel, Detour, Brevort); 0–5 mi inland
Sanilac	MI	Huron	1961, current		Lakeplains (Iosco, Melita); beach ridges (Croswell, Eastport); 2–5 Mi inland
Tuscola	MI	Huron	1986, current		Lakeplains (Essexville, Tappan); wave-washed till plains (Avoca, Londo); 1–5 mi inland
Cook	IL	Michigan	2012, current	1:12000	No general soil map
Lake	IL	Michigan	2005, current	1:12000	No general soil map
Lake	IN	Michigan	1972, current	1:15840	Lakeplain (Maumee, Bono, Whitaker); beach ridges and dunes (Oakville, Tawas); 2–9 mi inland
LaPorte	IN	Michigan	1982, current	1:12000	Beach ridges (Oakville, Morocco, Brems); lakeplains (Blount, Selfridge) against till; 4 mi inland
Porter	IN	Michigan	1981, current	1:12000	Beach ridges and dunes (Oakville, Maumee, Brems); patches of outwash (Door, Lydick); lakeplains (Whitaker, Milford, Del Rey); till on relict near-shore lakeplains (Blount, Morley, Pewamo); 5–9 mi inland
Allegan	MI	Michigan	1987, current		Beach ridges and dunes (Oakville, Morocco, Newton); lakeplains (Rimer, Pipestone, Adrian, Granby); 10–17 mi inland
Antrim	MI	Michigan	1978, current		Dunes and lakeplains (Deer Park, Roscommon); till-floored lakeplains (Emmet, Montcalm, Ensley, Tawas); 0.5–2 mi inland
Benzie, Manistee	MI	Michigan	2008, current		Dunes (Nordhouse, Spinks, Plainfield, Udipsamments); lakeplains (Kaleva, Covert, Pipestone, Benona, Adrian, Houghton); 4–10 mi inland
Berrien	MI	Michigan	1980, current		Beach ridges (Spinks, Oakville, Oshtemo); lakeplains (Morocco, Thetford, Granby, Pella, Kibbie, Brady, Gilford, Monitor); wave-washed till plain (Blount, Rimer); 0–10 mi inland
Charlevoix	MI	Michigan	1974, current		Dunes and beach ridges (Deer Park, Eastport); lakeplains (Detour, Kiva, Alpena, East Lake, Kalkaska, Mancelona); 1–5 mi inland
Delta	MI	Michigan	1995, current		Lakeplains (Rubicon, Dawson, Rousseau, Tawas, Cathro); glacial lake benches (Longrie, Summerville); wave-washed till (Ensley); 0–25 mi inland
Emmet	MI	Michigan	1973, current		Dunes and beach ridges (Deer Park, dune land); lakeplains (Alpena, Brimley, Bruce, Wainola, Thomas, Brevort, Iosco, Carbondale, Roscommon, Tawas); lake benches (Longrie, St. Ignace); 0–7 mi inland
Grand Traverse	MI	Michigan	1966, current		Lakeplains (Lupton, Roscommon); 0–2 mi inland
Leelanau	MI	Michigan	1973, current		Dunes (Deer Park, dune land); beach ridges (East Lake, Eastport, Lupton); 0–2 mi inland
Mason	MI	Michigan	1995, current		dunes (dune land, Nordhouse, Quartzipsamments, Typic Udipsamments, Plainfield, Coloma, Epworth, Entic Haplorthods); lakeplains (Covert, Pipestone, Saugatuck, Arkona, Wixom); 1–4 mi inland

(continued)

Table 1.3 (continued)

County	State	Great Lake	Date soil survey	Scale	Landforms/soil associations
Menominee	MI	Michigan	1989, current		Lakeplains (Lupton, Loxley, Cathro, Pickford, Wainola, Deford, Rousseau); 2–5 mi inland
Muskegon	MI	Michigan	1968, current		Dunes and beach ridges (Rubicon, Croswell, Deer Park); lakeplains (Au Gres, Roscommon, Granby, Allendale); 0.5–7 mi inland
Oceana	MI	Michigan	1996, current		Dunes (Epworth, Nordhouse, dune land, Typic Udipsamments, Entic Haplorthods); lakeplains (Plainfield, Coloma, Covert, Granby, Benona, Spinks); 0–2 mi inland
Ottawa	MI	Michigan	1972, current		Dunes (Deer Park, Rubicon); lakeplains (Rubicon, Granby, Croswell, Au Gres, Saugatuck, Bowers, Hettinger); 5–12 mi inland
Schoolcraft	MI	Michigan	2013, current		Dunes and beach ridges (Deford, Tawas, Deer Park); 0–4 mi inland
Van Buren	MI	Michigan	1986, current		Beach ridges and dunes (Oakville); lakeplains (Kingsville, Covert, Pipestone, Selfridge); 2–7 mi inland
Brown	WI	Michigan	1974, current		Glacial lakeplains (Roscommon, Tedrow, Carbondale, Cathro, Oshkosh, Shawano, Sisson, Allendale); 0–12 mi inland
Door	WI	Michigan	1978, current		Lakeplains (Cathro, Carbondale, Kiva, Markey, Deford); glacial lake benches (Longrie, Summerville); 0–2 mi inland
Kenosha, Racine	WI	Michigan	1970, current		Lakeplains (Granby, Hebron, Aztalan, Montgomery, Houghton, Palms); 2–5 mi inland
Kewaunee	WI	Michigan	1980, current		Dunes and lakeplains (Oakville, Wainola); glacial lake benches (Kolberg, Longrie, Namur); 0–0.5 mi inland
Calumet & Manitowoc	WI	Michigan	1980, current		Lakeplains (Nichols, Mundelein, Briggsville, Pella, Shiocton, Granby, Wasepi); till-floored lakeplains (Tedrow); 0–2 mi inland
Marinette	WI	Michigan	1991, current		Glacial lakeplains (Wainola, Deford); 2–9 mi inland
Milwaukee & Waukesha	WI	Michigan	1971, current		Old lakeplains (Martinton, Montgomery, Hebron, Saylesville); >11 mi inland
Oconto	WI	Michigan	1988, current		Lakeplains (Wainola); 0–4 mi inland
Ozaukee	WI	Michigan	1970, current		Old lakeplains (Hochheim, Sisson, Casco); >3 mi inland
Sheboygan	WI	Michigan	1978, current		Lakeplains (Mosel, Hebron, Oakville)
Mackinac	MI	Michigan, Huron	1997, current		Lake Michigan: beach ridges (Eastport, Esau); lakeplain (Leafriver, Croswell, Wainola, Wallace, Spot,); till plain (Angelica, Solona, Markey); glacial lake benches (St. Ignace, Alpena); 4 + mi inland; Lake Huron: beach ridges (Esau); lakeplain (Pickford, Rudyard, Finch, Allendale, Wakeley); till plain (Angelica, Solona, Satago, Battydoe, Markey); glacial lake benches (Alpena, Longrie, Shelter, Amadon, St. Ignace)
Cayuga	NY	Ontario	1971, current		Lakeplain (Williamson); within 10 mi of shoreline; wave-washed (?) till plain (Sodus, Ira) 10 + mi inland
Jefferson	NY	Ontario	1989, current		Lakeplain (Chaumont, Wilpoint, Guffin, Collamer, Niagara, Rhinebeck, Hudson, Vergennes, Kingsbury, Elmridge, Covington, Livingston, Carlisle, Palms, Wilette); bedrock lake benches (Galoo, Galway); extends 5 + mi inland
Monroe	NY	Ontario	1973, current		Lakeplain (Arkport, Collamer, Canandaigua, Niagara, Eel, Hilton, Hudson, Rhinebeck, Madalin, Schoharie, Odessa, Cayuga); extends 1–8 mi inland; backs against moraine
Orleans	NY	Ontario	1977, current		Beach ridges (Arkport); lakeplain (Galen, Elnora, Colonie, Junius, Cosad, Collamer, Niagara, Odessa, Churchville, Rhinebeck, Claverack, Canandaigua, Lakemont, Madalin, Barre); within 3 + mi of coast

(continued)

Table 1.3 (continued)

County	State	Great Lake	Date soil survey	Scale	Landforms/soil associations
Oswego	NY	Ontario	1981, current		Beach ridges (Oakville, Deerfield); lakeplains (Williamson, Amboy, Raynham, Canandaigua, Minoa, Lamson, Madalin, Fonda, Rhinebeck, Hudson, Adams, Windsor); basal till (Ira, Sodus); within 0–5 + mi of coast
St. Lawrence	NY	Ontario	2005, current		Lakeplain (Adjidaumo, Swanton, Deford); till-floored lakeplain (Naumburg); glacial lake benches (Quetico, Summerville); extends 0 to 2 + mi inland
Wayne	NY	Ontario	1978, current		Beach ridges (Alton, Phelps); lakeplain (Williamson, Elnora, Collamer, Minoa); extends 0.5–7 mi inland
Alger	MI	Superior	2013, current		Beach ridges (Deer Park); glacial lake bench (Sauxhead, Burt, Trout Bay, Shingleton, Chatham, Ruse); till-floored lake plain (Garlic, Blue Lake); 0–2 + miles inland
Baraga	MI	Superior	1988, current		Lakeplains (Carbondale, Greenwood, Au Gres, Croswell, Kinross, Rubicon, Grayling, Rousseau, Ocqueoc, Ontonagon, Nunica, Froberg); extends 0–1 mi inland
Chippewa	MI	Superior	1992, current		Beach ridges (Deer Park, Dawson, Au Gres); lakeplains (Pickford, Rudyard, Ontonagon, Kalkaska, Kinross, Alcona, Ingalls, Manistee, Fibre, Allendale); glacial lake benches (Shelter, Posen, Summerville);
Gogebic	MI	Superior	2010, current		Lake terrace in till (Flintsteel, Loggerhead); extends 0 mi inland
Houghton	MI	Superior	1991, current		Lakeplains (Ocqueoc, Rudyard, Froberg, Ontonagon, Lupton, Loxley, Roscommon, Halfaday, Au Gres); dissected lakeplains (Nunica, Fence, Alcona, Misery); water-worked till plain (Graveraet, Kalkaska, Keweenaw, Trimountain)
Isle Royale	MI	Superior	2012, current		
Keweenaw	MI	Superior	2006, current		Beach ridges and dunes (Dawson, Au Gres, Croswell, Deer Park, Rubicon, Deford); lakeplains (Arcadian); dissected lake terrace (Gay); glacial lake benches (Nipissing, Waiska); till-floored lake plain (Garlic, Paavola, Skanee)
Luce	MI	Superior	2006, current		Beach ridges and dunes (Deer Park); lakeplains (Loxley, Pickford, Annanias); till-floored lakeplains (Hendrie, Millecoquins, Rudyard, Pickford); beach ridges 1–4 mi inland from lake; lakeplain and till-floored lakeplain up to 28 mi inland from lake
Marquette	MI	Superior	2007, current		Beach ridges (Deer Park, Pence, Kalkaska, Deford); lakeplains (Alcona, Fence, Paquin, Carbondale, Tawas); till-floored lakeplains (Skanee, Gay, Garlic, Voelker); glacial lake benches (Waiska)
Ontonagon	MI	Superior	2010, current		Lakeplains (Liminga, Manido); dissected lake terrace (Algonquin, Negwegon); lake terrace in till (Flintsteel, Annalake, Loggerhead, Greenstone, Nonesuch);
Cook	MN	Superior	current		Not published
Lake	MN	Superior	current		Not published
St. Louis	MN	Superior	current		Not published
Ashland	WI	Superior	2006, current		No general soil map
Bayfield	WI	Superior	2006, current		No general soil map
Douglas	WI	Superior	2006, current		No general soil map
Iron	WI	Superior	2006, current		No general soil ma p
Apostle Islands Natl. Lakeshore	WI	Superior	2014		

a county on Lake Huron (Fig. 1.3). The map is composed of 17 map units and three soil series. The Clyde sand (gold color) occurs on Nipissing-aged sediments (now Tobico series); and the Clyde loam (gray) and Miami fine sandy loam (pink) occur on what is now recognized as wave-washed Port Huron till (Tappan-Belleville Association) from the Algonquin and Warren lake stages (Weesies 1980). The Web Soil Survey (2019) shows a similar distribution of soil series (not shown).

A soil map of Niagara County New York was published by Pippen and Rice (1901) and represents the first map of a county on Lake Ontario (Fig. 1.4). The map has 17 units and four soil series. The Dunkirk soil series was divided into eight soil types, with textures ranging from fine sand to clay. The coastline features the Dunkirk sand (light brown) and sandy loam (gold), a series that is still recognized along the southern Lake Ontario shore (Web Soil Survey 2019). The Collamer, Niagara, and Rhinebeck series were mapped on the Iroquois lake terrace in the 1972 county survey and are recognized on the shore today (not shown).

The earliest soil map on the Lake Erie shore was prepared for Erie County PA in 1910 (Maynadier and Bucher 1912) (Fig. 1.5). This map has 26 units and five soil series. The Dunkirk soil series was divided into 11 soil types, with textures ranging from gravel to clay. The Dunkirk gravelly loam (yellow), sandy loam heavy phase (dark gray), and sandy loam (pink) were mapped along the shore. The Presque Isle spit was mapped as dune sand. Although the Dunkirk series is still mapped in Erie County, the dominant soil series are the Rimer, Wauseon, and Berrien series (Web Soil Survey 2019).

As of 2019, the entire US Great Lakes coastline has been mapped at a scale of 1:24,000, except a small portion along the northwestern shore of Lake Superior.

On the Canadian Great Lakes shoreline, the earliest soil map was published in 1929 for Elgin County along the north shore of Lake Erie. Soil mapping was begun in earnest shortly after World War II, beginning in Durham (1946) and Prince Edward (1948) counties on Lake Ontario and followed by Essex County on Lake Erie in 1949 (Table 1.4). Soil survey coverage available in a digital GIS format is shown in Fig. 1.6. The most recent published soil survey for a county bordering a Great Lake in Canada was the revision of Chatham-Kent Counties in 1996.

The number of soil series recognized in the US Great Lakes Coastal Zone has increased linearly over the past 115 years (Fig. 1.7). Whereas only 15 soil series were mapped in 1900, more than 650 have been mapped by 2010.

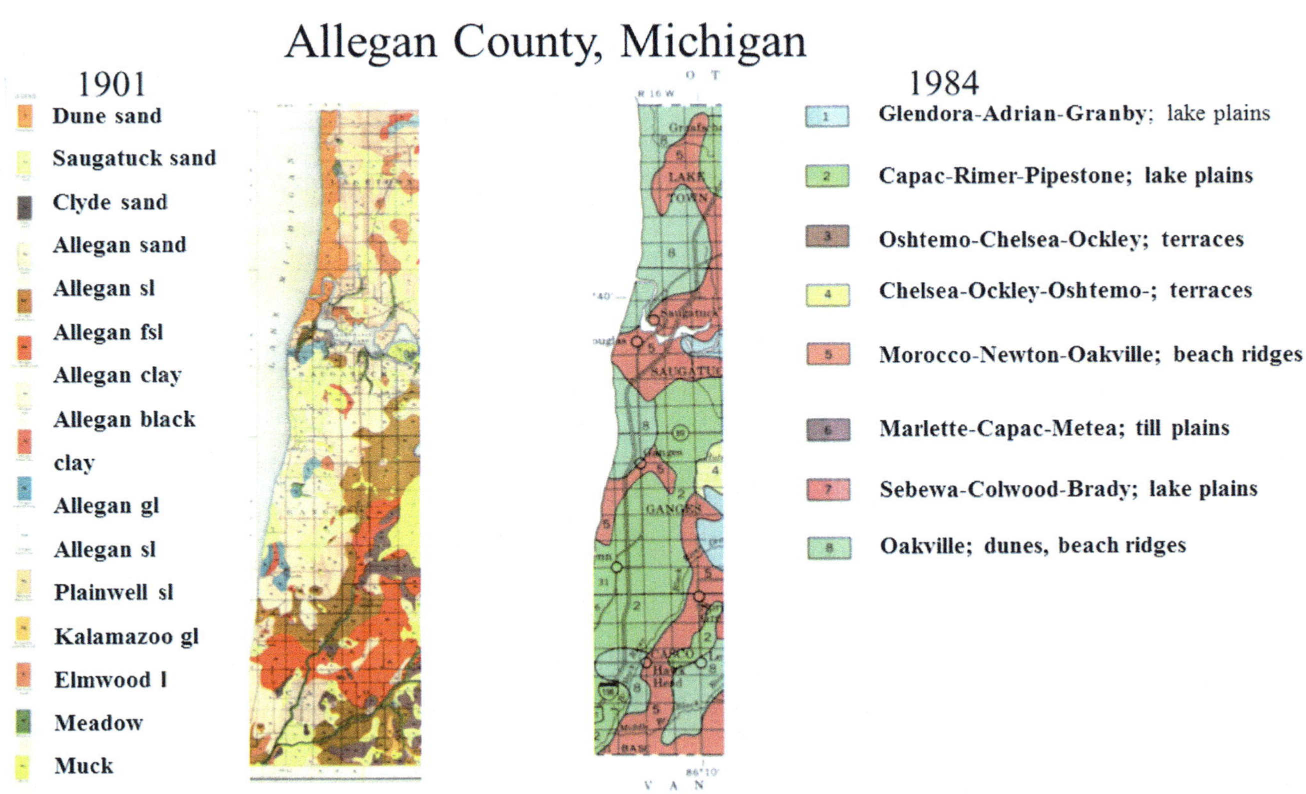

Fig. 1.1 A comparison of general soil maps of Allegan County, MI, on Lake Michigan in 1901 (Pippen and Rice 1901) and the general county soil map of 1984 (Knapp 1987)

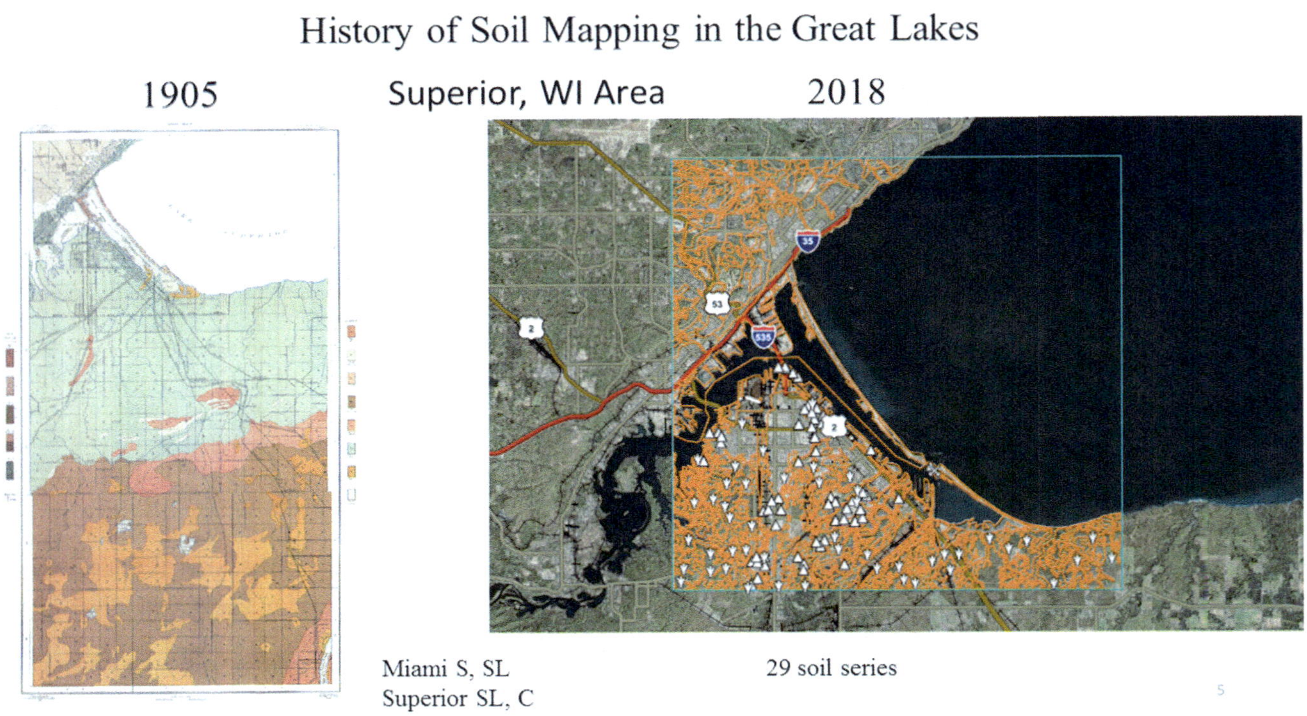

Fig. 1.2 A soil survey of the Superior, WI, area on Lake Superior in 1905 (Maynadier and Bucher 1912) and 2019 (Web Soil Survey)

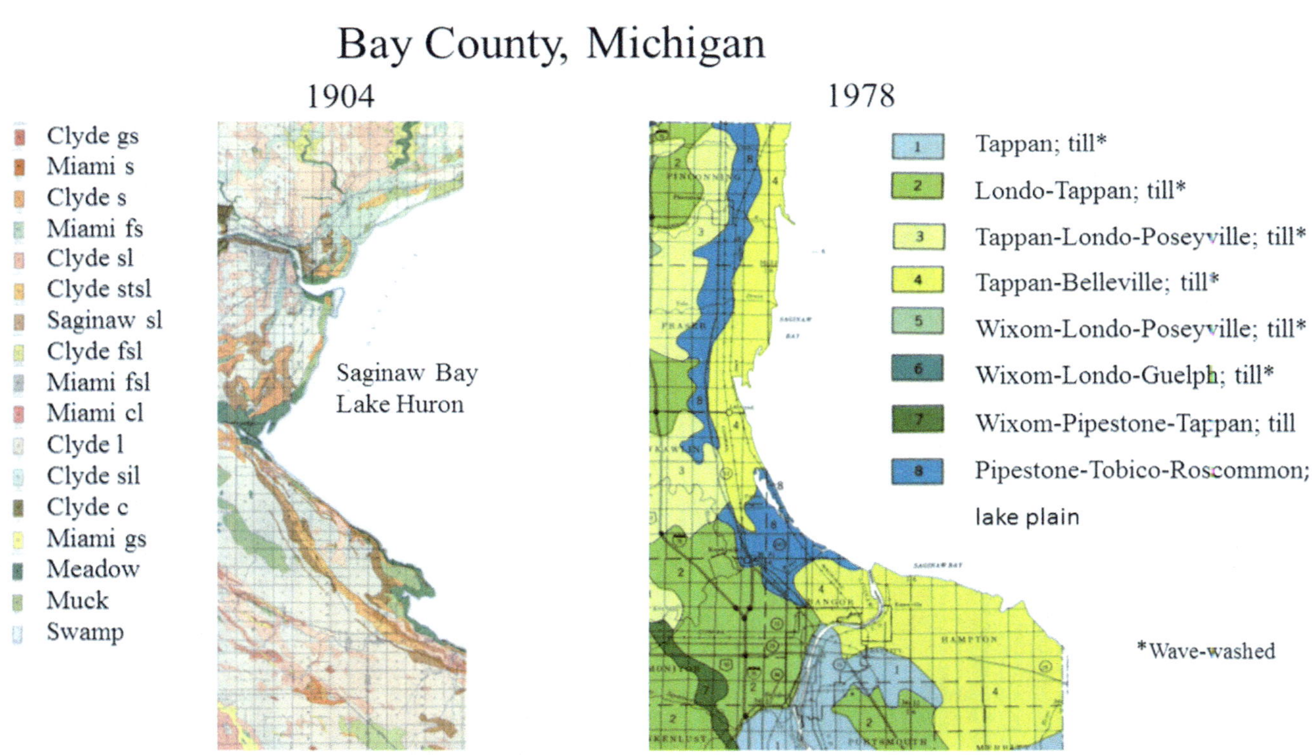

Fig. 1.3 A comparison of soil surveys in Bay County, MI, on Lake Huron for 1904 (McLendon and Carter 1904) and the general soil map for 1978 (Weesies 1980)

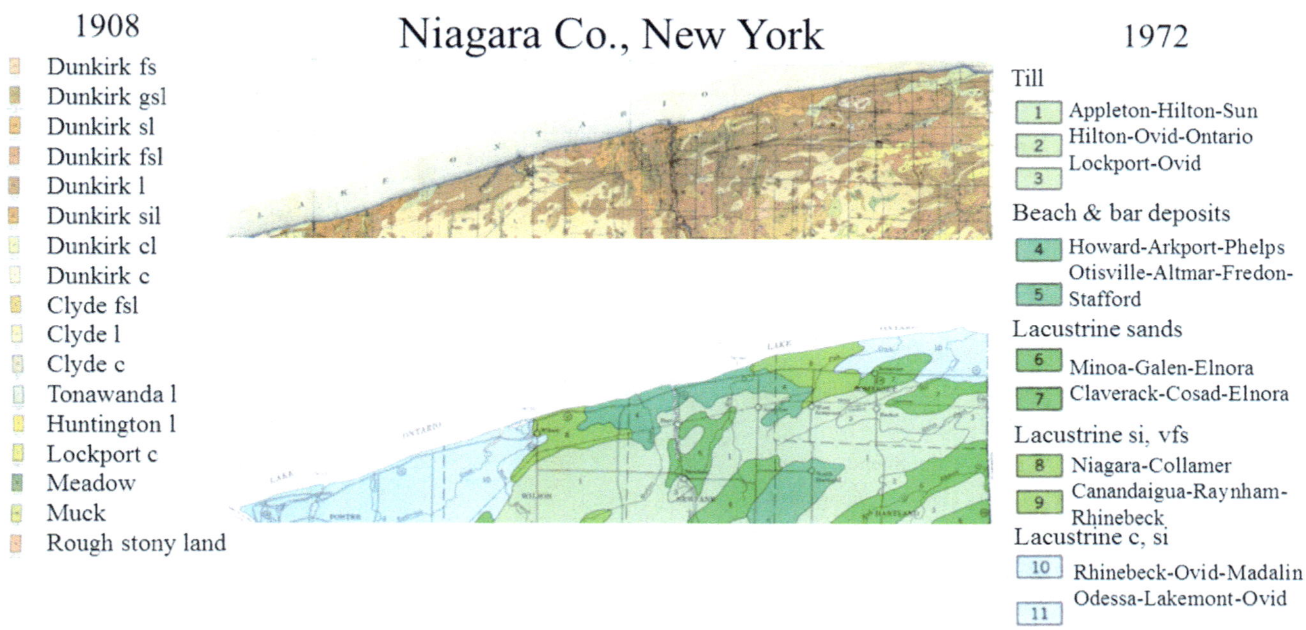

Fig. 1.4 A comparison of soil surveys in Niagara County, NY, on Lake Ontario for 1908 (Pippen and Rice 1901) and the general soil map for 1972 (Higgins et al. 1972)

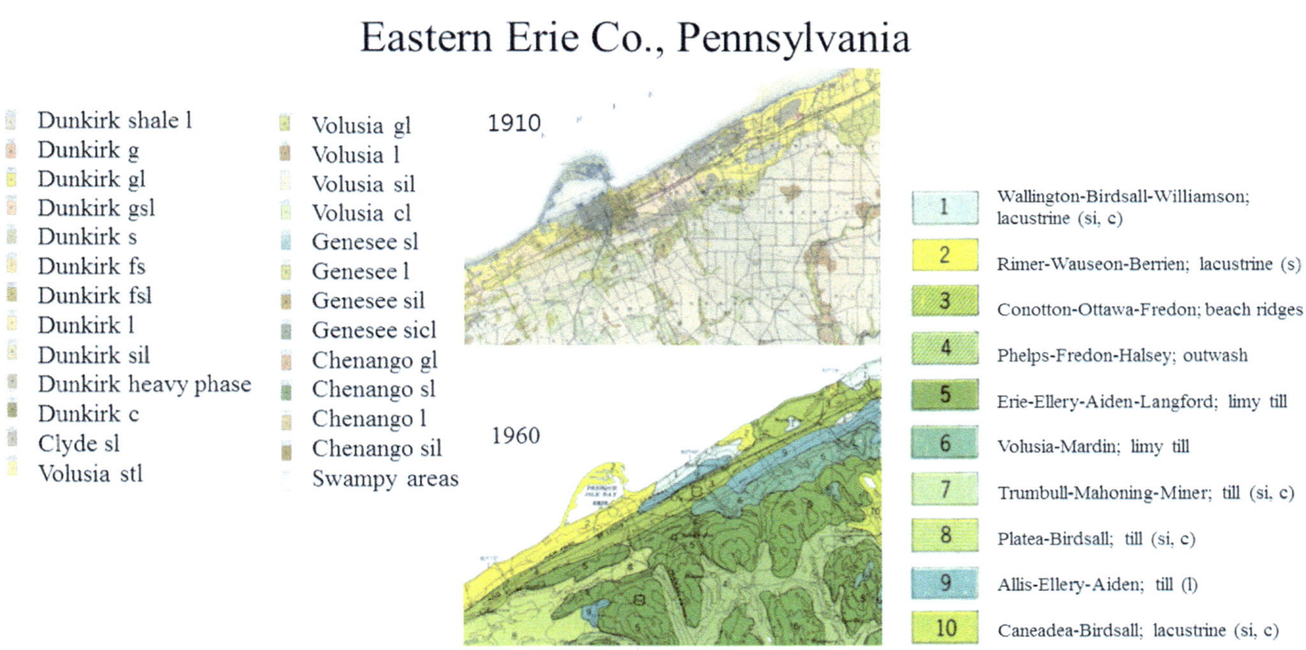

Fig. 1.5 A comparison of soil surveys for eastern Erie County, PA, from 1910 (Maynadier and Bucher 1912) and the general soil map of 1960 (Taylor et al. 1960)

By the beginning of 2018, 79% of the Great Lakes shoreline had been mapped for soils. Michigan is the lead state for 56% of the soil series identified in the GLCZ, followed by New York (15%), Wisconsin (10%), Ohio (8%), and Minnesota (4%) (Fig. 1.8). Michigan also has the greatest proportion of the US Great Lakes coastline at 63%, followed by Wisconsin (15%), New York (9%), Ohio (8%), and Minnesota (4%).

Previous summaries of soils of the Great Lakes focused on sand dunes (Albert 2000) and wetlands (Albert 2003). In

Table 1.4 County and district soil surveys for the Canadian Great Lakes Coastal Zone

County/District	Lake	Year	Scale	Map ID No.
Durham	ON	1946	1:126720	on09
Prince Edward	ON	1948	1:63360	on10
Essex	HU, ER	1949	1:63360	on11
Huron	HU	1952	1:63360	on13
Bruce	HU	1975	1:63360	on16
Grey	HU	1981	1:63360	on17
Peel	ON	1953	1:63360	on18
Lambton	HU	1957	1:63360	on22
Ontario	ON	1956	1:63360	on23
Manitoulin Island	HU	1959	1:63360	on26
Simcoe	HU	1962	1:63360	on29
Parry Sound	HU	1962	1:126720	on31
Lennox-Addington	ON	1963	1:63360	on36
Frontenac	ON	1965	1:63360	on39
Northumberland	ON	1974	1:63360	on42
Halton	ON	1971	1:63360	on43
Halton	ON	1971	1:63360	on43
Thunder Bay	SU	1981	1:50000	on48
Sault Ste. Marie-Blind River	HU	1983	1:50000	on50
Puskakwa	SU	1985	1:100000	on53
Norfolk	ER	1984	1:25000	on57
Niagara	ER, ON	1986	1:25000	on60
Elgin	ER	1992	1:50000	on63
Chatham-Kent	HU, ER	1996	1:63360	on64
Avg. range		1946–1996	1:25000–1:126720	

his regional landscape classification of Michigan, Minnesota, and Wisconsin, Albert (1995) recognized climate, bedrock geology, physiography, and vegetation as key components. However, brief descriptions were made of soils within each landscape map unit. Many of the case studies of soils along the Great Lakes shoreline utilized buried soils in dunes to identify periods of dune activity. An example is the paleosol buried in dunes at Mt. Baldy in the eastern part of the Indiana Dunes National Lakeshore on southern Lake Michigan (Fig. 1.9). The lower dune formed 3,000–4,000 year BP.

Some of the classic studies of soil chronosequences have been done on dunes and beach ridges along the Great Lakes shores (Olson 1958). Other studies of soil chronosequences in the Great Lakes region include those of Franzmeier and Whiteside (1963a, b), Cary et al. (1979), Berg (1980), Barrett and Schaetzl (1992), VandenBygaart and Protz (1995), Lichter (1998), Barrett (2001), and Howard et al. (2012).

One of the most famous soils in the GLCZ is the Two Creeks soil (Fig. 1.10). The soil underlies a buried spruce forest along the western Lake Michigan shore near Two Creeks, WI. The soil is developed on drift from the Cary glacial stage, buried by Valders drift, and dated at 11,850 year BP (Lee and Horn 1972).

1.3 Great Lakes Coastal Zone

The Laurentian Great Lakes contain nearly 17,000 km of coastline (includes islands) and are sometimes referred to as the North American "Third Coast" (Fig. 1.11). The Great Lakes contain 18% of the world and 90% of the US fresh surface water. The Great Lakes region supports 33 million people and is important to the US and Canada for shipping, fishing, recreation, forest products, power generation, and other land uses (Environmental Protection Agency 1995; Northeast-Midwest Institute and National Oceanic and

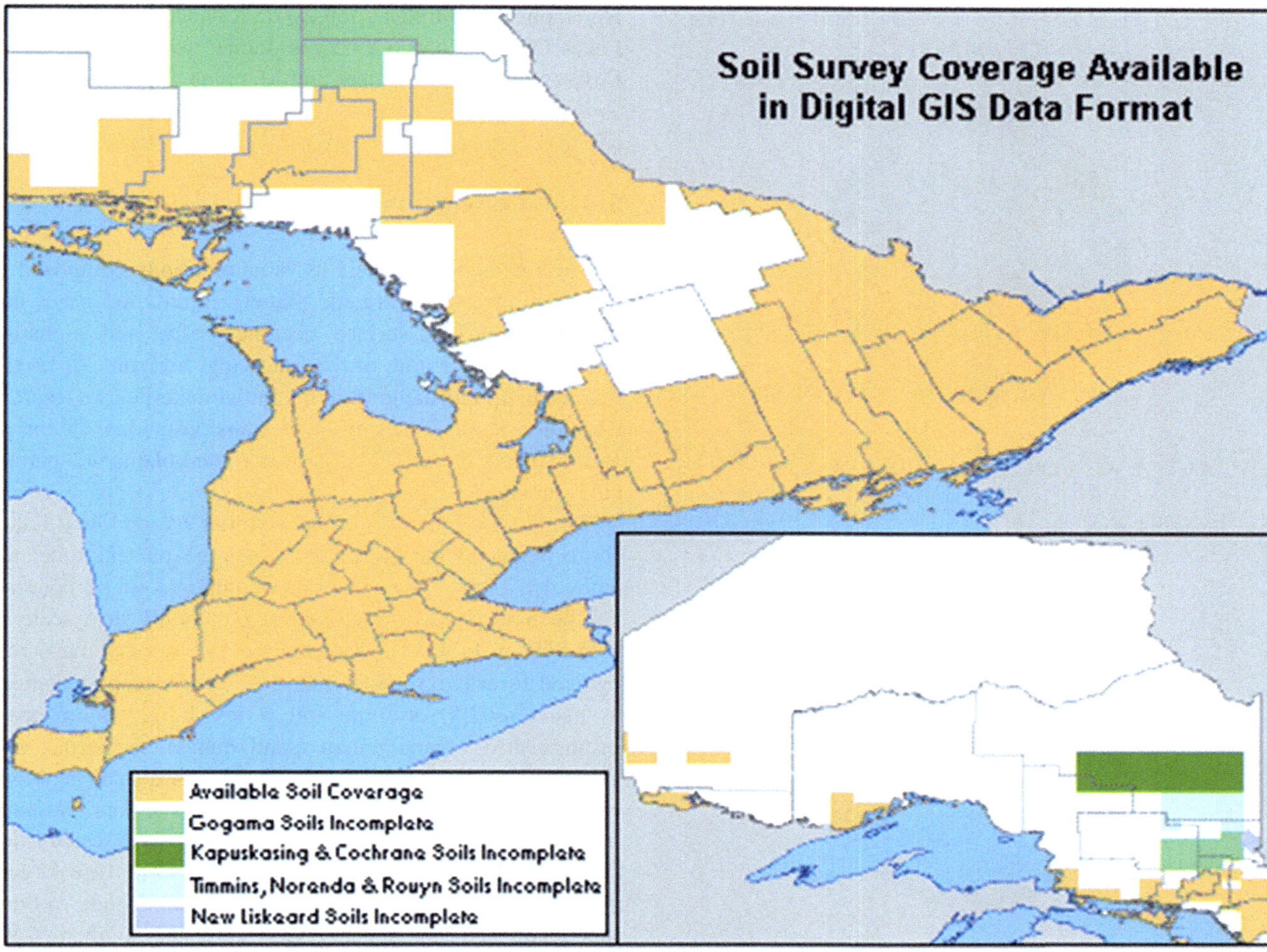

Fig. 1.6 Soil survey coverage available in digital GIS data format for the Canadian Great Lakes

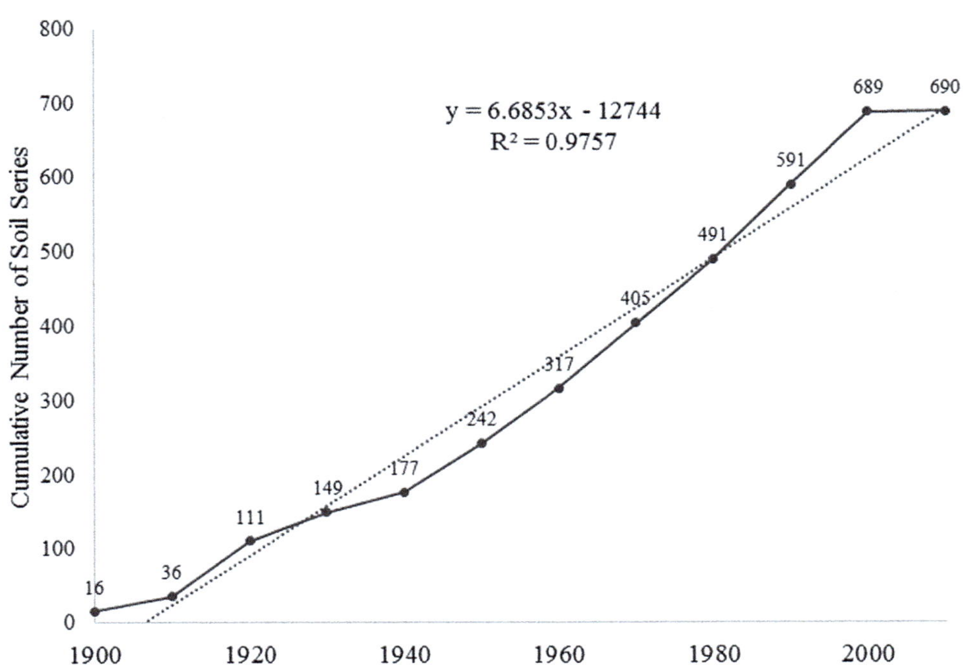

Fig. 1.7 History of soil series established in the Great Lakes Coastal Zone. The dashed line is the linear regression of the cumulative number of soil series against the mid-point of the decade

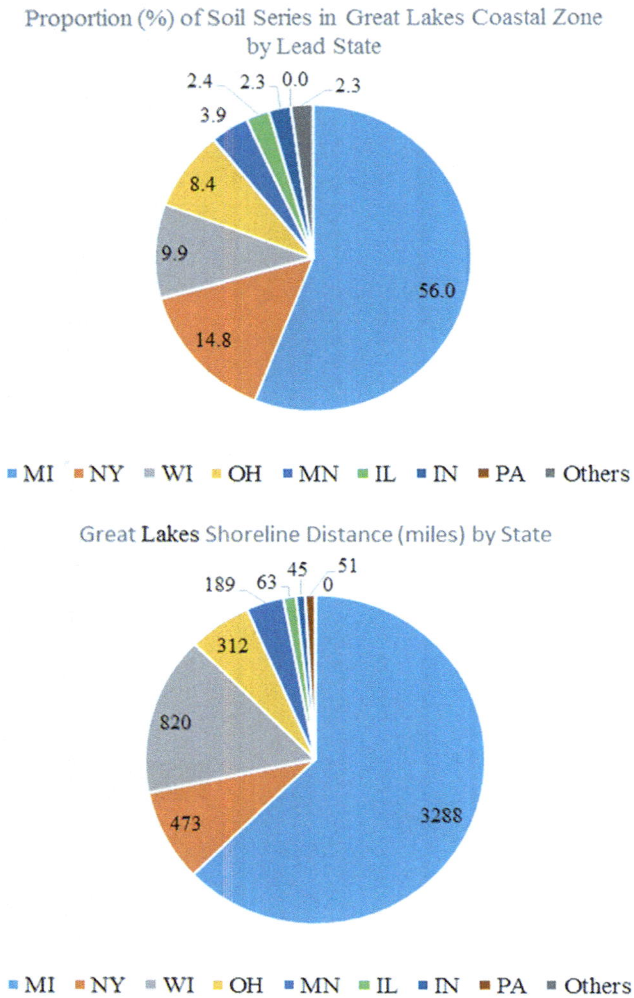

Fig. 1.8 Proportion (%) of soil series and shoreline length (miles) in the US Great Lakes Coastal Zone by state

Atmospheric Administration 2001; Kavcic 2016). The Great Lakes Coastal Zone (GLCZ) is defined by the Great Lakes Environmental Assessment and Mapping Project (GLEAM) as all land and water within 5 km of the shoreline.

1.4 Conclusions

The soil is recognized here as "a natural body comprised of solids (minerals and organic matter), liquid, and gases that occurs on the land surface, occupies space, and is characterized by one or both of the following: horizons, or layers, that are distinguishable from the initial material as a result of additions, losses, transfers, and transformations of energy and matter or the ability to support rooted plants in a natural environment."

The first soil surveys of US counties on the Great Lakes shore began in the early 1900s. At the present time, the entire US Great Lakes shoreline composed of 82 counties has been mapped in detail, most commonly at a scale of 1:15,840 or 1:24,000. Whereas the first county maps recognized from two to six soil series, current maps recognize as many as 100 or more soil series. Early soil mappers distinguished among eolian (sand dunes), lacustrine, and glacial parent materials. Frank Leverett, a glacial geologist who conducted research along the Great Lakes shoreline from 1902 to 1939, cooperated with the newly established Bureau of Soils. However, Great Lakes soils were largely ignored until the late 1950s with J.S. Olson's study of dune succession and soil development along the southern Lake Michigan shore. Studies beginning in the 1990s examined soil chronosequences and buried soils in dunes.

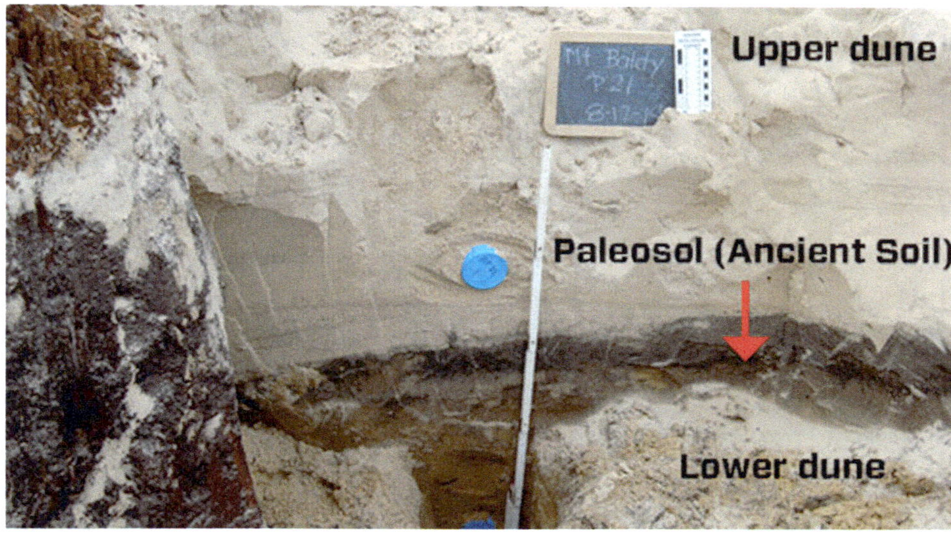

Fig. 1.9 An example of a buried Spodosol in dunes on Mt. Baldy, Indiana Dunes National Lakeshore (photo by W. Monaghan)

Fig. 1.10 The Two Creeks buried forest bed and buried Spodosol near Two Creeks, WI, on Lake Michigan (photo by Tom Wilch, Albion College)

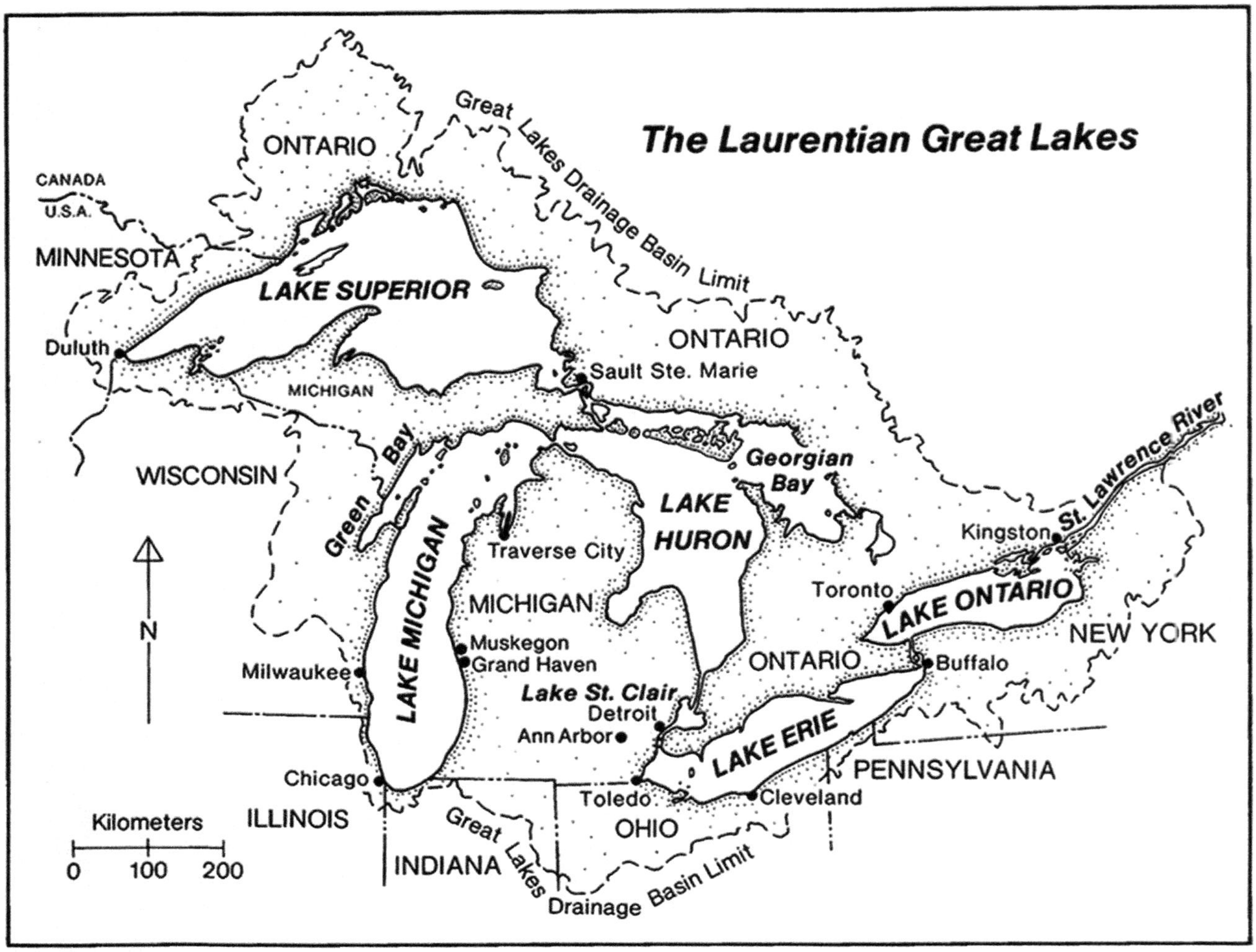

Fig. 1.11 The Laurentide Great Lakes with key place names and watershed boundaries

References

Albert, D. A. 1995. *Regional landscape ecosystems of Michigan, Minnesota, and Wisconsin: a working map and classification*. U.S. Forest Service, North Central For. Exp. Stn., Gen. Tech. Rep. NC-179.

Albert, D. A. 2000. *Borne of the Wind: an Introduction to the Ecology of Michigan Sand Dunes*. Michigan Natural Features Inventory, 63.

Albert, D. A. 2003. *Between land the lake: Michigan's Great Lakes Coastal Wetlands*. Mich. Nat. Features Inventory, Mich. State Univ. Ext., Ext. Bull. E-2902, 96.

Barrett, L. R. (2001). A strand plain soil development sequence in northern Michigan, USA. *CATENA, 44,* 163–186.

Barrett, L. R., & Schaetzl, R. J. (1992). An examination of podzolization near Lake Michigan using chronofunctions. *Canadian Journal of Soil Science, 72,* 527–541.

Berg, R. C. (1980). Use of stepwise discriminant analysis to assess soil genesis in a youthful sandy environment. *Soil Science, 129,* 353–365.

Cary, S. J., McDowell, P. F., & Graumlich, L. J. (1979). Soils and surficial geology of four Apostle Islands. *The Wisconsin Academy of Sciences, Arts and Letters, 67,* 14–30.

Environmental Protection Agency. 1995. The Great Lakes: an environmental atlas and resource book (3rd edn). EPA 9-5-B-95-001.

Franzmeier, D. P., & Whiteside, E. P. (1963a). A chronosequence of Podzols in northern Michigan. I. Ecology and description of pedons. *Michigan State University Agricultural Experiment Station, Quarterly Bulletin, 46,* 2–20.

Franzmeier, D. P., & Whiteside, E. P. (1963b). A chronosequence of Podzols in northern Michigan. II. Physical and chemical properties. *Michigan State University Agricultural Experiment Station, Quarterly Bulletin, 46,* 21–36.

Higgins, B. A., P. S. Puglia, R. P. Leonard, Yoakum, W. A. Wirtz. 1972. *Soil survey of Niagara County, New York*. US Dep. Agr., Soil Conserv. Serv., Cornell Univ. Agr. Exp. Stn., US Gov. Print. Office, Washington, D.C., 208.

Howard, J. L., Clawson, C. R., & Daniels, W. L. (2012). A comparison of mineralogical techniques and potassium adsorption isotherm analysis for relative dating and correlation of late Quaternary soil chronosequences. *Geoderma, 179–180,* 81–95.

Kavcic, R. 2016. Connecting across borders: a special report on the Great Lakes and St. Lawrence regional economy. http://www.cglslgp.org/media/1818/2016-cglslgp-bmo-economic-report.pdf. Accessed 03 April 2017.

Knapp, B. D. 1987. *Soil survey of Allegan County, Michigan.* US Dep. Agr., Soil Conserv. Serv., Mich. Agric. Exp. Stn., 188.

Lee, G. B., & Horn, M. E. (1972). Pedology of the Two Creek section, Manitowoc County, Wisconsin. *Wisconsin Academy of Sciences, Arts, and Letters, 60,* 183–199.

Leverett, F. 1902. *Glacial formations and drainage features of the Erie and Ohio basins.* U.S. Geol. Surv. Monog. 41. 802.

Leverett, F. 1910. *Surface geology of the northern peninsula of Michigan.* Mich. Geol. & Biol. Surv., Publ. 7, Geol. Ser. 5, 113.

Leverett, F., and C. F. Schneider. 1912. *Surface geology and agricultural conditions of the southern peninsula of Michigan.* Mich. Geol. & Biol. Surv., Lansing, MI. 185.

Leverett, F., and F. B. Taylor. 1915. *The Pleistocene of Indiana and Michigan and the history of the Great Lakes.* U.S. Geol. Surv. Vol. III, 529.

Leverett, F. 1929. *Moraines and shorelines of the Lake Superior region.* U.S. Geol. Surv. Prof. Pap. 154-A, 72.

Lichter, J. (1998). Rates of weathering and chemical depletion in soils across a chronosequence of Lake Michigan sand dunes. *Geoderma, 85,* 255–282.

Maynadier, G., W. J. Geib, L. Schoenmann, and F. L. Musback. 1912. *Soil survey of the Bayfield Area, Wisconsin.* U.S. Dept. Agr., Bureau of Soils, Govt. Print. Office, Washington, D.C., 31.

Maynadier, G. B., and F. S. Bucher. 1912. *Soil survey of Erie County, Pennsylvania.* US Dep. Agr., Bureau of Soils, 49.

McLendon, W. H., and M. E. Carr. 1904. *Soil survey of the Saginaw Area, Michigan.* US Dep. Agr., Bureau of Soils, 37.

Northeast-Midwest Institute and National Oceanic and Atmospheric Administration. 2001. Revealing the economic value of protecting the Great Lakes. http://www.nemw.org/wp-content/uploads/2015/06/GLEconVal.pdf. Accessed 13 Jan 2018.

Olson, J. S. (1958). Rates of succession and soil changes on southern Lake Michigan sand dunes. *Botanical Gazette, 119,* 125–170.

Pippen, E. O, and T. D. Rice. 1901. *Soil survey of Allegan County, Michigan.* US Dep. Agr., Bureau of Soils, 124.

Pippen, E. O., G. B. Jones, W. J. Geib, O. L. Ayrs, C. W. Mann. 1908. *Soil survey of Niagara County, New York.* US Dep. Agr., Bureau of Soils, Govt. Print. Off., Washington, D.C., 56.

Soil Survey Staff. 2014. Keys to Soil Taxonomy (12th edit). U.S. Dep. Agric., Nat. Resour. Conserv. Serv.

Taylor, D.C., Beard, J.S., and Anderson, J., et al. (1960). *Soil survey of Erie county, Pennsylvania.* US Department of Agriculture, Soil Conservation Service. Web Soil Survey. 2019. https://websoilsurvey.nrces/usda.gov

VandenBygaart, A. J., & Protz, R. (1995). Soil genesis on a chronosequence, Pinery Provincial Park, Ontario. *Canadian journal of soil science, 75,* 63–72.

Weesies, G. A. 1980. *Soil survey of Bay County, Michigan.* US Dep. Agr., Soil Conserv. Serv., Mich. Agr. Exp. Stn., 117.

2 Soil-Forming Factors of the Great Lakes Coastal Zone

2.1 General

Soils result from the interaction of climate, organisms (plants, animals, and humans), relief (topography), parent material, and time—all of which have been influenced by humankind. Each of these factors varies considerably across the Great Lakes region, which extends 1,560 km (935 miles) west to east, 760 km (455 miles) north to south, and has a coastal zone elevation range of 74–305 m (243–1,000 ft.).

2.2 Climate

There are considerable climate variations not only due to latitude and longitude but also to the lakes themselves. The Great Lakes act as heat sinks and moderate the regional temperatures, with cooler temperatures in the summer and milder temperatures in the winter. In the Great Lakes Coastal Zone (GLCZ), the mean annual air temperature ranges between 0 and 5 °C in the north and 5–10 °C in the south. The mean annual precipitation ranges between 400 and 800 mm throughout most of the basin, but is between 800 and 1,200 mm along southern Lakes Michigan, Erie, and Ontario. The Great Lakes receive large amounts of snow, particularly in the snow-belt regions (Fig. 2.1). Snowfall exceeds 500 cm/year on the Keweenaw Peninsula in Lake Superior and is around 380 cm/year on the southeastern shores of the other four Great Lakes.

The plant hardiness zone is 3 along the north shore of Lake Superior, 4 along the west and south shore of Lake Superior and the north shore of Lake Huron, and 5 and 6 around Lakes Michigan, Erie, and Ontario (Fig. 2.2). The effect of the large water bodies on regional climate is readily visible in the sharp changes in plant hardiness zone with proximity to the lakes. Hinkel and Nelson (2012) reported near-shore sites along Lake Superior to be 1–2 °C cooler than sites 5 km inland in spring and summer and 1 °C warmer in winter. The mean daily temperature differed by as much as 11 °C among 26–30 study sites, but the average difference was 2.5–3.0 °C.

2.3 Vegetation

There are four broad terrestrial provinces in the Great Lakes region (Fig. 2.3). The Boreal Forest is limited to the north shore of Lake Superior and includes forest cover types comprised predominantly of white spruce (*Picea glauca*), black spruce (*P. mariana*), balsam fir (*Abies balsamifera*), jack pine (*Pinus banksiana*), paper birch (*Betula papyrifera*), and trembling aspen (*Populus tremuloides*) (Fig. 2.4a). The Laurentian Mixed (Great Lakes-St. Lawrence) Forest encompasses the shores of southern Lake Superior, the northern half of Lake Michigan, and all but the southern tip of Lake Huron. The mixed forest cover types are composed of red pine (*Pinus resinosa*), eastern white pine (*P. strobus*), eastern hemlock (*Tsuga Canadensis*), yellow birch (*Betula alleghaniensis*), maples (*Acer* spp.), and oaks (*Quercus* spp.) (Fig. 2.4b). The Midwest Broadleaf (Deciduous; Carolinian) Forest occurs along the shores of southern Lake Michigan and Huron and the shores of Lakes Erie and Ontario shores; it is comprised of beech (*Fagus grandifolia*), maples (*Acer* spp.), hickories (*Carya* spp.), and oaks (*Quercus* spp.) (Fig. 2.4c). A fourth province is the Prairie Parkland which nearly reaches the southern shore of Lake Michigan (Fig. 2.4d). The areas with a yellow pattern are cultivated.

From an analysis of the soil series descriptions, the dominant vegetation types in the GLCZ are temperate mixed coniferous and broad-leaved forest (43%) and temperate broad-leaved forest (30%), followed by lowland mixed forest (21%), lowland shrubs (2.5%), prairie (1.6%), and oak- or pine-savanna (1.4%) (Fig. 2.5).

The vegetation zones of the upper Great Lakes region have not remained stable since recession of the glaciers. Rather they have responded to natural and human-caused changes in the climate. Many of these changes have occurred as a result of shifts in the tension zone, a zone reflecting a

J. G. Bockheim, *Soils of the Laurentian Great Lakes, USA and Canada*,
https://doi.org/10.1007/978-3-030-52425-8_2

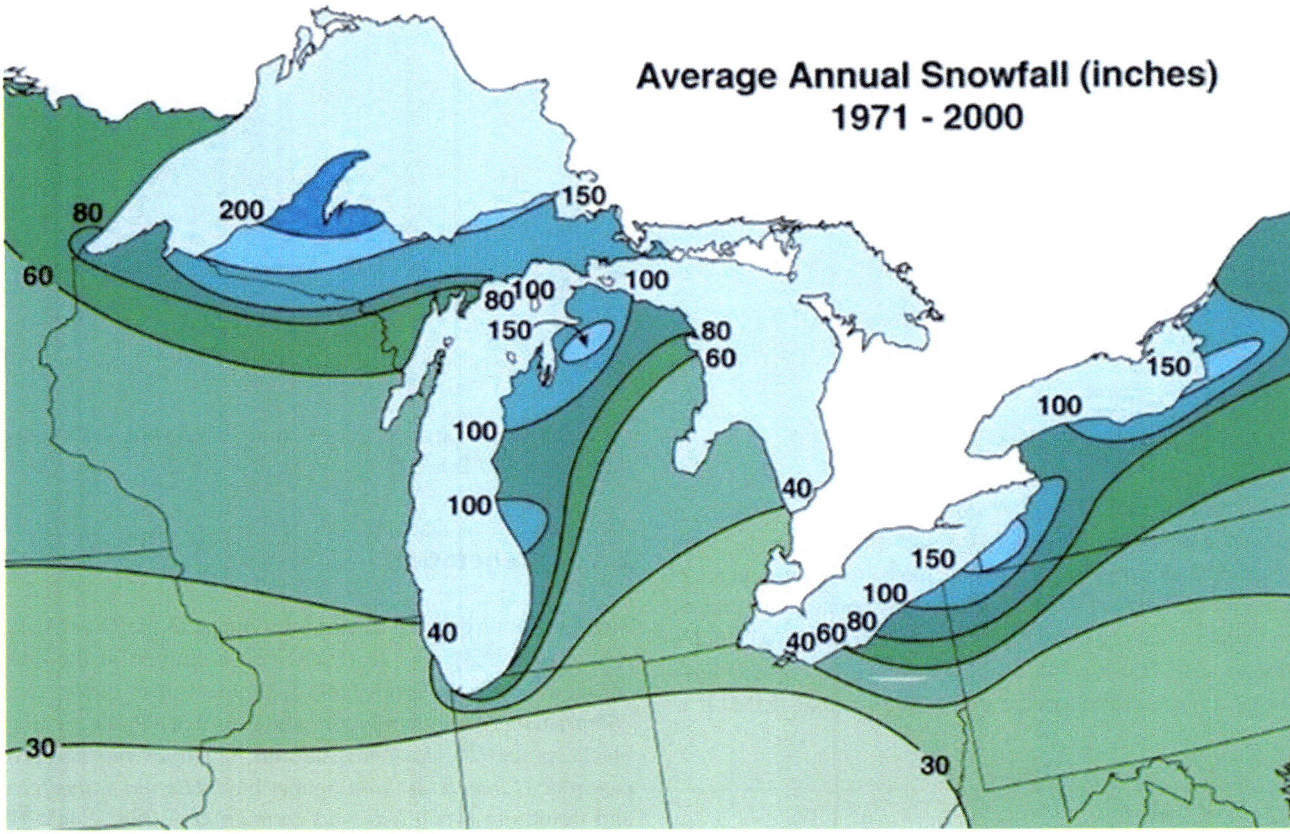

Fig. 2.1 Average annual snow distribution in the Great Lakes Coastal Zone. Lake-effect snow occurs in the eastern portion of each of the Great Lakes (http://serc.carleton.edu/download/images/35582/great_lakes_profile.jpg)

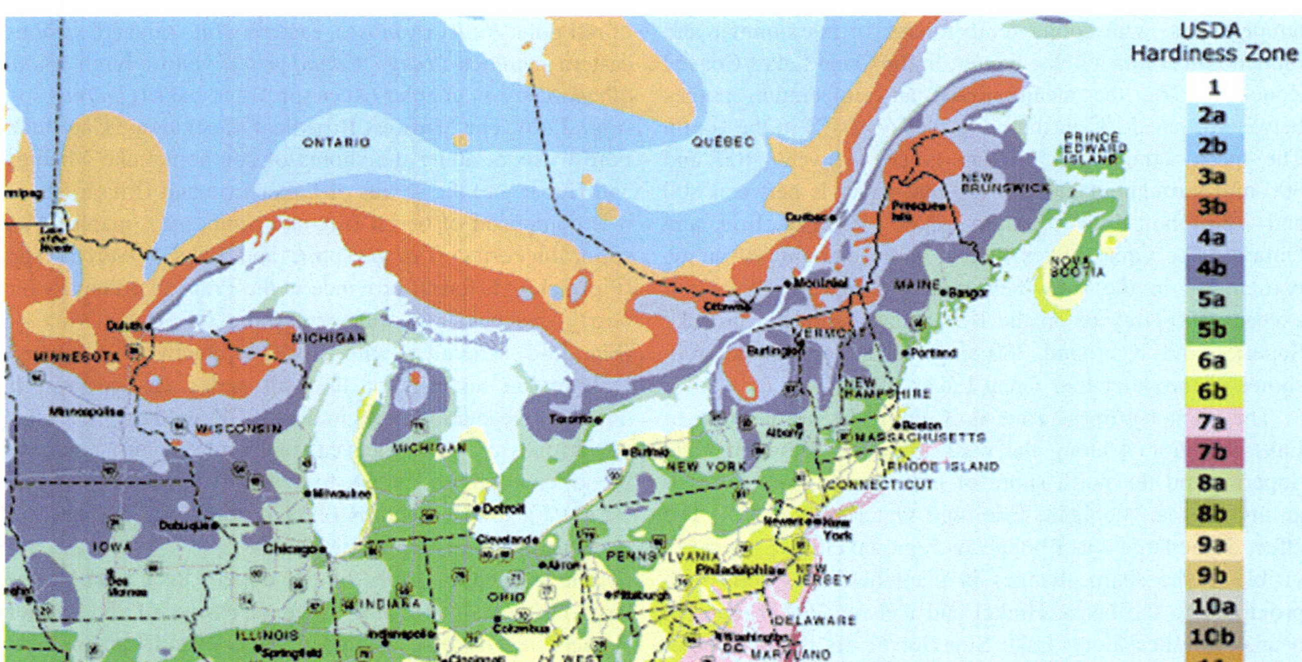

Fig. 2.2 USDA Plant Hardiness Zones in the Great Lakes Coastal Zone (2012). The zones are arranged from 1, with the lowest annual extreme minimum temperatures, to 11, with the highest annual extreme minimum temperatures

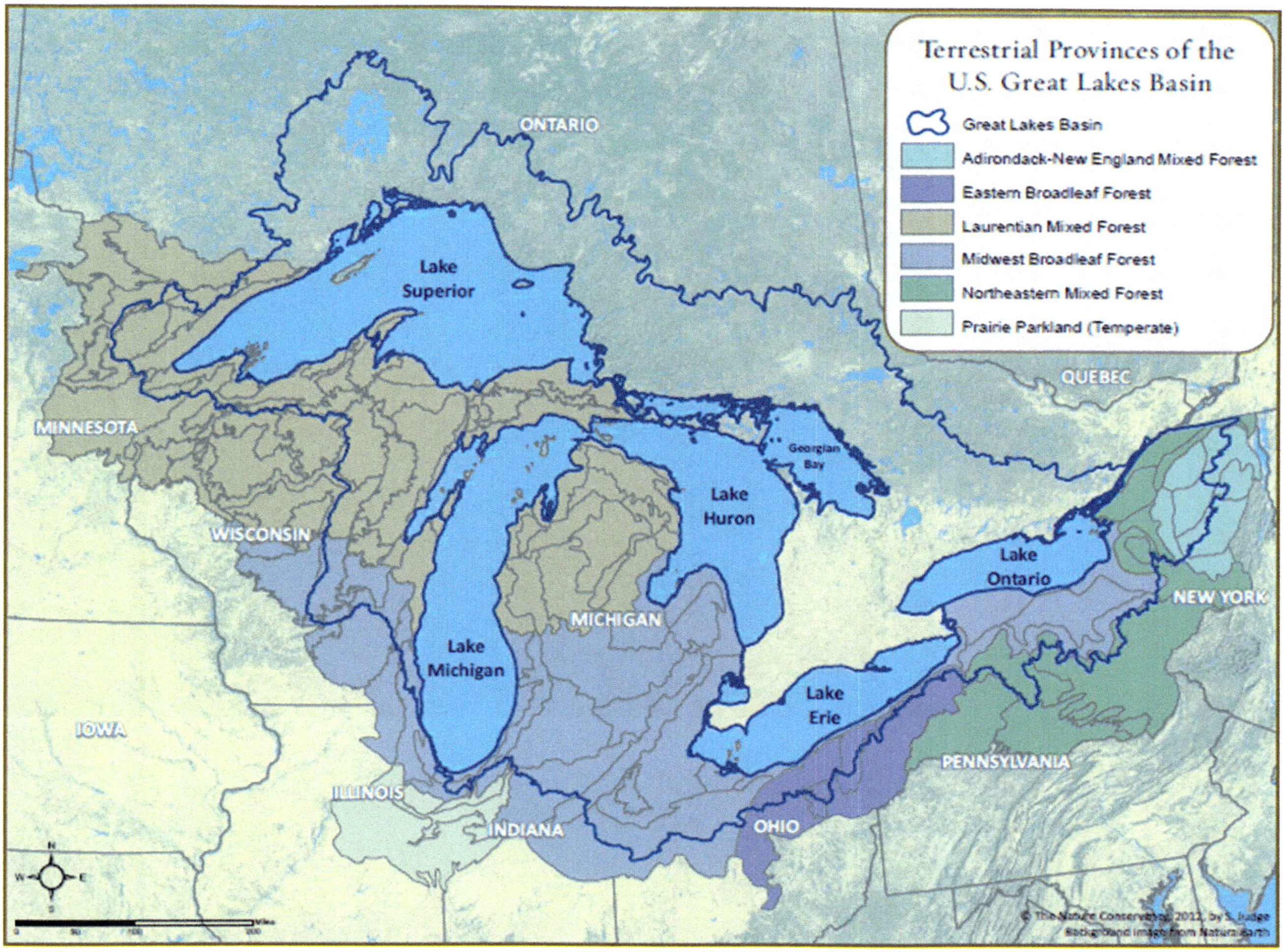

Fig. 2.3 Terrestrial provinces on the US Great Lakes basin (*Source* Great Lakes Inform; http://serc.carleton.edu/download/images/35582/great_lakes_profile.jpg)

transition or mixture of floristic elements. The tension zone illustrated in Fig. 2.6 shows separation of the mixed forest (M) and primarily broad-leaved forest (D) across the Great Lakes states and provinces.

Coastal wetlands are important in the Great Lakes basin in terms of biodiversity, nesting and migratory wildfowl habitat, and maintenance of water levels and quality (Albert 2003). Wetlands comprise about 50% the Great Lakes shoreline (Fig. 2.7). Some wetland types are include Lowland Mixed Coniferous and Broad-leaved Forest (A), Lowland Conifer Forest (B), and Non-forested Wetland (C) (Fig. 2.8).

2.4 Relief and Elevation

The greatest elevations in the Great Lakes basin occur in the Lake Superior and Lake Huron basins and are shown in the darkest pattern of Fig. 2.9. The ranges in elevations give an idea about the slopes present in the Great Lakes basin. Some of the steepest relief occurs on the Bayfield and Keweenaw Peninsulas and Nipigon area of Lake Superior, the eastern and southern shores of Georgian Bay (Lake Huron), the southern and eastern shores of Lake Erie, and the eastern shore of Lake Ontario.

a

b

Fig. 2.4 **a** Boreal forest near Thunder Bay, ON, in the Lake Superior Coastal Zone. The dominant species are spruce, fir, pines, birch, and aspen (photo by J. Bockheim). **b** Temperate mixed coniferous and broad-leaved forest in the Huron Mountains of Lake Superior's south shore (photo by A.E. Hartemink). The dominant species are pine, spruce, maple and birch, cont.
c Temperate broad-leaved forest in the Huron Mountains along the south shore of Lake Superior (photo by J. Bockheim). **d** Prairie near the Indiana Dunes National Lakeshore (photo by Alex Zaideman)

Fig. 2.4 (continued)

c

d

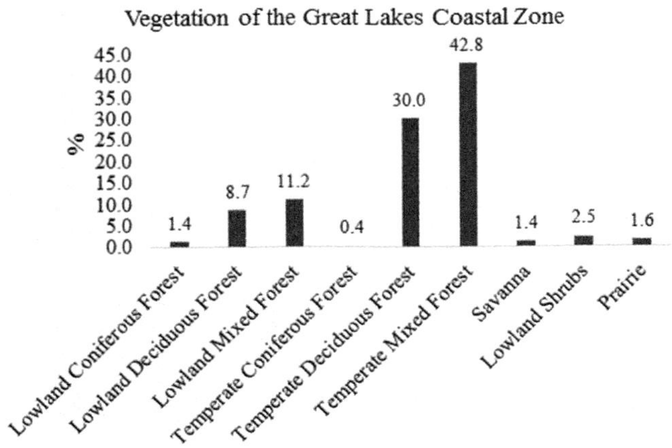

Fig. 2.5 Distribution of vegetation in the US Great Lakes Coastal Zone based on the areal distribution of soil series and vegetation descriptions in Official Soil Series Descriptions

2.5 Bedrock Geology

Nearly all of the Lake Superior basin and the north shore of Georgian Bay are part of the Canadian Shield and are composed of granites and gneisses (Fig. 2.10). Sedimentary rocks comprise the Michigan Basin, which is a geologic basin centered on the Lower Peninsula but including all of the Lakes Michigan and most of the Lake Huron and Lake Erie basins. The Michigan Basin is a depression containing limestones, dolomitic limestones, sandstones and shales.

The Niagara Escarpment, a long escarpment or cuesta composed of dolomitic limestone, extends from the Genesee River along the south shore of Lake Ontario to the northwest along the Bruce Peninsula and Manitoulin Island in Lake Huron, and around the northern and western shores of Lake Michigan (Fig. 2.11). A 435-mile (725-km) stretch of the escarpment from Niagara Falls to the tip of the Bruce Peninsula composes one of six UNESCO World Biosphere Reserves along the Great Lakes shores.

2.6 Surficial Geology

(*Note* a word about geologic dating: rocks and other materials are dated by a variety of techniques. Most of the dates included here were determined from the radiocarbon method. Therefore, I follow the convention of using "years Before Present," or kyr (thousands of years) and Ma (millions of years) BP. Some publications provide calendar (uncorrected) dates, i.e., cal yr BP. More recent publications have used Thermoluminescence, TL; Optically Stimulated Luminescence, OSL; Accelerator Mass Spectrometry, AMS; and other dating techniques.)

The entire Great lakes basin was glaciated during the Port Huron advance of the late Wisconsinan (Fig. 2.12). The pre-existing topography of the basin played an important role in directing the flow of glacial ice. The Des Moines lobe in the west passed in a NW–SE direction; the lobes over Lakes Michigan and Huron moved nearly N–S; and Lake Ontario received ice directed by the Appalachian Mountains from E–W.

The Laurentide ice left predominantly eroded landforms in the northern Lake Superior and Lake Huron basins and deposited till, outwash, and glaciolacustrine materials in the southern Great Lakes basin (Fig. 2.13). Evidence of glacial erosion include striations, scoured bedrock, U-shaped valleys, roches moutonnées, over-deepenings, and other features. Roche moutonnées refer to rock formations with an asymmetric erosional form as a result of abrasion on the "stoss" (upstream) side of the rock and plucking on the "lee" (downstream) side. Glacial depositional features include moraines, kames, drumlins, eskers, and outwash fans. A moraine is a landform comprised of unconsolidated glacial debris. Kames are steep-sided mounds of sand and gravel deposited by a melting ice sheet. Drumlins are elongated hills that reflect erosion or deposition by a glacier that parallel the direction of ice movement at the time of formation. Eskers are long, winding ridges of stratified sand and gravel formed within ice-walled tunnels by steams which flowed within and under glaciers. Outwash fans are fan-shaped bodies of sediments deposited by braised streams from a melting glacier.

There were at least six glaciations in the Great Lakes basin over the past 20 kyr, including from oldest to youngest, the Valparaiso, Tinley-Defiance, Lake Border, Port Huron, Greatlakean, and Marquette advances. The deposits are named are type localities based on association with a particular lobe of ice advancing into the basin. By 9.5 ky BP, Laurentide ice had left the Great Lakes basin.

During the Laurentide glaciations, the ice mass depressed the land; the land rebounded isostatically following recession of the ice. Figure 2.14 shows that the land was up to 980 ft. (300 m) lower in the northern Lake Superior basin 13,000 years ago than at present due to the mass of the Laurentide ice. The diagram shows that the "hinge-line," or 0 feet of uplift passes through the middle of Lake Michigan, the southern tip of Lake Huron, and the northwestern portion

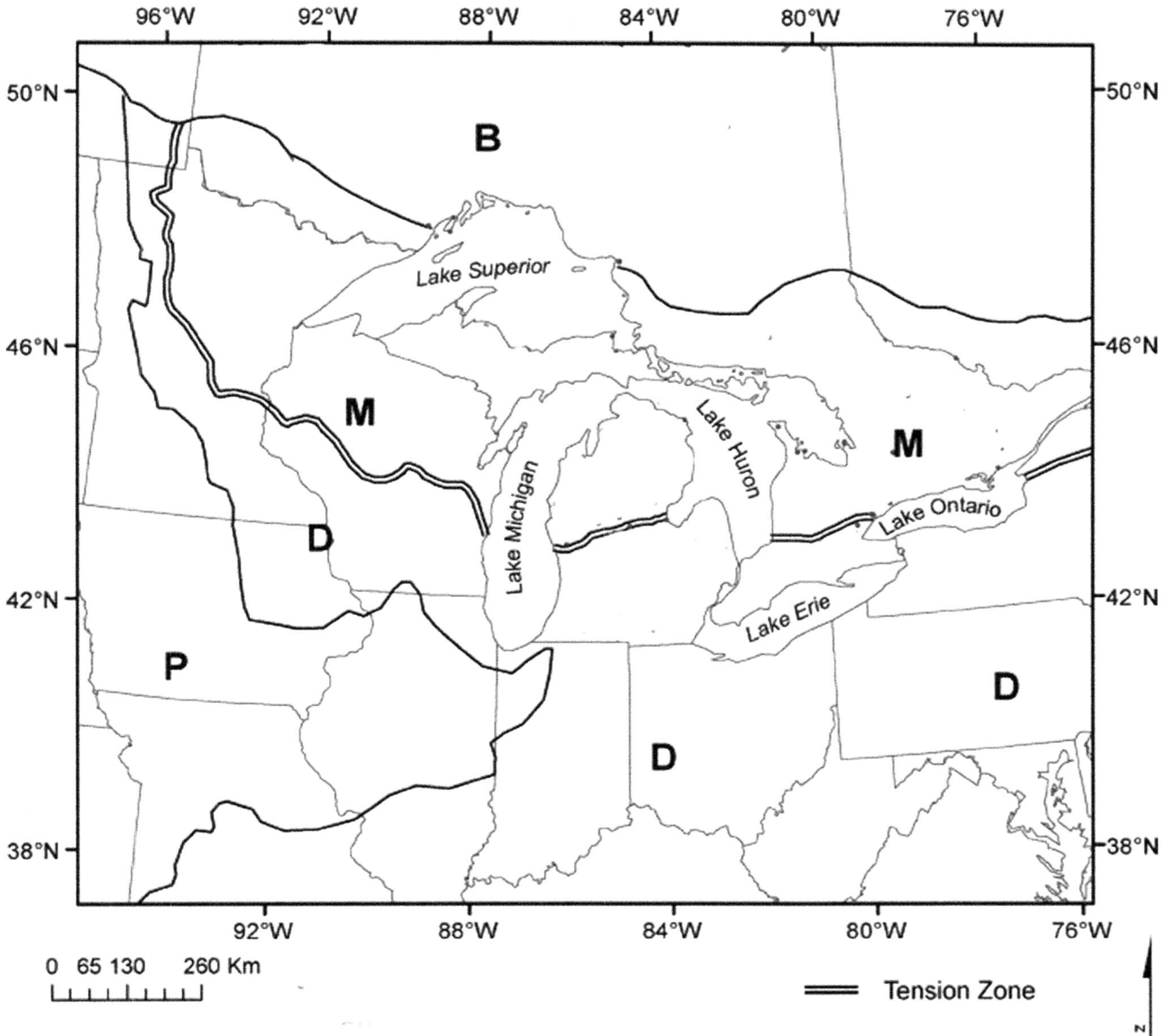

Fig. 2.6 Vegetation tension zone across the Great Lakes region: B = boreal forest; M = mixed broad-leaved and coniferous forest; D = broad-leaved forest; P = prairie (Hupy 2013)

of Lake Erie. Areas to the south of the hinge line on Lake Erie have experienced permanent depression of up to 65 ft. (20 m). Glacio-isostatic adjustment continues today, necessitating adjustment of vertical datums of the Great Lakes every 25–35 years as part of the requirements permitting hydropower water diversions, navigation, and other uses.

During and following the Laurentide ice fluctuations, glacial and postglacial lakes formed in the Great Lakes basin. Figure 2.15 shows shorelines of prominent postglacial lakes in the Great Lakes watershed and their spillways and outlets. Erosional evidence of former lake levels include headlands, spillways, inlet lakes, sea caves, and sea stacks. Headlands are promontories with a sheer drop along a coastline caused by intense wave action. Spillways represent former river valleys where water passed between lakes. Inlet (bay-mouth) lakes are formed by stream incision during low lake stands followed by drowning of these old river valleys during high stands. Sea caves (cavettos) are formed by wave action in a lake or the ocean; and sea stacks are columns of resistant bedrock left from intense wave action along a shoreline.

Evidence of catastrophic erosion from outbursts of glacial Lake Agassiz include boulder fields, closed basins, dry cataracts, plunge pools, and giant potholes. Boulder fields

Fig. 2.7 Coastal wetlands occupy about 25% of the Great Lakes Coastal Zone (US Environmental Protection Agency)

are areas composed of rounded and sub-rounded boulders from intense water erosion. Closed (endorheic) basins were hollowed out by rapidly flowing water and have no apparent outlet. Dry cataracts are former waterfalls; plunge pools are basins carved into bedrock at the bottom of these waterfalls. Potholes are deep circular holes in bedrock formed by water erosion by the rotation of stones in an eddy.

Depositional evidence of former shorelines include beach ridges, raised beaches, sand spits, tombolos, drowned forests, and varved lakes. Beach ridges are ridges running parallel to a shoreline that are formed by wave action. Raised beaches are former shorelines that occur near or above the modern lake or ocean level due to isostatic rebound. Sand spits are narrow point of sandy land projecting into the sea or a lake. Tombolos are bars joining an island to a mainland. Drowned forests result from changes in lake level due to flooding. Varves (rhythmites) are pairs of annual layers of silt and clay of contrasting color that represent the deposit of a single year (summer and winter) in a lake.

One of the problems in interpreting past shorelines of the Great Lakes region is the manner in which the data were recorded. Some investigators record elevations of wave-cut bluffs, which result from storm-wave maximum heights; others record wave-cut benches, and still others report the surfaces or base or base of wave-built terraces. The distance from the top of wave-cut bluff and the base of the wave-built terrace may be 100 ft. (30 m) or more. Schaetzl et al. (2002) recommend that former shorelines be measured using a differential global positioning system (DGPS) at the surface of the boulder lag at the base of the wave-cut bluff.

The timing of the changes in former levels of the Great Lakes is based on hundreds of dates collected from former lake terraces. From these dates, a chronology of the Great Lakes basin can be established (Table 2.1). From the work of Hough (1958), modified by Prest (1970), and depicted by Clark et al. (2012), the oldest lakes identified in the Great Lakes basin existed 13.8 kyr BP and include Glacial Lake Oshkosh in front of the Green Bay lobe, Glacial Lake

Fig. 2.8 **a** Lowland mixed coniferous and broad-leaved forest near Ford River, MI on the northwest shore of Lake Michigan. The dominant species are cedar, spruce, birch, and aspen (photo by J. Bockheim). **b** Lowland conifer forest near Terrace Bay, ON, on the north shore of Lake Superior (photo by J. Bockheim). **c** Open (non-forested) wetland on the north shore of Lake Superior (photo by J. Bockheim)

a

b

Fig. 2.8 (continued)

c

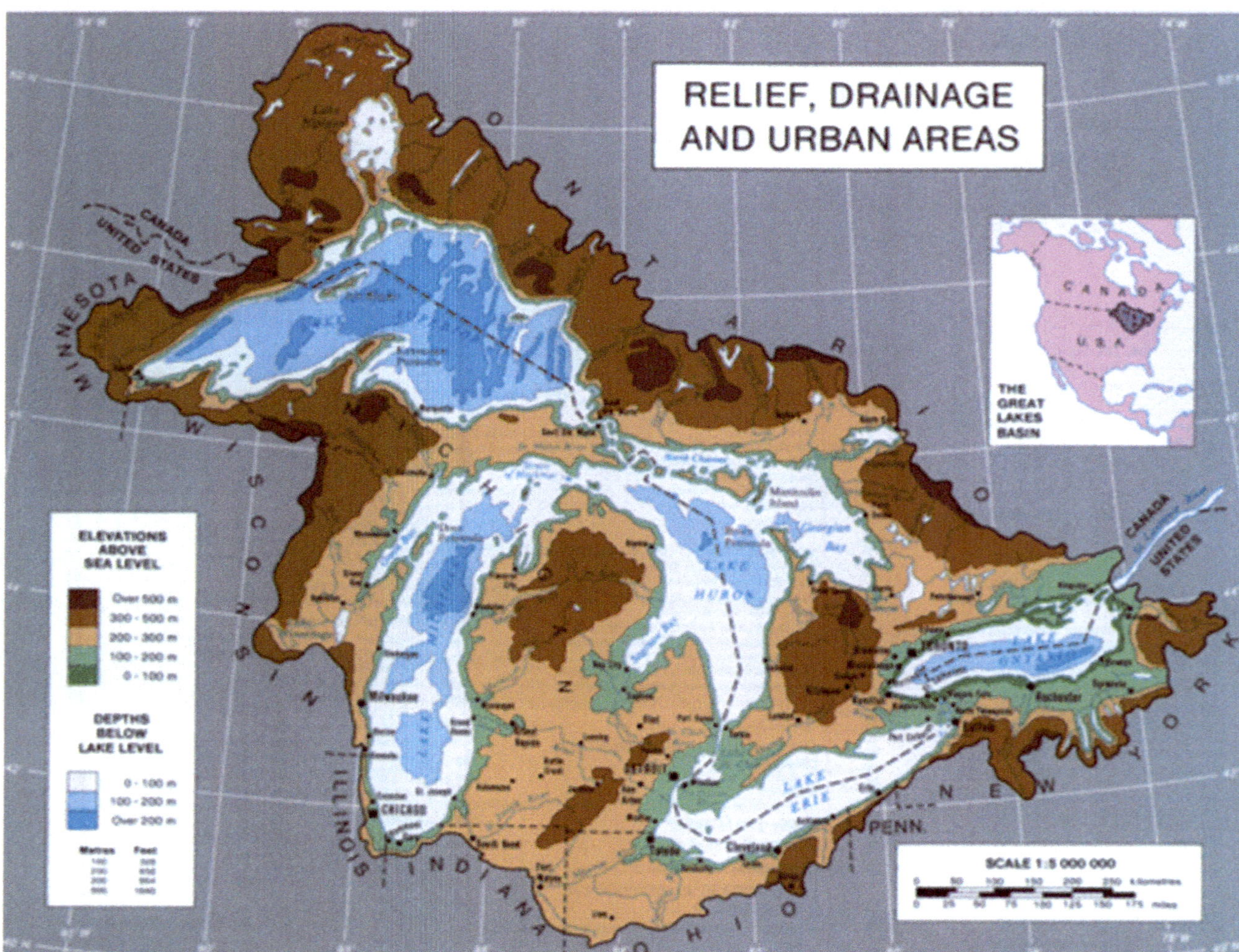

Fig. 2.9 Relief and drainage in the Great Lakes region (*Source* Great Lakes Coalition; https://www.greatlakescoalition.org/water-levels)

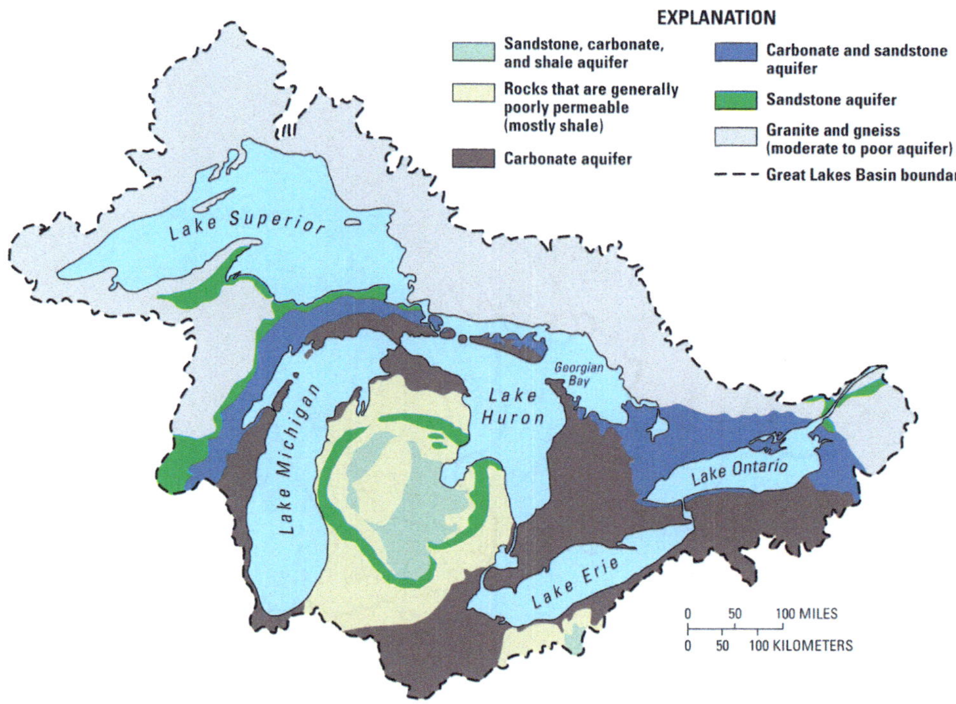

Fig. 2.10 Bedrock geology map of the Great Lakes region (*Source* Bornhorst 2016)

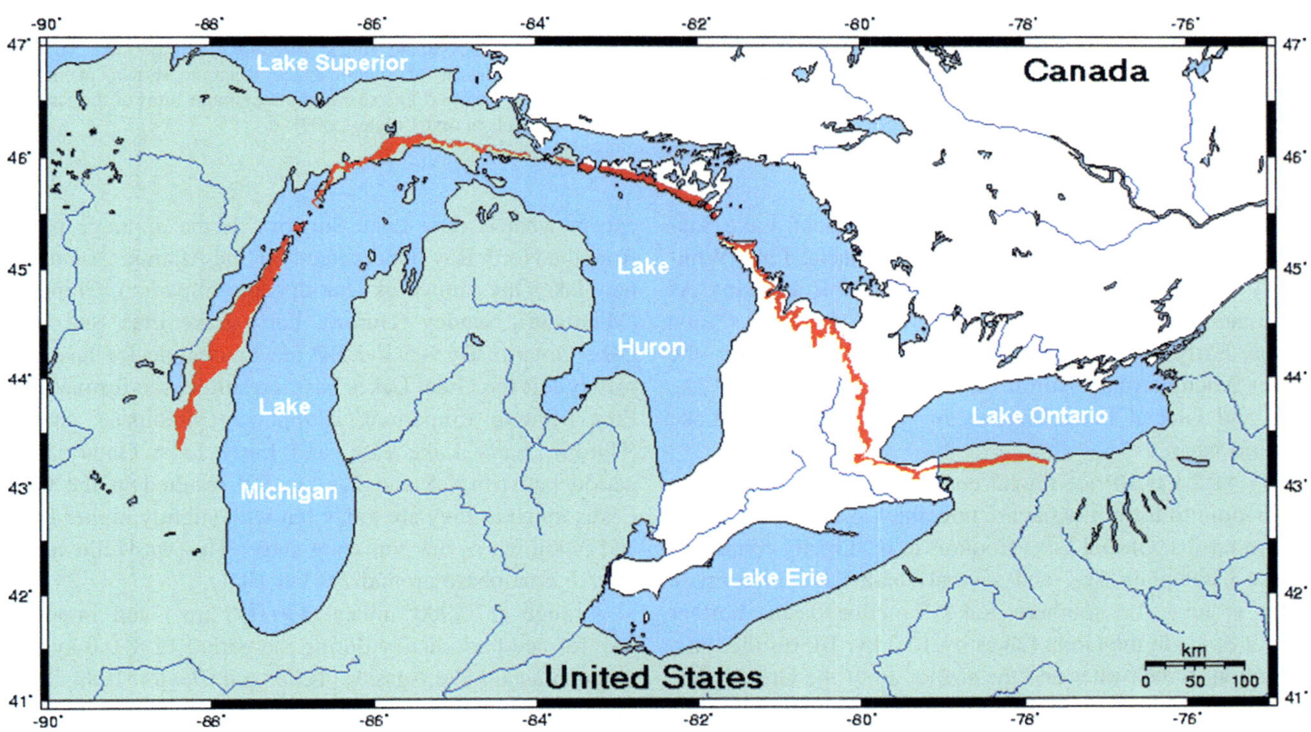

Fig. 2.11 The Niagara Escarpment is shown in red (*Source* http://www.aquarius.geomar.de/omc/make_map.html)

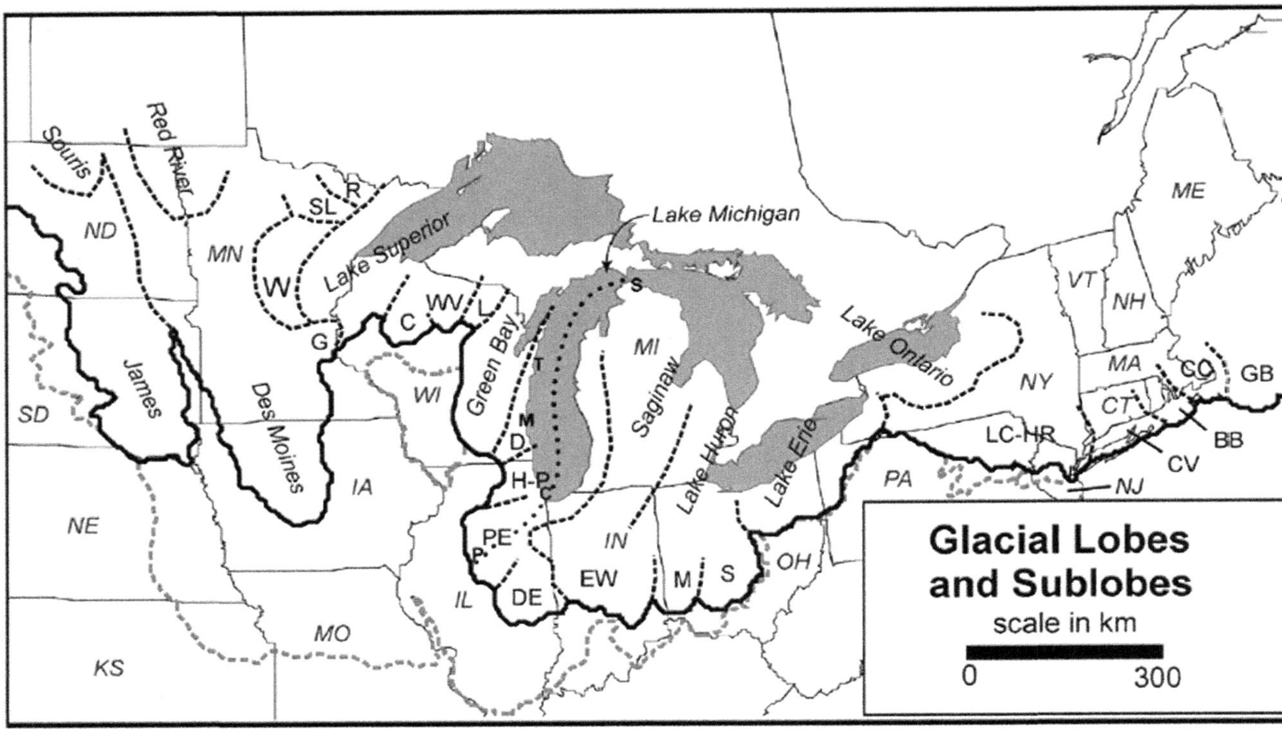

Fig. 2.12 Glacial lobes and sublobes of the southern Laurentide Ice Sheet in north central USA during the late Wisconsinan Glaciation: G = Grantsburg; W = Wadena; SL = St. Louis; R = Rainey; C = Chippewa; WV = Wisconsin Valley; L = Langlade; D = Delevan; H-P = Harvard-Princeton; PE = Peoria; DE = Decatur; EW = East White; M = Miami; S = Scioto; LC = Lake Champlain; HR = Hudson River; CV = Connecticut Valley; BB = Buzzards Bay; CC = Cape Cod; GB = Georges Bank. Dotted line along axis of Lake Michigan lobe shows location of time-distance diagram shown in original reference. Light dashed line shows the maximum limit of the ice sheet (*Source* Mickelson and Colgan 2003)

Chicago in what is now Lake Michigan, Glacial Lake Saginaw in front of the Saginaw lobe, and Glacial Lake Whittlesey in what are now the Lakes Huron and Erie basins. As the Laurentide ice retreated during the Two Creeks Low Phase, Early Lake Algonquin formed in what are now the Lakes Michigan and Huron basins, along with Early Lake Erie and Glacial Lake Iroquois in what is now the Lake Ontario basin.

By 11.2 kyr BP ice retreat enabled the huge Main Lake Algonquin to form in what are now the Lakes Michigan and Huron basins, Glacial Lake Iroquois to be directly connected to the Champlain Sea, and several small lakes to form in what is now the southern Lake Superior basin. Further retreat of ice in the Great Lakes by 10.2 kyr BP resulted in a reduction in meltwater and the size of all of the Great Lakes except what is now Lake Superior as the drainage passed from the North Bay outlet directly in the Atlantic Ocean. The new lakes are known as Duluth (now Superior), Chippewa (Michigan), Stanley (Huron), Early Lake Erie, and Early Lake Ontario. By 8.5 kyr BP the Laurentide ice has completely left the Great Lakes basin, resulting the formation of Post Minong (Superior), Chippewa (Michigan), Stanley (Huron), Early Lake Erie, and Early Lake Ontario. The period between 7.5 and 5.3 kyr BP resulted in the Great Lakes much as they are today but with slightly higher levels and is known as the Nipissing stage. This was followed by the Algoma phase around 2.9 kyr BP.

A huge (170,000^2 miles; 440,000 km^2) and important lake formed periodically during the period 12.8–9.0 kyr BP, called Glacial Lake Agassiz (Teller and Clayton 1983; Teller

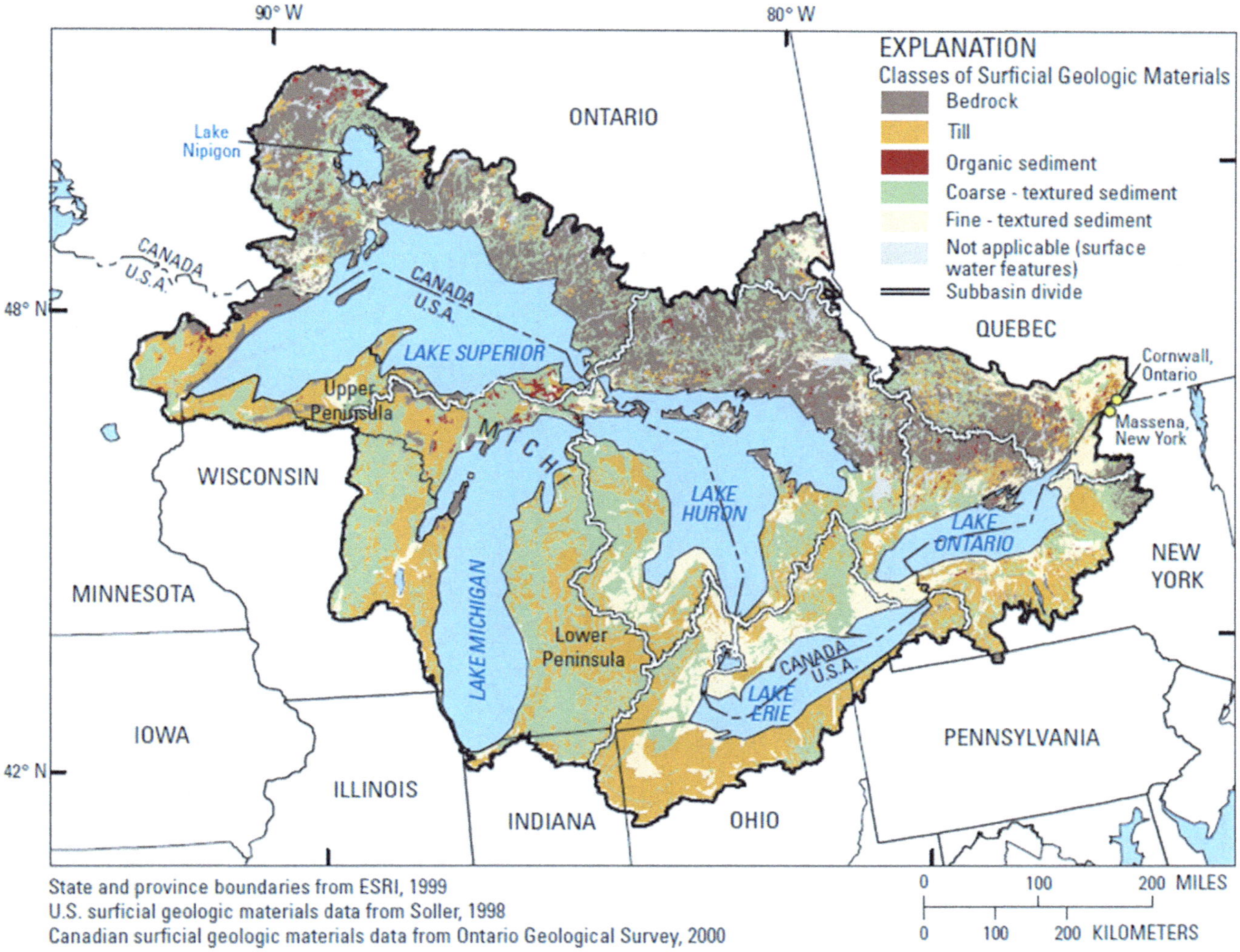

Fig. 2.13 Sources of surficial geologic materials in the Great Lakes region (*Source* Neff et al. 2005; USGS; http://www.aquarius.geomar.de/omc/make_map.html)

et al. 2002). This lake drained in various directions but often drained catastrophically into the Lake Superior basin and through the Great Lakes and St. Lawrence River into the Atlantic Ocean by what are called "outbursts".

The glacial and post-glacial lake features described herein occur within 5 km of the current lake shores. However, there is geological evidence that shoreline features occur up to 100 miles (165 km) from the modern shorelines. In areas featuring negative isostatic rebound, particularly along the south shore of Lake Erie, paleo-shoreline features may be buried beneath the modern lakes.

2.7 Land Use and Population

Although forest is still prevalent in the Superior and Huron basins, clearing for agricultural and urbanization has reduced the forest cover to 50% or less in the Ontario, Michigan, and Erie basins. Agriculture accounts for 58% of the Erie shore, 36% of the Michigan shore, and 28% of the Ontario shore (Fig. 2.16). Twenty percent of the shorelines of Lakes Michigan and Ontario are urbanized. The total population of the Great Lakes shoreline is over 35 million, with the

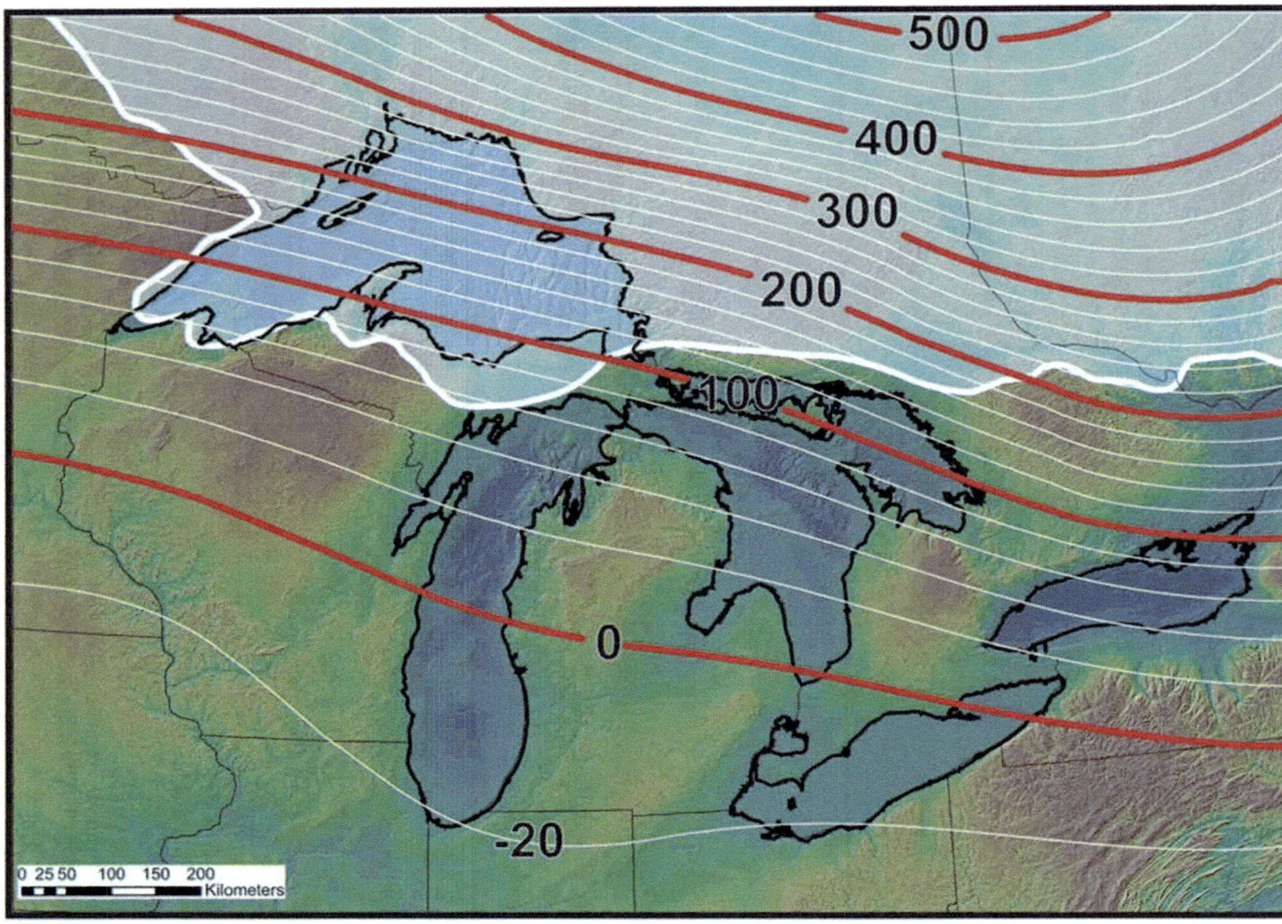

Fig. 2.14 Isostatic rebound (in meters) of the land following recession of the Laurentide ice sheet in the Great Lakes region (Clark and Befus 2009)

greatest numbers along the Erie (12 million), Michigan (10 million), and Ontario (8 million) shores.

2.8 Conclusions

Because the Great Lakes watershed occupies a large area (522,000 km^2), there is considerable variation in the soil forming factors, including climate, vegetation, relief, parent materials, and the duration of soil formation. The Great Lakes exert a modifying effect on summer and winter air temperatures and on snowfall amounts within 50 km of the shoreline. Four broad vegetation types in the Great Lakes Coastal Zone include boreal forest, temperate mixed forest, temperate broad-leaved forest, and native prairie. A floristic tension zone cuts across central portions of Lake Michigan, Huron, and Ontario. Wetlands occupy about 25% of the shoreline. The northern portion of the Great Lakes shoreline is underlain by Precambrian granitic and metamorphic rocks and the southern portion by carbonate rocks. The Niagara Escarpment follows the southern Lake Ontario shoreline, bisects Georgian Bay from the main body of Lake Huron, and connects the Garden Peninsula and Door Peninsula in Lake Michigan. The entire Great Lakes basin was glaciated during the late Wisconsinan. The Great Lakes basin features exposed bedrock, sandy outwash and lacustrine sediments, and clayey till in roughly equal

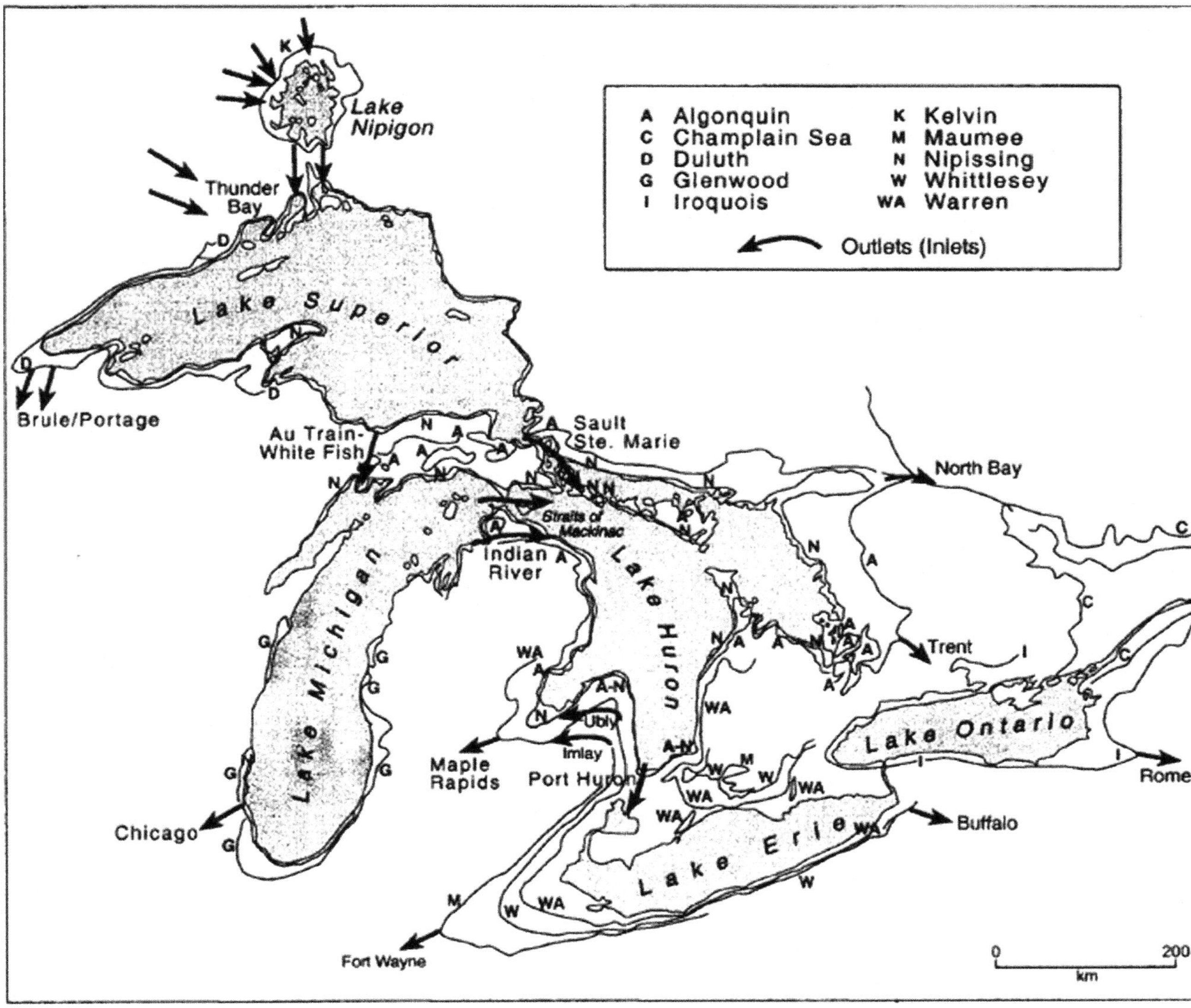

Fig. 2.15 Locations of shorelines of prominent proglacial lakes in the Great Lakes basin and their spillways and outlets (*Source* Larson and Schaetzl 2001; after Karrow 1984)

areas. Glacioisostatic recovery ranges from −20 to 300 m in the Great Lakes basin. Shorelines of prominent proglacial lakes include the Algonquin, Champlain Sea, Duluth, Glenwood, Iroquois, Maumee, Nipissing, Whittlesey, and Warren lake stages. Urbanization has left a strong imprint on the southeastern Lake Michigan shore, Lake St. Claire, the western and south-central Lake Erie shore, and the northwestern Lake Ontario shore.

Table 2.1 Glacial and post-glacial lakes of the Great Lakes basin

Dates (kyr BP)	Superior	Michigan	Huron	Erie	Ontario
13.8	[ice]	Chicago (Glenwood)	Maumee	Maumee	[ice]
			Arkona	Arkona	
			Whittlesey	Whittlesey	
			Warren	Wayne	
		Chicago (Calumet)		Warren	
				Grassmere	
			Lundy	Lundy	
12.4	[ice]	Early Lake Algonquin (Toleston)	Early Lake Algonquin	Early Lake Erie	Iroquois (Earliest)
11.2	Duluth	Main Lake Algonquin	Main Lake Algonquin	Early Lake Erie	Iroquois (Main)
10.2	Duluth	Chippewa	Stanley	Early Lake Erie	Early Lake Ontario
	[Sub-Duluth	[Wyebridge	[Wyebridge		[Frontenac
	Highbridge	Battlefield	Penetang		Sydney
	Moquah	Fort Brady]	Cedar Point		Belleville
	Washburn		Payette		Trenton]
	Manitou		Sheguiandah		
	Beaver Bay		Korah]		
	Minong				
	Dorion				
	Houghton]				
8.5	Post-Minong	Chippewa	Stanley		Early Lake Ontario
5.3	Nipissing	Nipissing	Nipissing		
2.9	Algoma Sault	Algoma	Algoma		

Sources Hough 1958; Calkin and Feenstra 1985; Cowan 1985; Farrand and Drexler 1985; Larsen 1987; Clark et al. 2012. Bracketed lake stages are ordered chronologically but do not necessarily correlate across columns

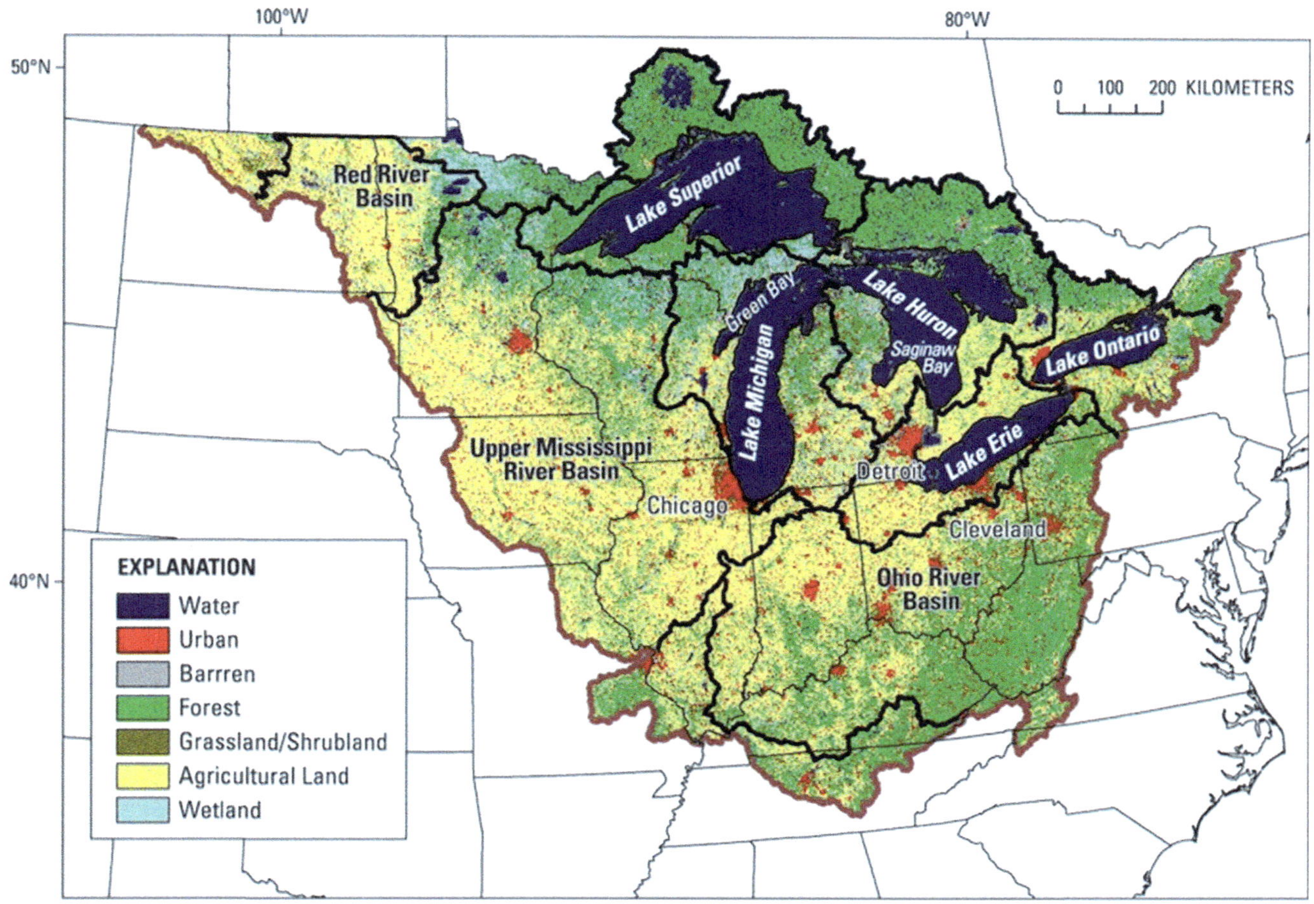

Fig. 2.16 Land use in the Great Lakes and adjoining basins (Robertson and Saad 2011; open access)

References

Albert, D.A. 2003. *Between Land the Lake: Michigan's Great Lakes Coastal Wetlands*. Mich. Nat. Features Inventory, Mich. State Univ. Ext., Ext. Bull. E-2902. 96.

Bornhorst, T.J. 2016. *An overview of the geology of the Great Lakes basin*. A.E. Seaman Mineral. Museum, Web Publ. 1, p. 8. (http://www.museum.mtu.edu/sites/default/AESMM_Web_Pub_1_Great_Lakes_Geology/_0.pd.

Calkin, P.E. and B.H. Feenstra. 1985. Evolution of the Erie-basin Great Lakes. In *Quaternary evolution of the great Lakes*, eds. P.F. Karrow and P.E. Calkin, 149–170. Canada: Geological Association of Canada (Spec. Pap. 30).

Clark, J.A., and Befus, K.M. 2009. *Shifting shorelines: modeling the 20,000-year history of the Great Lakes*. ArcUser (summer), 24–26.

Clark, J.A., K.M. Befus, and G.R. Sharman. 2012. A model of surface water hydrology of the Great Lakes, North America during the past 16,000 years. *Physics and Chemistry of the Earth* 53–54: 61–71.

Cowan, W.R. 1985. Deglacial Great Lakes shorelines at Sault Ste. Marie, Ontario. In *Quaternary evolution of the great Lakes*, eds. P. F. Karrow and P.E. Calkin, 33–37. Canada: Geological Association of Canada (Spec. Pap. 30).

Farrand, W.R., C.W. Drexler and C.W. 1985. Late Wisconsinan and Holocene history of the Lake Superior basin. In *Quaternary evolution of the Great Lakes*, eds. P.F. Karrow and P.E. Calkin, 17–32. Canada: Geological Association of Canada (Spec. Pap. 30).

Hinkel, K.M., and F.E. Nelson. 2012. Spatial and temporal aspects of the lake effect on the southern shore of Lake Superior. *Theoretical and Applied Climatology* 109: 415–428.

Hough, J.L. 1958. *Geology of the Great Lakes*. Univ. of Illinois Press. 313 pp.

Hupy, C.M. 2013. Mapping ecotone movements: Holocene dynamics of the forest tension zone in central Lower Michigan, USA. *Physical Geography* 33 (5): 473–490.

Karrow, P.F. 1984. Quaternary stratigraphy and history, Great Lakes-St. Lawrence region. *QuaternaryStratigraphy of Canada—A Canadian Contribution to IGCP Project* 24, Geological Survey Canada Paper84–10, pp. 137–153.

Larsen, C.E. 1987. Geologic history of Glacial Lake Algonquin and the upper Great Lakes. *U.S. Geological Survey Bulletin* 1801: 36.

Larson, G., and Schaetzl, R. 2001. Origin and evolution of the Great Lakes. *Journal of Great LakesResearch* 27: 518–546.

Mickelson, D.M., and Colgan, P.M. 2003. The southern Laurentide ice sheet. *Developments in Quaternary Sciences* 1:1–7

Neff, B.P., Piggott, A.R., and Sheets, R.A. 2005. Estimation of shallow ground-water recharge in the Greatlakes basin. US Geological Survey Scientific Investigations Report 2005–5284.

Prest, V.K. 1970. Quaternary Geology of Canada. In *Geology and economic minerals of Canada*, ed Douglas, R.J.W. Geol. Surv. Canada. Econ. Geol. 676–764.

Robertson, D.M., and Saad, D.A. 2011. Nutrient inputs to the Laurentian Great Lakes by source and watershed estimated using

Sparrow watershed models. *Journal of the American Water Resources Association* 47:1011–1033.

Schaetzl, R.J., S.A. Drzyzga, B.N. Weisenborn, K.A. Kincare, X.C. Lepczyk, K. Shein, C.M. Dowd, and J. Linker. 2002. Measurement, correlation, and mapping of glacial Lake Algonquin shorelines in northern Michigan. *Annals of the Association of American Geographers* 92: 399–415.

Teller, J.T., and Clayton, L. (eds.). 1983. Glacial Lake Agassiz. Geol. Assoc. Can. Spec. Pap. 26. p. 451.

Teller, J.T., D.W. Leverington, and J.D. Mann. 2002. Freshwater outbursts to the oceans from glacial Lake Agassiz and their role in climate change during the last glaciation. *Quaternary Science Reviews* 21: 879–887.

3 Soil Taxonomic Systems Used in the Great Lakes Coastal Zone

3.1 Soil Horizons

Genetic soil horizons are important role in describing and classifying soils. A soil horizon is a layer that lies more or less parallel to the land surface that results from the interplay of the soil-forming factors described in the previous chapter. Table 3.1 lists the master horizons and suffix symbols that are used for soils in the Great Lakes region in the US and Canadian soil classification systems. These symbols will be referred to in the text.

3.2 Approaches to Soil Classification

During the first half of the twentieth century, soil classification systems were based on poorly understood soil-forming processes, including soil classification systems used in the USA from 1927 until the late 1950s (Marbut 1927; Baldwin et al. 1938). However, starting with the "Seventh Approximation" (Soil Survey Staff 1960) and culminating with *Soil Taxonomy* (ST) (Soil Survey Staff 1975, 1999), soils in the USA and in countries adopting ST were classified with quantitative properties, particularly morphological properties, delineated as diagnostic epipedons and horizons. Soil-forming processes were de-emphasized and kept in the background. A similar approach was used by the FAO-UNESCO (1974) and in the World Reference Base (WRB) for Soil Resources (FAO 1998).

The movement away from an emphasis on soil processes was predicated on the assumption that soil properties are more readily quantifiable than soil processes and that soil processes occur simultaneously in a given soil, reinforcing or contradicting one another (Simonson 1959). It was also assumed that polygenesis likely has occurred in most, if not all soils, making genetic interpretations difficult. As soil-forming factors change, soil-forming processes change, resulting in a change in soil taxa. An additional criticism of soil classification systems based on soil processes is that they often contain insufficient taxa to satisfactorily delineate global soils.

One of the creators of ST, Smith (1983, p. 43) emphasized: "The genesis per se, cannot be used to define soil taxa and meet this objective. The processes that go on can rarely be observed or measured. Nevertheless, the genesis of soils is extremely important both to the taxonomy of soils and to the mapping in the field. Genesis is important to the classification partly because it produces the observable or measureable differences that can be used as differentiae. Genesis does not appear in the definitions of the taxa but lies behind them."

3.3 Soil Taxonomy

3.3.1 Hierarchical Levels

Soil Taxonomy (the full title is *Soil Taxonomy: a Basic System of Soil Classification for Making and Interpreting Soil Surveys*) was initially published in 1975 and revised in 1999. The system is hierarchical and includes six levels from broadest to narrowest: order, suborder, great group, subgroup, family, and series (Table 3.2). The system is based on diagnostic surface (epipedons) and subsurface horizons.

3.3.2 Diagnostic Horizons

Eight epipedons are defined on the basis of color, organic carbon concentration, thickness, base saturation, presence of andic properties, and evidence for human disturbance. In the Great Lakes region, five of the epipedons are present, including the ochric (thin or light colored and low organic C), mollic (mineral enriched with organic C $\geq$ 18 cm thick, dark-colored, abundant organic C and bases), histic (organic $\geq$ 30 cm thick), umbric (same as mollic except with low base saturation), and folistic (organic horizon $\geq$ 15 cm thick that is saturated for less than 30 days). The ochric

J. G. Bockheim, *Soils of the Laurentian Great Lakes, USA and Canada*,
https://doi.org/10.1007/978-3-030-52425-8_3

Table 3.1 Genetic soil horizons used in the US and Canadian soil classification systems for the Great Lakes region

US	Definition	Canadian equivalent
Master horizon		
O	Organic layer ≥ 17% organic C	O (also L, F, H)
A	Mineral horizon near surface that is darkened by organic matter accumulation	A
E	Mineral horizon subject to evaluation of clay, Fe, Al, OM, etc.	e
B	Mineral horizon subject to enrichment of OM, Fe, Al, clay, etc.	B
C	Mineral horizon relatively unaffected by pedogenesis	C
R	Consolidated bedrock	R
Suffix symbols		
a	Highly decomposed	h
b	Buried soil horizon	b
c	Concretions or nodules	cc
d	Densic layer	–
e	Organic material of intermediate decomposition	m
g	Strong gleying	g
h	Illuvial organic matter	h
i	Slightly decomposed organic matter	f
k	Accumulation of secondary carbonates	ca, k
m	Pedogenic cementation	c
p	Layer mixed by cultivation, logging, etc.	p
s	Illuvial Al, Fe, with organic C	f
ss	Slickensides (shiny surfaces from shrinking and swelling	ss
t	Translocation of silicate clay	t
w	Development of color or structure	m
x	Fragipan character	x
–	Juvenile development of a particular horizon	j
–	Horizon disrupted by physical or faunal processes	u
–	Vertical cracks	v

epipedon is the most frequently occurring diagnostic surface horizon (83%), followed by the mollic and histic (8% each), and umbric (1%) (Table 3.3). The histic epipedon is the thickest surface horizon at 82 cm, followed by the umbric (29 cm), mollic (28 cm), and ochric (20 cm).

Diagnostic subsurface horizons are defined primarily on the basis of the accumulation of specific weathering products, such as silicate clays, Ca, Si, Fe and Al oxyhydroxides; cementation; bleaching; mixing; and human disturbances. Of the 20 diagnostic subsurface horizons, eight are present in Great Lakes soils, including from most to least, the argillic horizon (accumulation of clay), albic horizon (bleaching of sand and silt grains), cambic horizon (minimal development of color and/or structure), spodic horizon (accumulation of Fe and Al oxyhydroxides complexed with organic C), and glossic horizons (tonguing in E and Bt in degrading argillic horizons) (Table 3.3). Less common are soils with fragipans (reversibly cemented based on soil moisture content), and ortstein (cemented by Fe oxyhydroxides). The fragipan tends to be the thickest diagnostic subsurface horizon at 60 cm, followed by the argillic (51 cm), cambic (47 cm), glossic (33 cm), ortstein (26 cm), and albic (24 cm). Soils of the Great Lakes contain one epipedon and 1.4 ± 1.2 subsurface horizons. Some soil series contain as many as five diagnostic subsurface horizons, including the Frohling, Gogebic, Skanee, Tokiahok, and Yalmer.

3.3.3 Constructing a Soil Name

Soil series bear a name that often recognizes its type location, e.g., Gogebic from Lake Gogebic in the Upper Peninsula of Michigan. Taxonomic names are constructed from abbreviations taken generally from the first vowel and

Table 3.2 A comparison of hierarchical soil classes and diagnostic horizons in Soil Taxonomy (2014) and The Canadian Soil Classification system (1998)

Soil Taxonomy (2014)	The Canadian System of Soil Classification (1998)
Soil classes	
Order	Order
Suborder	Great group
Great group	Subgroup
Subgroup	
Family	[Family]
Series	Series
Diagnostic horizons[a]	
Folistic	[Folisol]
Histic	Organic (Fibric, Mesic, Humic)
Mollic	Chernozemic
Ochric	–
Umbric	–
Albic	[Ae, Ahe horizons]
Argillic	[Bt horizon]
Cambic	[Bm horizon]
Fragipan	Fragipan
Glossic	–
Ortstein	Ortstein
Spodic	Podzolic B

[a]Includes only those reported in the GLCZ

Table 3.3 Thickness and occurrence of diagnostic horizons in 330 of the most common soil series in the Great Lakes Coastal Zone

Horizon	Number	Frequency (%)	Thickness (cm)	St. Dev
Surface				
Ochric	273	83	20	11
Mollic	27	8	28	15
Histic	27	8	82	58
Umbric	3	1	29	8
Subsurface				
Albic	103	31	14	11
Argillic	129	39	51	32
Cambic	87	26	47	26
Spodic	84	25	32	18
Glossic	53	16	33	34
Fragipan	17	5	60	30
Ortstein	6	2	26	17

the following consonants of order names. For the six soil orders found in the Great Lakes region, **alf** is used for Alfisols, the base-enriched soils with an argillic horizon; **ent** is used for the Entisols, which lack a diagnostic subsurface horizon; **ept** is used for Inceptisols, which feature minimal soil development and often contain a cambic horizon; **od** is used for Spodosols, the acid forest soils enriched in Al and Fe complexed with humus; **oll** is used for Mollisols, which contain a thick, dark-colored, base-rich surface horizon; and **ist** is used for Histosols, the organic soils.

The formative elements listed in the upper part of Table 3.4 are used to construct suborder names. An Aqualf is a wet Alfisol; a Psamment is a sandy Entisol, and a Hemist is a Histosol containing organic materials of intermediate decomposition. The formative elements listed in the middle of Table 3.4 are used to build upon the suborder names and

Table 3.4 Formative elements for classifying soils of the Great Lakes at the suborder, great-group, and subgroup levels

Formative element	Meaning	Alfisols	Entisols	Inceptisols	Spodosols	Mollisols	Histosols
Suborder							
Aqu-	Wet	X	X	X	X	X	
Fluv-	Young river sediments		X				
Fol-	Organic-rich (dry)						X
Hem-	Organic-rich (middle decomp.)						X
Orth-	"Normal"		X		X		
Psamm-	Sandy		X				
Ren-	From limestone					X	
Sapr-	Organic-rich (highly decomp.)						X
Ud-	Sufficient soil moisture	X		X		X	
Great group							
Argi-	Translocated clay					X	
Dur-	Cemented by Si				X		
Dystr-	Acidic			X			
Endo-	Wet from below	X	X	X	X	X	
Epi-	Wet from above	X	X	X	X	X	
Eutr-	Basic			X			
Fluv-	Young river sediments		X				
Fragi-	Fragipan	X		X	X		
Gloss-	Glossic horizon	X					
Hapl-, Hap-	"Normal"	X			X	X	X
Hum-	Organic-rich in mineral horizon			X			
Psamm-	Sandy		X				
Quartzi-	Quartz-rich sand		X				
Ud-	Sufficient soil moisture		X				X
Subgroup							
Aeric	More aeration than Typic subgroup	X	X	X			
Alfic	Argillic horizon				X		
Alfic Oxyaquic	Alfic and water saturated but not reduced				X		
Aqualfic	Alfic and Aquic				X		
Aquic	Wetter than Typic subgroup	X	X	X		X	
Aquollic	Aquic and Mollic	X					
Aquultic	Aquic and Ultic	X					
Arenic	50–100 cm sandy textured surface	X					
Arenic Oxyaquic	Arenic and Oxyaquic	X					
Argic	Argillic horizon				X		
Cumulic Vertic	Thickened epipedon and cracks					X	
Dystric	Lower base saturation percentage			X			
Entic	Weakly developed				X	X	
Fluvaquentic	River sediments and poorly drained			X		X	X
Fluventic	River sediments			X			
Fragiaquic	Fragipan and Aquic	X					
Fragic	Fragipan				X		

(continued)

Table 3.4 (continued)

Formative element	Meaning	Alfisols	Entisols	Inceptisols	Spodosols	Mollisols	Histosols
Glossaquic	Glossic horizon and Aquic	X					
Glossic	Glossic horizon	X					
Haplic	Minimum horizon development	X					
Hemic	Intermediate OM decomposition						X
Histic	Histic epipedon			X	X		
Inceptic	Weakly developed	X				X	
Lamellic	Lamellae	X	X	X	X		
Limnic	Limnic layer						X
Mollic	Mollic-like horizon	X	X	X			
Mollic Oxyaquic	Mollic and water saturated but not reduced	X					
Oxyaquic	Water saturated but not reduced	X	X	X	X	X	
Oxyaquic Vertic	Oxyaquic and vertic	X					
Psammentic	Sandy	X					
Spodic	Spodic-like horizon		X				
Terric	Mineral substratum within 1 m						X
Typic	"Normal"	X	X	X	X	X	X
Udollic	Intergrade With Udoll	X					
Ultic	Ultisol-like				X		
Vertic	Cracking	X		X		X	

construct great-group names. An Endoaqualf features a high water table; a Quartzipsamment is derived from quartz-rich sand; and a Haplohemist is a "normal" Hemist. The formative elements in the last section of Table 3.4 are used to construct subgroups, such as Aeric Endoaqualfs, Spodic Quartzipsamments, and Terric Haplohemists.

3.4 The Canadian System of Soil Classification

The Canadian system of soil classification (Soil Classification Working Group 1998) also is hierarchical and has four levels, including order, great group, subgroups, and series (Table 3.1). The Canadian system, which applies only to Canada, has 10 orders. Six of these orders occur in the Great Lakes region, including Brunisols (weakly developed B horizon 5–10 cm thick), Gleysols (saturated with water), Luvisols (accumulation of translocated clay), Organic soils ($\geq$17% organic C; $\geq$40 cm thick), Podzols (accumulation of Fe and Al oxhydroxides often with organic C), and Regosols (minimally developed soils).

The Canadian soil classification system has a limited number of diagnostic horizons, and only four (Chernozemic, Fragipan, Ortstein, and Podzolic B) occur in the GLCZ (Table 3.1). Genetic horizons are used in place diagnostic horizons to distinguish among some soil taxa, e.g., Ae, Ahe, Bt, and Bm.

3.5 Linking the US and Canadian Soil Taxonomic Systems

Both the US and Canadian approaches to soil classification are natural, i.e., are based on measurable soil properties. Whereas the US system is global and attempts to classify soils throughout the world, the Canadian soil classification system is intended to classify only the soils of Canada. To provide a comprehensive overview of the soils of the Great Lakes Coastal Zone (GLCZ), I attempted to merge the two systems.

Difficulties in comparing the two soil classification systems centered primarily on two Canadian orders—the Gleysols and the Brunisols (Table 3.5). Gleysols include

Table 3.5 Comparison of soil taxa on the Canadian and US Great Lakes shore

Canadian soil subgroup[a]	No. soil series	US Great Group equivalent[b]
Brunisolic Gray Brown Luvisol	41	Hapludalfs-Glossudalfs
Gleyed Gray Brown Luvisol	20	Endoaqualfs, Epiaqualfs
Gleyed Brunisolic Gray Brown Luvisol	19	Endoaqualfs, Epiaqualfs
Orthic Gray Brown Luvisol	8	Hapludalfs
Orthic Gray Luvisol	7	Hapludalfs
Gleyed Gray Luvisol	5	Endoaqualfs, Epiaqualfs
Podzolic Gray Luvisol	2	Glossudalfs
Podzolic Gray Brown Luvisol	1	Glossudalfs
Gleyed Podzolic Gray Brown Luvisol	1	Glossaqualfs
Orthic Melanic Brunisol	20	Eutrudepts, Hapludolls
Gleyed Melanic Brunisol	5	Endoaquepts, Epiaquepts
Gleyed Eluviated Melanic Brunisol	4	Endo-, Epiaquepts; Endo-, Epiaquolls
Eluviated Melanic Brunisol	3	Eutrudepts, Hapludolls
Orthic Eutric Brunisol	1	Eutrudepts, Hapludolls
Orthic Sombric Brunisol	1	Dystrudepts
Orthic Humo-Ferric Podzol	6	Haplorthods
Gleyed Humo-Ferric Podzol	4	Endoquods, Epiaquods
Orthic Regosol	1	Udipsamments
Fibric Mesisol	1	Haplohemists
Orthic Humic Gleysol	46	Endoaquepts, Epiaquepts, Endoaquolls, Epiaquolls
Regosolic Humic Gleysol	4	Endo-, Epiaquepts; Endo-, Epiaquolls
Humic Luvic Gleysol	3	Endoaquepts, Epiaquepts
Rego Gleysol	2	Endoaquepts, Epiaquepts

[a]Soil Classification Working Group (1998)
[b]Soil Survey Staff (2014)

Aquolls, Aquepts, and Aquents saturated by a water table from below (Endo-) or from above (Epi-) in *Soil Taxonomy*. Brunisols include Udolls and Udepts with good drainage and Aquolls and Aquepts with imperfect drainage.

3.6 Conclusions

More than three-quarters (79%) of the soils of the Great Lakes Coastal Zone have been identified, classified, and mapped. *Soil Taxonomy* (US) and *The Canadian System of Soil Classification* have been to classify the soils of the region. Whereas ST is a global system, the Canadian approach is strictly national. However, both schemes are natural systems that delineate soils based on their properties rather than their presumed genesis. Both systems are hierarchical, with ST having six levels and the Canadian system having four. Both approaches map soil series and soil associations. ST uses diagnostic horizons for taxonomic levels; the Canadian system employs genetic and some diagnostic horizons. Whereas ST uses formative elements to build soil taxonomic names, the Canadian system employs two levels of word modifiers. The two systems are somewhat comparable. Difficulties in comparing the two soil classification systems center primarily on two Canadian orders—the Gleysols and the Brunisols. Gleysols include Aquolls, Aquepts, and Aquents saturated by water from below (Endo-) or from above (Epi-) in ST. Brunisols include Udolls and Udepts under good drainage and Aquolls and Aquepts under imperfect drainage.

References

Baldwin, M., C.E. Kellogg, and J. Thorp. 1938. *Soil classification. Soils and men,* 979–1001. U.S. Dep. Agric. Yearbook. Washington, DC: U.S. Govt. Print. Office.

FAO. 1974. *Key to soil units for the new soil map of the world.* Rome: Legend 1 FAO.

FAO, ISRIC, and ISSS. 1998. *World reference base for soil resources*. World Soil Resour. Rep. 84, Food & Agric. Organ. United Nations, Rome.

Marbut, C.F. 1927. A scheme for soil classification. In *Proceedings of the 1st International Congress on Soil Science and Communication*, vol. 5, 1–31.

Simonson, R.W. 1959. Outline of a generalized theory of soil genesis. *Soil Science Society of America, Proceedings* 23: 152–156.

Smith, G.D. 1983. Historical development of soil taxonomy-background. In *Pedogenesis and soil taxonomy: 1. Concepts and interactions*, ed. L.P. Wilding, 23–49. Amsterdam: Elsevier.

Soil Classification Working Group. 1998. *The Canadian system of soil classification*, 3rd ed. Agriculture and Agri-Food Canada Publ. 1646.

Soil Survey Staff. 1960. *Soil classification, a comprehensive system, 7th approximation*. Washington, DC: U.S. Govt. Print. Office.

Soil Survey Staff. 1975. *Soil taxonomy: A basic system of soil classification for making and interpreting soil surveys*. Agric. Handbook No. 436. Washington, DC: U.S. Govt. Print. Office.

Soil Survey Staff. 1999. *Soil taxonomy: A basic system of soil classification for making and interpreting soil surveys*, 2nd ed. Agric. Handbook 436. USDA, Natural Resources Conserv. Serv. US Dept.

Soil Survey Staff. 2014. *Keys to soil taxonomy*, 12th ed. U.S. Dep. Agric., Nat. Resour. Conserv. Serv.

4 Soil Taxonomic Structure and Factors Affecting Soil Distribution in the Great Lakes Coastal Zone

4.1 Soil Taxonomic Structure in the US Great Lakes Coastal Zone

There are six orders, 16 suborders, 39 great groups, 140 subgroups, 507 families, and 691 soil series in the US GLCZ. Of the 691 soil series in the US GLCZ, 28% are Alfisols 26% are Spodosols, 19% are Inceptisols, 12% are Entisols, 11% are Mollisols, and 4% are Histosols (Table 4.1). Ten great groups account for 65% of the soil series, including Haplorthods (16%), Hapludalfs (12%), Glossudalfs (7.1%), Endoaquepts (5.2%), Endoaquolls (4.6%), Eutrudepts (4.6%), Endoaqualfs (4.5%), Udipsamments (4.2%), Haplosaprists (3.5%), and Epiaquepts (3.5%) (Fig. 4.1).

Nearly a third (28%) of the soil series follow the central concept of a taxon, i.e., Typic and Haplic subgroups (Fig. 4.2). Nearly half (45%) of the soil series are extragrades in that they deviate from the central concept of a taxon. Drainage is the key factor in this deviation, accounting for 30% of the soil series. More than one-quarter (27%) of the soil series are intergrades, which are soils that have properties linking them with other soil orders. These orders are most commonly Mollisols, Alfisols, and Entisols, but include a few Histosols and Spodosols.

The distribution of soil families suggests that 71% of the soil series are in loamy (coarse-loamy, fine-loamy, and loamy) and sandy (sandy and sandy-skeletal) particle-size classes, 81% are in the mixed soil-mineral class, about half are in the frigid soil-temperature class (the other half is mesic), and about half are in active and superactive CEC activity classes (Fig. 4.3). Aquic soils (Hemists, Saprists, Aqualfs, Aquods, Aquepts, Aquolls, and Aquents) account for 42% of the soil series (Table 4.2).

4.2 Soil Taxonomic Structure in the Canadian Great Lakes Coastal Zone

A total of 205 soil series occur along the mapped portions of the Canadian Great Lakes coastal zone; more than three-quarters of the soil subgroups are Orthic Humic Gleysols (22%), Brunisolic Gray Brown Luvisols (20%), Gleyed Gray Brown Luvisols (10%), and Orthic Melanic Brunisols (10%) (Fig. 4.4). A large portion of the unmapped Canadian GLCZ is provisionally mapped as Humo-Ferric Podzols (Great Lakes Forestry Centre). Most of these soils would be classified as Haplorthods in ST. For this reason, the proportion of Luvisols (51%) in the Canadian GLCZ is substantially greater the proportion of Alfisols (28%) in the US GLCZ; and the proportion of Gleysols (27%) in the Canadian GLCZ is larger than the proportion of Aquents, Aquepts, and Aquolls (23%) in the GLCZ. Whereas Histosols occupy 4.0% of the US GLCZ, Organic soil series comprise less than 1% of the Canadian GLCZ. However, landforms identified as "marshes" and "bogs" in Canada may contain Histosols, as well as wet mineral soils. Whereas 5.4% of the US GLCZ features Psamments, Orthic Regosols compose less than 1% of the Canadian GLCZ.

In the US GLCZ, 42% of the soils are aquic, i.e., are either Hemists or Saprists or occur in Aqu-suborders of mineral soils (Table 4.2). In the Canadian GLCZ, 55% of the soil series are Gleysols or in Gleyed subgroups. In the US GLCZ, the largest proportion (42%) of the soil series are in loamy particle-size classes and 30% are in sandy classes; 81% of the soil series have a mixed mineralogy; 48% of the series are in superactive and active cation-exchange-capacity activity classes; and the proportion of soils are equally

J. G. Bockheim, *Soils of the Laurentian Great Lakes, USA and Canada*,
https://doi.org/10.1007/978-3-030-52425-8_4

Table 4.1 Taxonomic structure of soils along the shoreline of the US Great Lakes

Order	Suborder	Great groups	No. soil series	%
Alfisols	Aqualfs	Endoaqualfs	31	4.5
		Epiaqualfs	17	2.5
		Fragiaqualfs	1	0.1
		Glossaqualfs	5	0.7
	Udalfs	Fragiudalfs	1	0.1
		Fraglossudalfs	1	0.1
		Glossudalfs	49	7.1
		Hapludalfs	86	12.4
		Subtotal	191	27.6
Entisols	Aquents	Endoaquents	5	0.7
		Epiaquents	4	0.6
		Fluvaquents	5	0.7
		Psammaquents	17	2.5
	Fluvents	Udifluvents	3	0.4
	Orthents	Udorthents	14	2.0
	Psamments	Quartzipsamments	3	0.4
		Udipsamments	29	4.2
		Subtotal	80	11.6
Histosols	Folists	Udifolists	1	0.1
	Hemists	Haplohemists	5	0.7
	Saprists	Haplosaprists	24	3.5
		Subtotal	30	4.3
Inceptisols	Aquepts	Endoaquepts	36	5.2
		Epiaquepts	24	3.5
		Fragiaquepts	3	0.4
		Humaquepts	20	2.9
	Udepts	Dystrudepts	16	2.3
		Eutrudepts	27	3.9
		Fragiudepts	4	0.6
		Subtotal	130	18.8
Mollisols	Aquolls	Argiaquolls	10	1.4
		Endoaquolls	32	4.6
		Epiaquolls	7	1.0
	Rendolls	Haprendolls	4	0.6
	Udolls	Argiudolls	11	1.6
		Hapludolls	13	1.9
		Subtotal	77	11.1
Spodosols	Aquods	Duraquods	6	0.9
		Endoaquods	23	3.3
		Epiaquods	13	1.9
		Fragiaquods	6	0.9
	Orthods	Durorthods	6	0.9
		Fragiorthods	16	2.3
		Haplorthods	113	16.4
		Subtotal	183	26.5
		Total	691	100.0

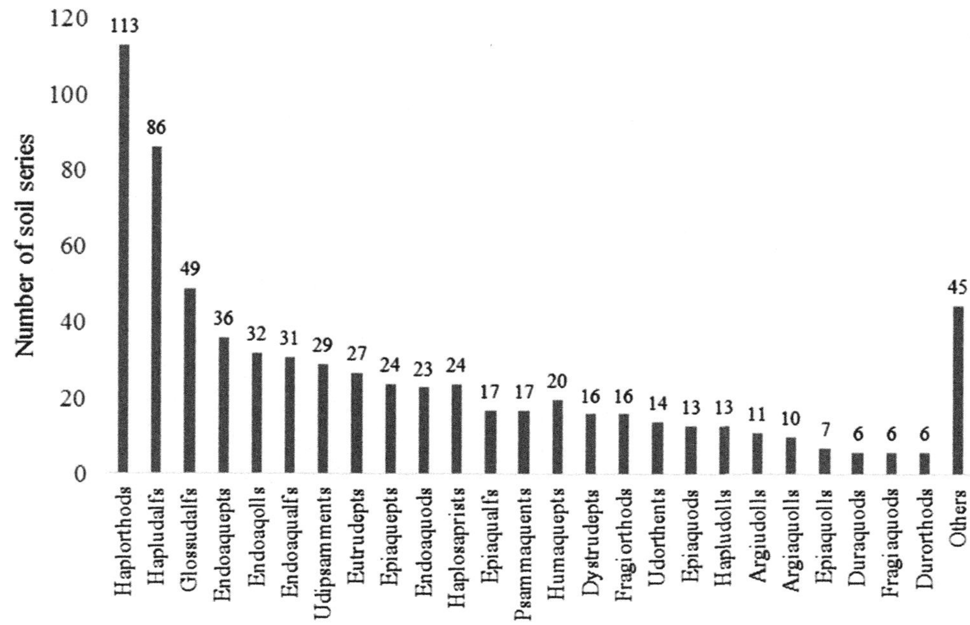

Fig. 4.1 Distribution of soil series by great group in the US Great Lakes Coastal Zone

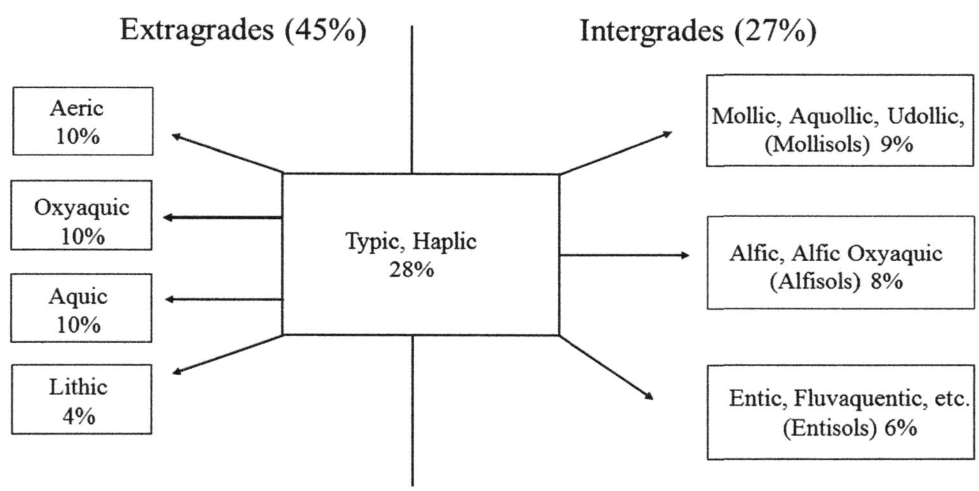

Fig. 4.2 The distribution of soil subgroups by central concept (Typic, Haplic), extragrades (variation within a taxon), and intergrades (linkage of the taxon with other soil orders)

divided into mesic and frigid soil-temperature classes (Fig. 4.3). In the Canadian GLCZ, the largest proportion of the soil series are in the moderately fine particle-size class (20%), followed by the medium (18%), fine (16%), very coarse (14%), and other classes (Fig. 4.5).

There are 16 soil series with the same name in the US and Canadian GLCZ, eight of which are classified similarly. These include the Eastport (Orthic Regosol/Udipsamments), Granby (Orthic Humic Gleysol/Endoaquolls), Guelph (Brunisolic Gray Brown Luvisol/Glossudalfs), Niagara (Gleyed Gray Brown Luvisol/Endoaqualfs), Ontario (Brunisolic Gray Brown Luvisol/Hapludalfs), Rubicon (Gleyed Humo-Ferric Podzol/Haplorthods), Toledo (Orthic Humic Gleysol/Endoaquolls), and Wauseon (Orthic Humic Gleysol/Epiaquolls) soil series.

A large portion of the soils in the Canadian GLCZ are calcareous, including 100% of those derived from glaciolacustrine materials, 70% from till, 39% from lacustrine deposits, and 32% from outwash (Fig. 4.6).

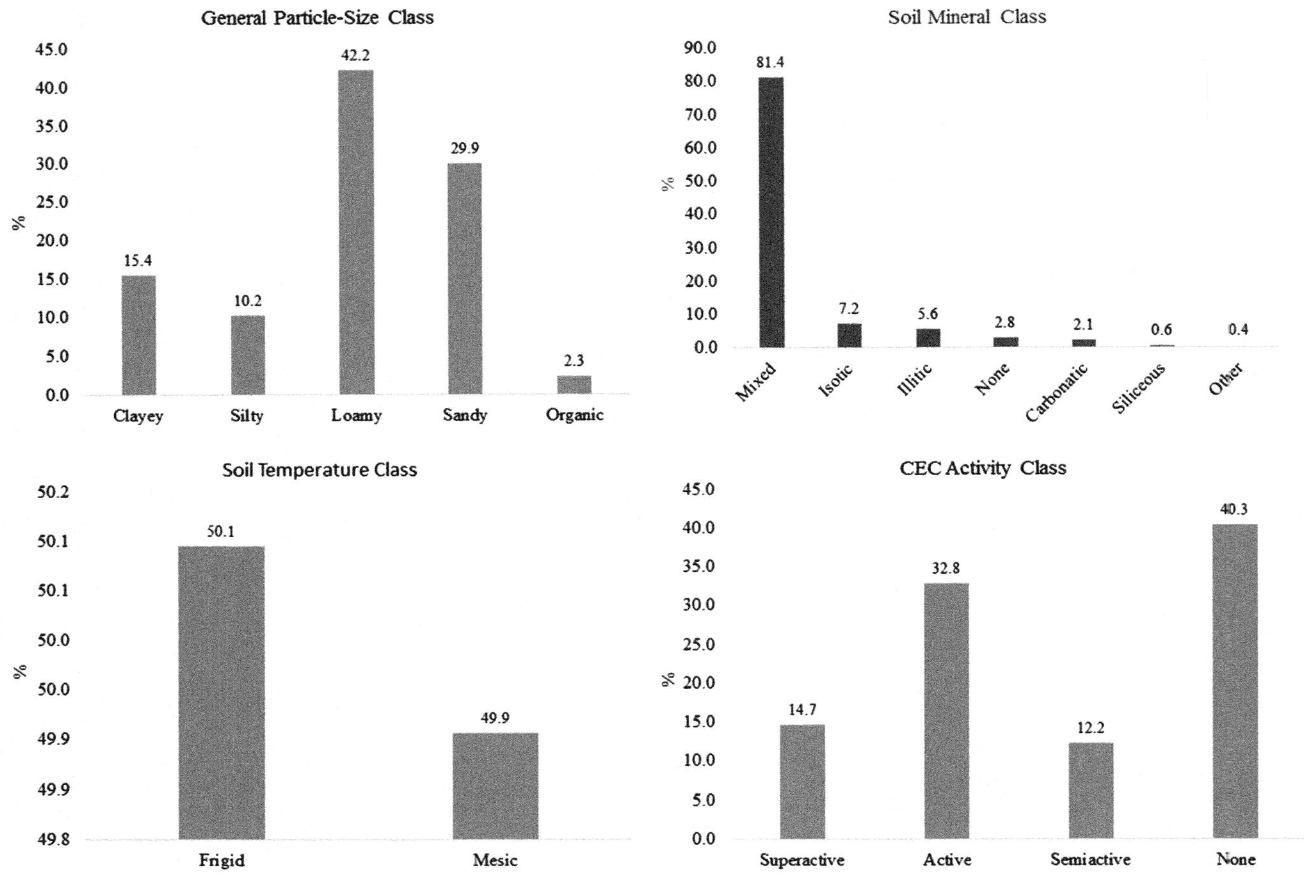

Fig. 4.3 Proportion of soil series by family class, including general particle-size class, soil mineral class, soil temperature class, and cation-exchange capacity activity class

4.3 Factors Affecting Soil Distribution in the Great Lakes Coastal Zone

Four soil-forming factors appear to have special importance with regards to the distribution of soil taxa in the Great Lakes Coastal Zone, including climate, vegetation, parent material, and time.

4.3.1 Climate and Soil Distribution

The Great Lakes Tension Zone was illustrated in Fig. 2.6. Not only does this tension zone separate mixed broad-leaved and coniferous forest from broad-leaved forest, but also it separates the mesic and frigid soil-temperature classes (Hole 1976; Bockheim and Schliemann 2014). Figure 4.7 shows that the frigid soil temperature zone contains dominantly Haplorthods, Endoaquods, Fragiorthods, Fragiaquods, Durorthods, and a large portion of the Glossudalfs. In contrast, the mesic soil-temperature zone contains primarily Hapludalfs, Endoaquepts, Endoaquolls, and Endoaqualfs.

4.3.2 Vegetation and Soil Distribution

There is a correspondence between ecosystem type and soil great group in the Great Lakes Coastal Zone (Fig. 4.8). Lowland Deciduous Forest, Lowland Mixed Forest, and Lowland Shrubs are contained in Endo- and Epi-great groups of Aquolls, Aquepts, and Aquents and in the Haplosaprist and Haplohemist great groups. Temperate Deciduous Forest is underlain primarily by Alfisols (52%) and Spodosols (15%). In contrast, Temperate Mixed Forest contains 25% Alfisols and 24% Spodosols. There were only three soil series supporting prairie vegetation, two of which were Argiudolls and one an Argiaquolls.

4.3.3 Landforms, Parent Material and Soil Distribution

The dominant landforms in the US coastal zone are lake plains-terraces (26%), till plains (26%), and glacial lake plains (16%) (Fig. 4.9). The dominant landforms on the Canadian portion of the GLCZ include moraines (36%), lake

Table 4.2 Aquic-gleyed soils in the Great Lakes Coastal Zone

US GLCZ	No. soil series	%
Aqualfs	54	7.8
Aquents	31	4.5
Aquept	83	12.0
Aquods	48	6.9
Aquolls	49	7.1
Hemists	5	0.7
Saprists	24	3.5
Subtotal	294	43
Non-aquic	397	57
Total	691	100
Canada		
Gleysols	44	30.6
Gleyed Luvisols	22	15.3
Gleyed Brunisols	8	5.6
Gleyed Podzols	2	1.4
Subtotal	76	53
Non-gleyed	68	47
Total	144	100

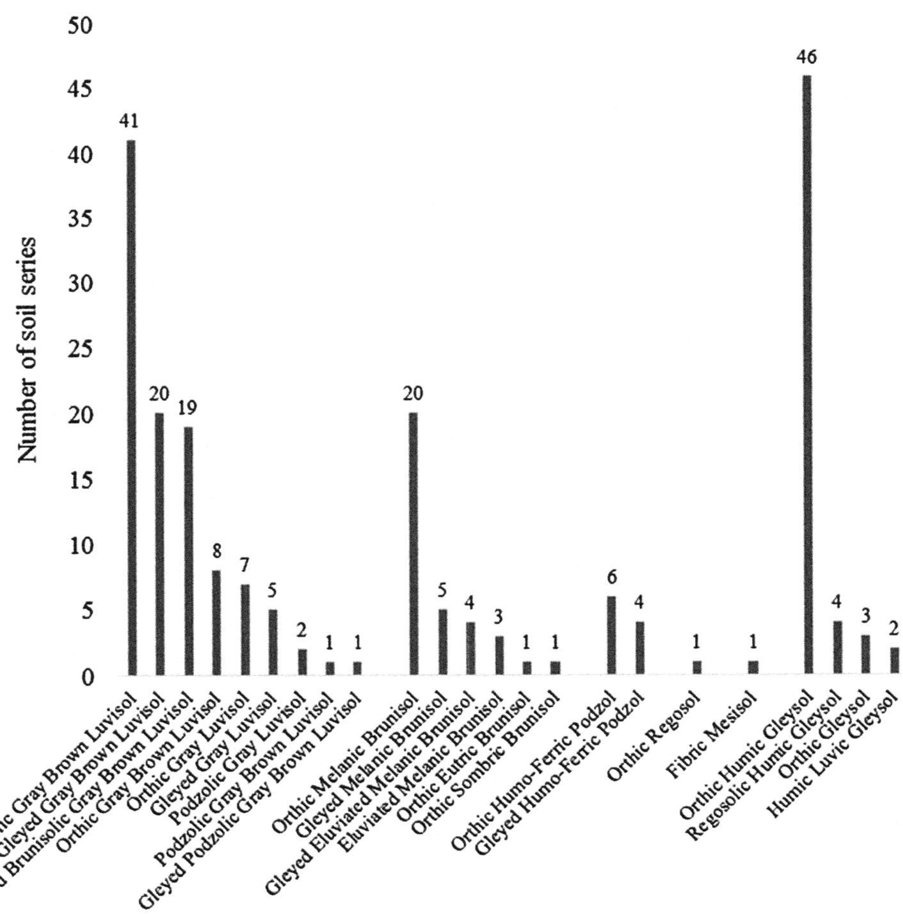

Fig. 4.4 Distribution of soil series by subgroup in the Canadian Great Lakes Coastal Zone

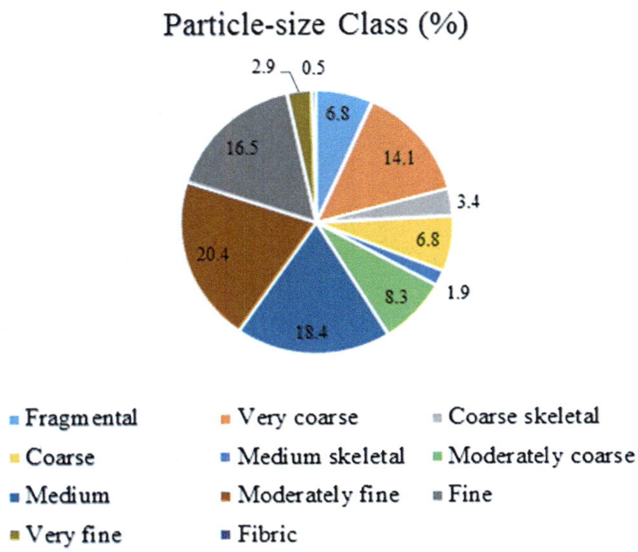

Fig. 4.5 Distribution of soil series in the Canadian GLCZ by particle-size class

plains (24%), glacial lake plains (17%), and outwash plains (14%) (Fig. 4.10).

In the US portion of the GLCZ, till accounts for 32% of the coastal zone parent materials, followed by glaciolacustrine (19%), lacustrine (18%), and outwash (16%) (Fig. 4.11). Although the mapping is incomplete, the proportions of parent materials in the Canadian coastal zone are comparable to those of the US, with 33% mapped as till, 24% as lacustrine, 19% as glaciolacustrine, and 14% as outwash (Fig. 4.6).

The major kinds of parent materials found in the Great Lakes CZ are illustrated in Fig. 4.12. Figure 4.12a, taken near Kingsville, OH, on Lake Erie, shows 2 m of lacustrine deposits from the Warren lake stage (orange colored) over gray, low-lime till. Figure 4.12b shows till containing well-rounded clasts near Meaford, ON. Outwash derived from limestone is shown in Fig. 4.12c. Figure 4.12d shows a classic sequence of alluvial sediments covered by lake sediments from the Iroquois stage, known as the Scarborough Bluffs, near Toronto, ON, on the western shore of Lake Ontario. Figure 4.12e shows dunes near Manistique, MI, on the north shore of Lake Michigan. Figure 4.12f shows limestone residuum exposed in a quarry near Moran, MI, on the north shore of Lake Michigan.

The waxing and waning of the paleo-Great Lakes have created interesting landforms and deposits. In some cases, the lakes washed pre-existing till causing a planation of the surface often accompanied by a concentration of boulders whereby the fines have been winnowed away (Fig. 4.13). The lakes also may have cut benches into the local bedrock, with limestone and sandstone being particularly erodible. In many cases glaciolacustrine or lacustrine sediments were deposited in thicknesses ranging from a decimeter to hundreds of meters.

Some examples of soil series on different parent materials are given in Fig. 4.14. The Toledo and Elnora soil series are derived from clay and sandy glaciolacustrine materials, respectively. The Elnora soil series is a relict longshore bar commonly found on the shorelines of Lakes Erie and Ontario. The Blount soil series, which occurs in the Lake Michigan and Lake Huron coastal zones, represents till that was wave-worked during the Algonquin lake stage 11.2–12.4 ky BP. The Augustana soil series is derived from a thin layer of glaciolacustrine sediments overlying till. The Summerville soil series contains loamy glaciolacustrine sediments over limestone bedrock on a glacial lake bench common to Lakes Michigan, Huron, and Ontario. The Deer Park soil series, common to Lakes Superior, Michigan, and Huron, occurs on excessively drained beach ridges and

Fig. 4.6 Distribution of soil series in the Canadian GLCZ by parent material

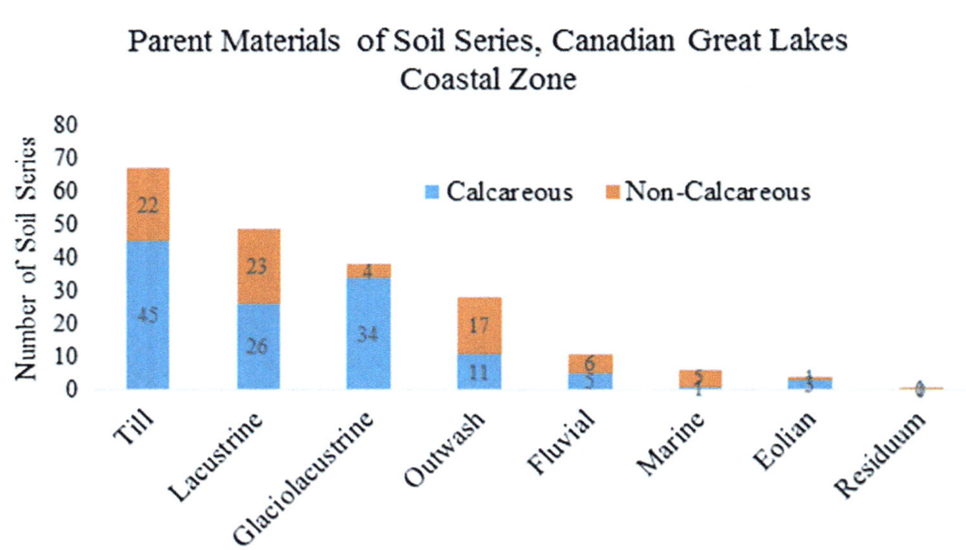

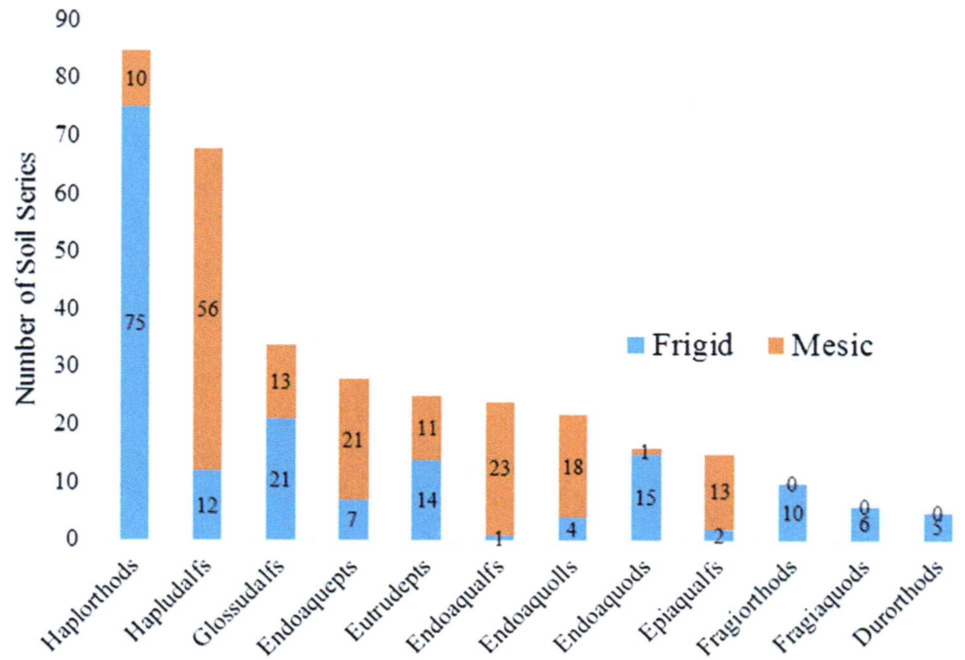

Fig. 4.7 Distribution of soil series by great group and soil-temperature class in the US Great Lakes Coastal Zone

stabilized dunes. The Oakville soil series is common on dunes along the shorelines of Lakes Michigan, Erie, and Ontario.

4.3.4 The Time Factor and Soil Distribution

Soils of the GLCZ occur in five age classes related to dated lake stages and drift units. Nearly half (46%) of the soil series are derived from lake-stage events at 11.2 kyr BP, i.e., from the Duluth (Lake Superior), Main Algonquin (Lakes Superior, Michigan, and Huron); 28% began forming during lake-stage events at 13.8 kyr BP, i.e., from the Glenwood, Calumet, Maumee (Lake Michigan), and Whittlesey and Warren (Lakes Huron and Erie) (Fig. 4.15).

Whereas Entisols are most common on the mid-Holocene Algoma and Nipissing surfaces, Spodosols and Inceptisols are dominant on Algonquin and Duluth surfaces of approximately 11.2 kyr in age, and Alfisols and Mollisols are most common on Glenwood, Whittlesey, and Warren surfaces of approximately 13.8 kyr in age (Fig. 4.16).

These data enable the determination of the time required to develop different diagnostic horizons. The argillic horizon, the most common diagnostic horizon, may form in 11.2 kyr, but is most common in soils that are 13.8 kyr in age (Fig. 4.17). The spodic horizon, the second most common diagnostic horizon in Great Lakes soils, appears to require 11.2 kyr to form. The glossic, fragipan, and ortstein horizons all seem to require 11.2 kyr to develop. The mollic and cambic horizons are most common in 11.2 kyr and 13.8 kyr soils, but are somewhat common on soils of Nipissing (5.3 kyr BP) age.

These trends are borne out by published soil chronosquences along the Great Lakes shoreline (Table 4.3). On coarse-textured (sands and loamy sands), quartz-rich beach ridges, dunes, outwash, and lake sediments, Psammaquents persist under wet conditions and Udipsamments and Quartzipsamments under dry conditions throughout the late Wisconsinan and Holocene (Fig. 4.13). However, on sandy loam and loamy fine sand outwash, lake sediments, and till, Udorthents evolve to Dystrudepts by 5.3 kyr, become weakly developed Spodosols on materials dated at ~7 kyr, and form strongly developed Spodosols at 11.2 kyr (Figs. 4.17 and 4.18).

4.4 Conclusions

There are six orders, 16 suborders, 39 great groups, 140 subgroups, 507 families, and 691 soil series in the US GLCZ. Of the 691 soil series in the US GLCZ, 28% are Alfisols 26% are Spodosols, 19% are Inceptisols, 12% are Entisols, 11% are Mollisols, and 4% are Histosols. Nine great groups account for 60% of the soil series, including Haplorthods (16%), Hapludalfs (12%), Glossudalfs (7.1%), Endoaquepts (5.2%), Endoaquolls (4.6%), Endoaqualfs (4.5%), Eutrudepts (3.9%), Haplosaprists (3.5%), and Epiaquepts (3.5%). The distribution of soil families suggests that 71% of the soil series are in loamy (coarse-loamy, fine-loamy, and loamy) and sandy (sandy and

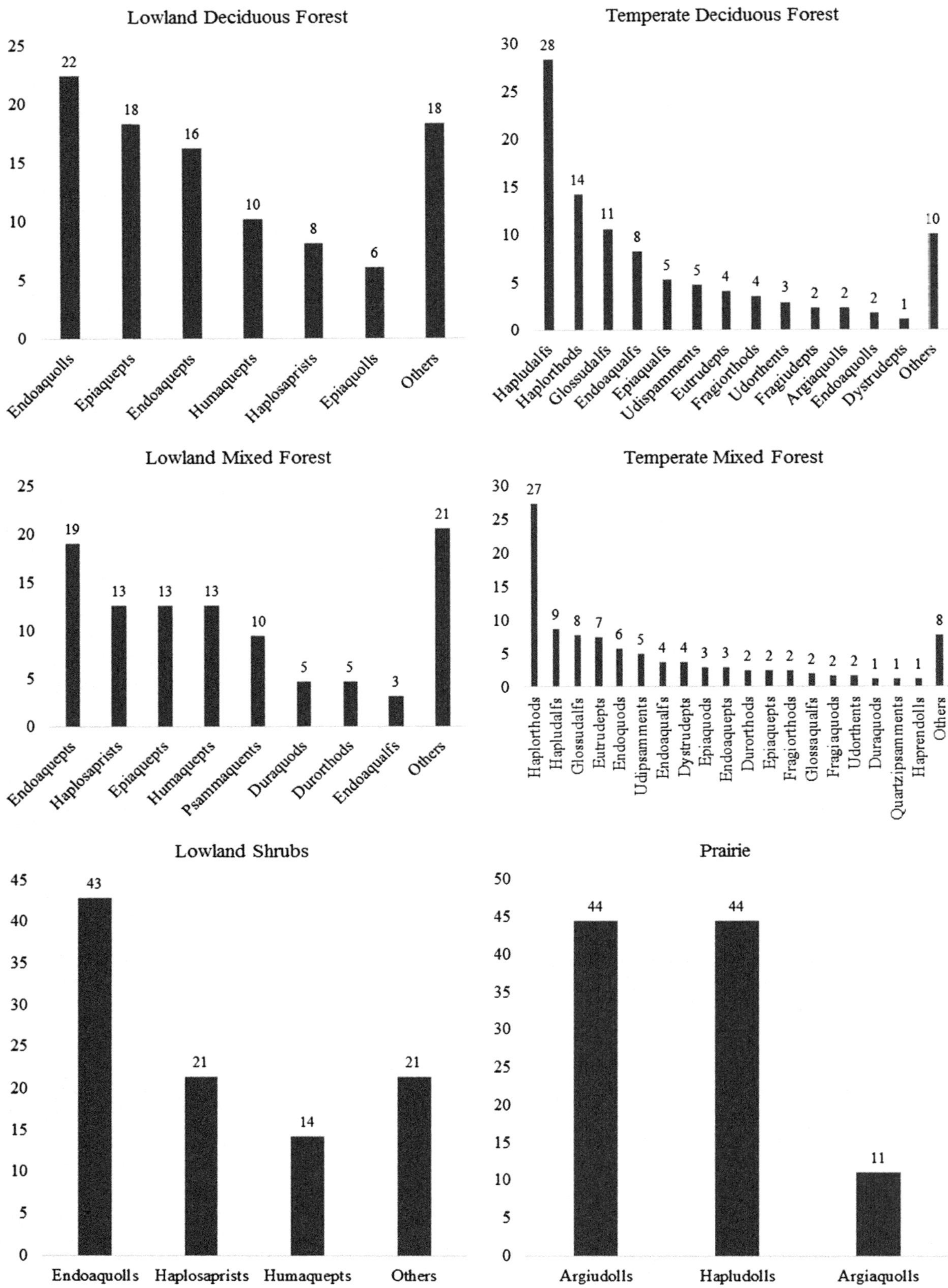

Fig. 4.8 Distribution of soil series by great group and vegetation type in the US Great Lakes Coastal Zone

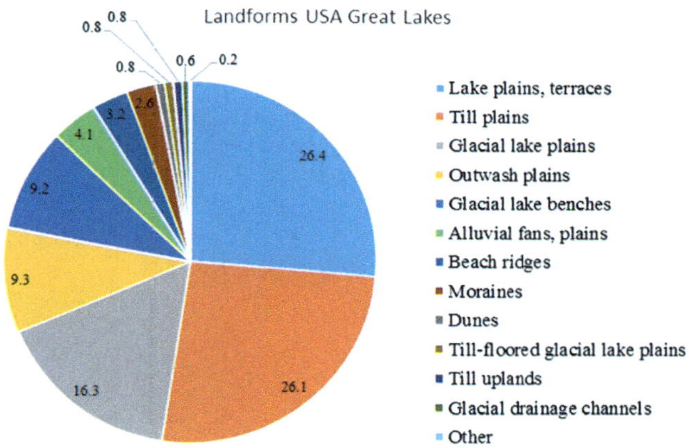

Fig. 4.9 Distribution of soil series in the US GLCZ by landform

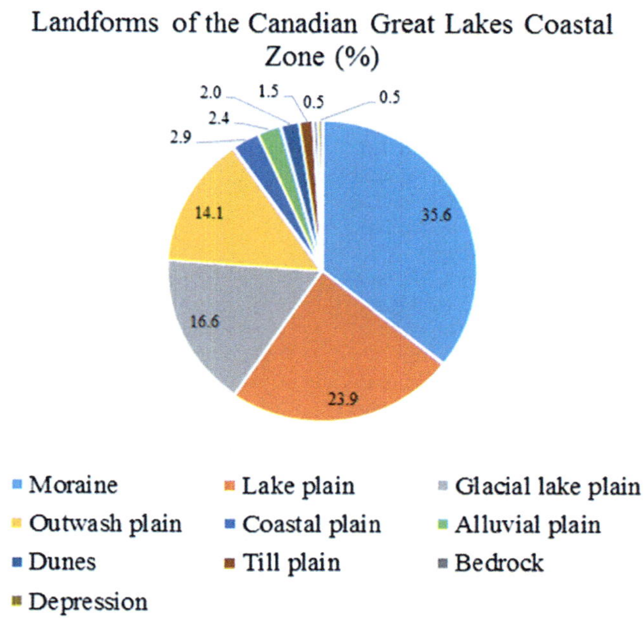

Fig. 4.10 Distribution of soil series in the Canadian GLCZ by landform

sandy-skeletal) particle-size classes, 81% are in the mixed soil-mineral class, about half are in the frigid soil-temperature class (the other half is mesic), and about half are in active and superactive CEC activity classes. Aquic soils (Histosols and Aqualfs, Aquods, Aquepts, Aquolls, and Aquents) account for 42% of the soil series.

A total of 204 soil series occur along the mapped portions of the Canadian Great Lakes coastal zone; more than three-quarters of the soil subgroups are Orthic Humic Gleysols (22%), Brunisolic Gray Brown Luvisols (20%), Gleyed Gray Brown Luvisols (10%), and Orthic Melanic Brunisols (10%). A large portion of the unmapped Canadian GLCZ is provisionally mapped as Humo-Ferric Podzols, suggesting that the distribution of soil orders in ST is comparable in the US and Canadian GLCZ.

Soil great groups (ST) are strongly related to soil climate, vegetation type, parent material-geomorphic surface, and age of material. Spodosols are common in areas with a frigid soil-temperature class, temperate mixed vegetation, sandy lacustrine and outwash parent materials; Alfisols are common in areas with a mesic soil-temperature class, temperate broad-leaved vegetation, till parent materials.

Fig. 4.11 Distribution of soil series in the US GLCZ by parent material

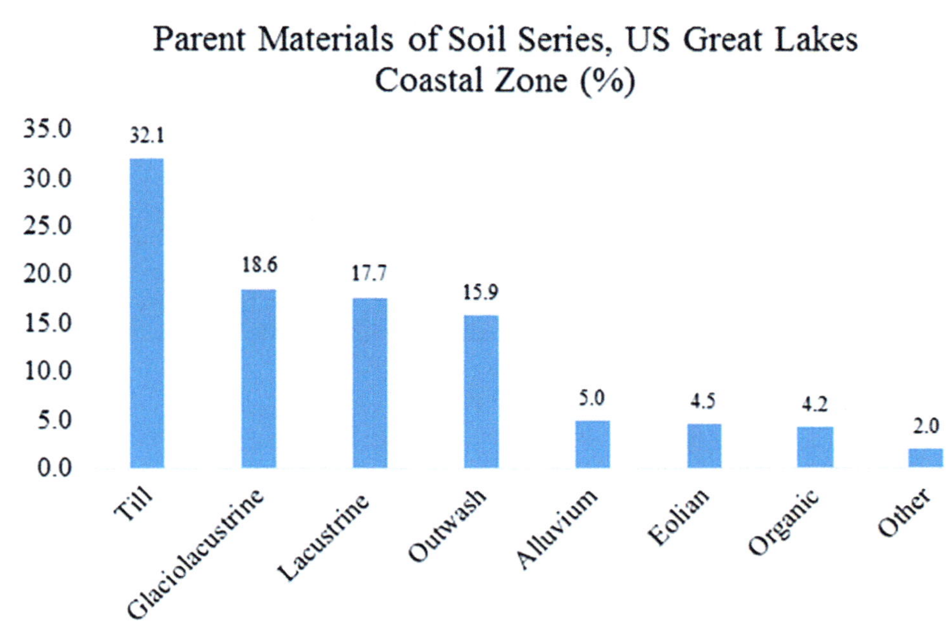

Fig. 4.12 **a** An example of lacustrine deposits with a soil (orange colored sediments) over till (gray deposits) near Kingsville, OH, on Lake Erie (photo by J. Bockheim). **b** Till with well-rounded and polished clasts near Meaford, ON, on Lake Huron (photo by J. Bockheim). **c** Outwash derived from limestone near Moran, MI, on the north shore of Lake Michigan (photo by J. Bockheim). **d** Scarborough Bluffs near Toronto, ON, on eastern Lake Ontario showing alluvial deposits (photo by https://goo.gl/images/V4cnyA). **e** Sand dunes near Manistique, MI, on the northwest shore of Lake Michigan (photo by J. Bockheim). **f** Residuum from limestone near Moran, MI, on the north shore of Lake Michigan (photo by J. Bockheim)

a

b

Fig. 4.12 (continued)

Fig. 4.12 (continued)

e

f

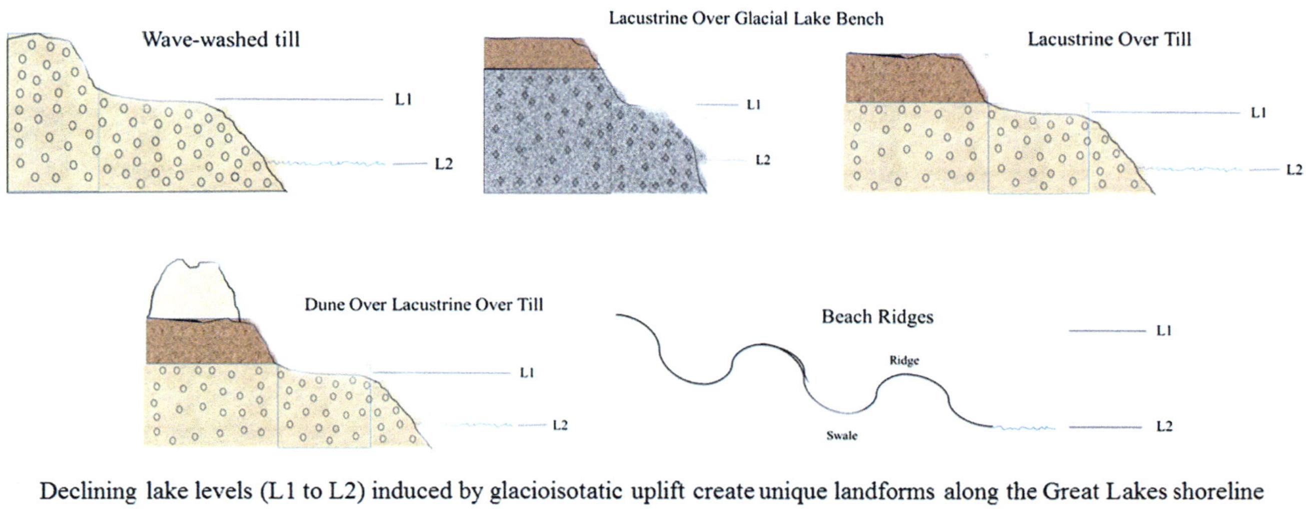

Fig. 4.13 Declining lake levels (L1 to L2) induced by glacioisostatic uplift create unique landforms along the Great Lakes shoreline

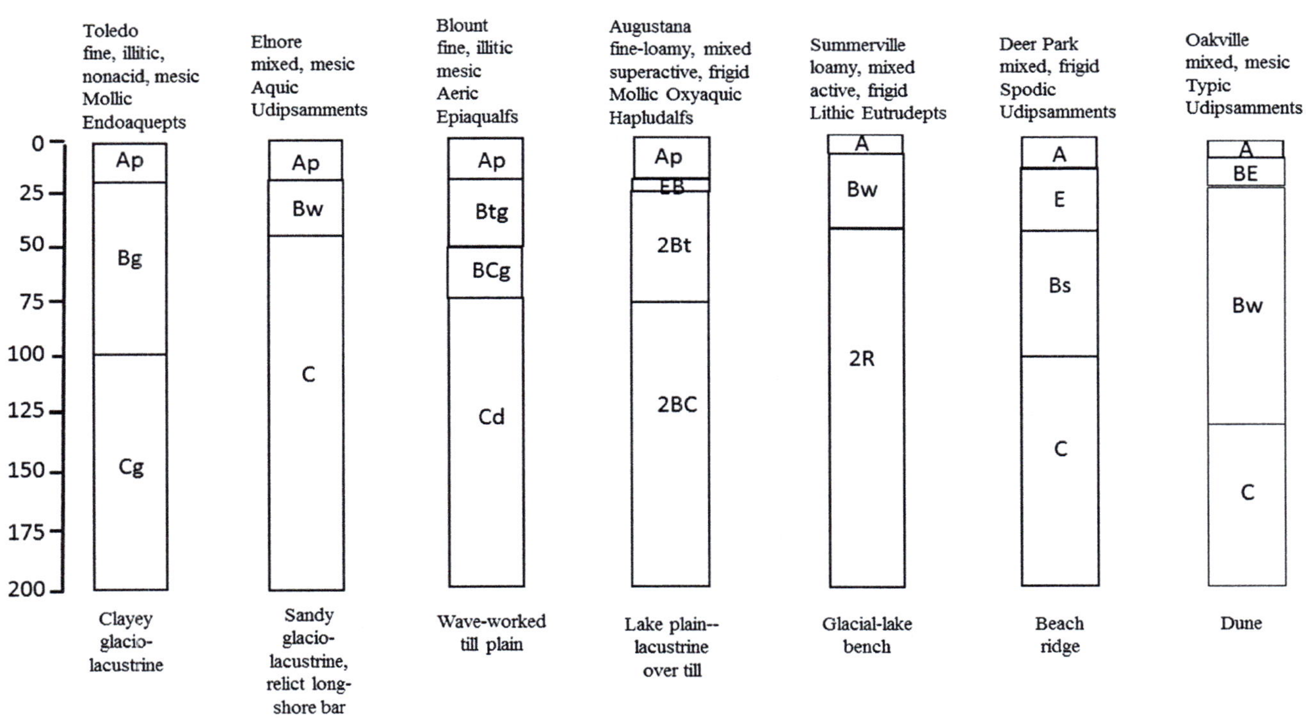

Fig. 4.14 Examples of soil series on different parent materials in the Great Lakes Coastal Zone

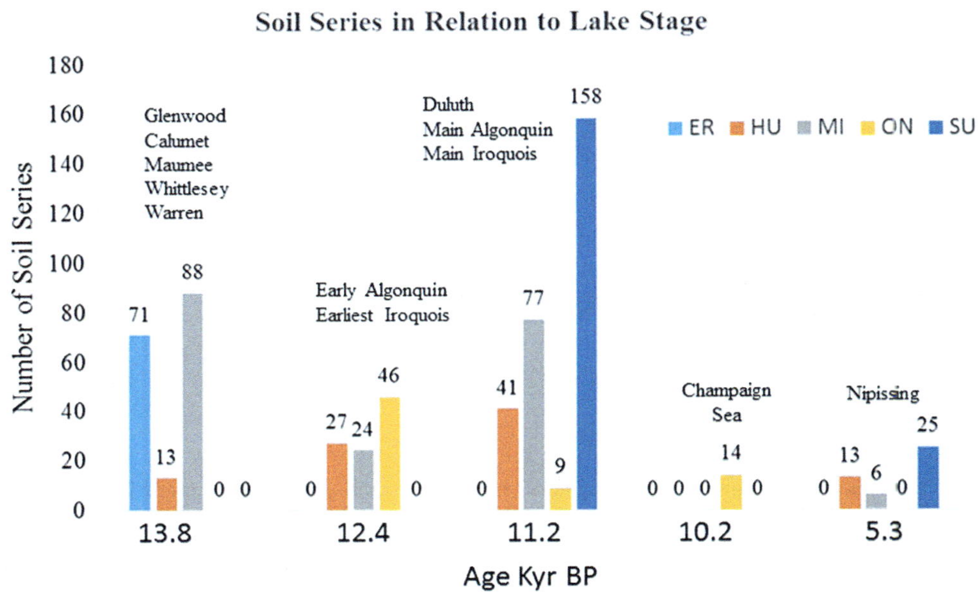

Fig. 4.15 Distribution of soil series by lake stage-age class in the US Great Lakes Coastal Zone

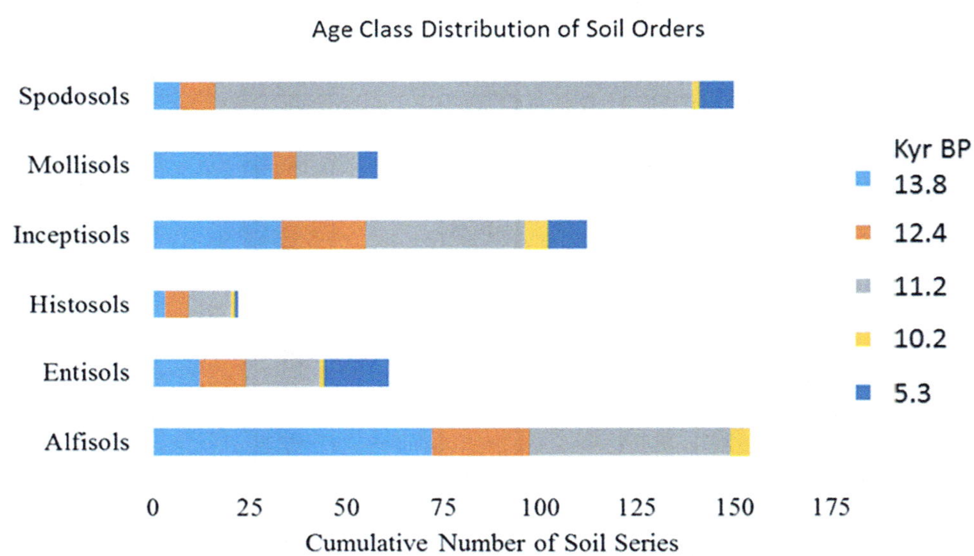

Fig. 4.16 Distribution of soil orders by age class in the US Great Lakes Coastal Zone

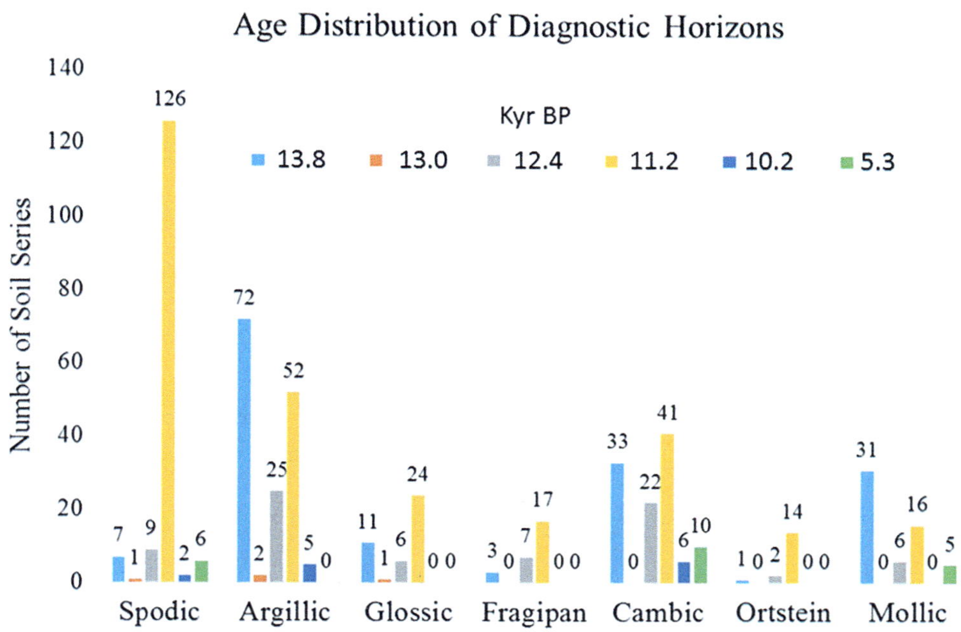

Fig. 4.17 Age distribution of diagnostic horizons in the US Great Lakes Coastal Zone

Table 4.3 Characteristics of soil chronosequences from the Laurentian Great Lakes

Location	Landform/composition	Soil taxa	Age range (kyr BP)	Properties	Model	References
NE Lake Michigan	Raised beaches, sandy	Typic Udipsamments, Entic Haplorthods, Typic Haplorthods	3–11	Profile OC, Fe, Al	Linear	Barrett and Schaetzl (1992)
NE Lake Michigan	Beach ridges, sandy	Typic Udipsamments, Spodic Udipsamments	0.01–5.0	Extr. Fe, Al, OC	Linear, log	Barrett (2001)
SW Lake Huron	River terraces, beach ridges, variable	Fluvaquents, Hapludalfs	<4.5–14.5	K adsorption	Log	Howard et al. (2012)
NE Lake Michigan	Dune-capped beach ridges, sandy	Typic Udipsamments, Spodic Udipsamments	0.0025–4.15	Depth to B horizon, C, N, etc.	Log, linear, exponential	Lichter (1998)
S Lake Michigan	Dunes	–	0.01–12	$CaCO_3$, %N, CEC, ex. acid., ex. bases, % si	Log	Olson (1958)
S Lake Huron	Beach ridges, sandy	Udipsamments, Dystrudepts	0.1–4.7	B chroma and thickness, Si, C, OM, $CaCO_3$, pH, depth to B	Linear, log	VandenBygaart and Protz (1995)
SW Lake Michigan	Dunes, beach ridges, sand	Udipsamments	0.01–3.5	A thickness, % clay in B	Discriminant analysis	Berg (1980)
NE Lake Michigan	Raised beaches, sandy	Typic Udipsamments, Entic Haplorthods, Typic Haplorthods				Franzmeier and Whiteside (1963a, b)

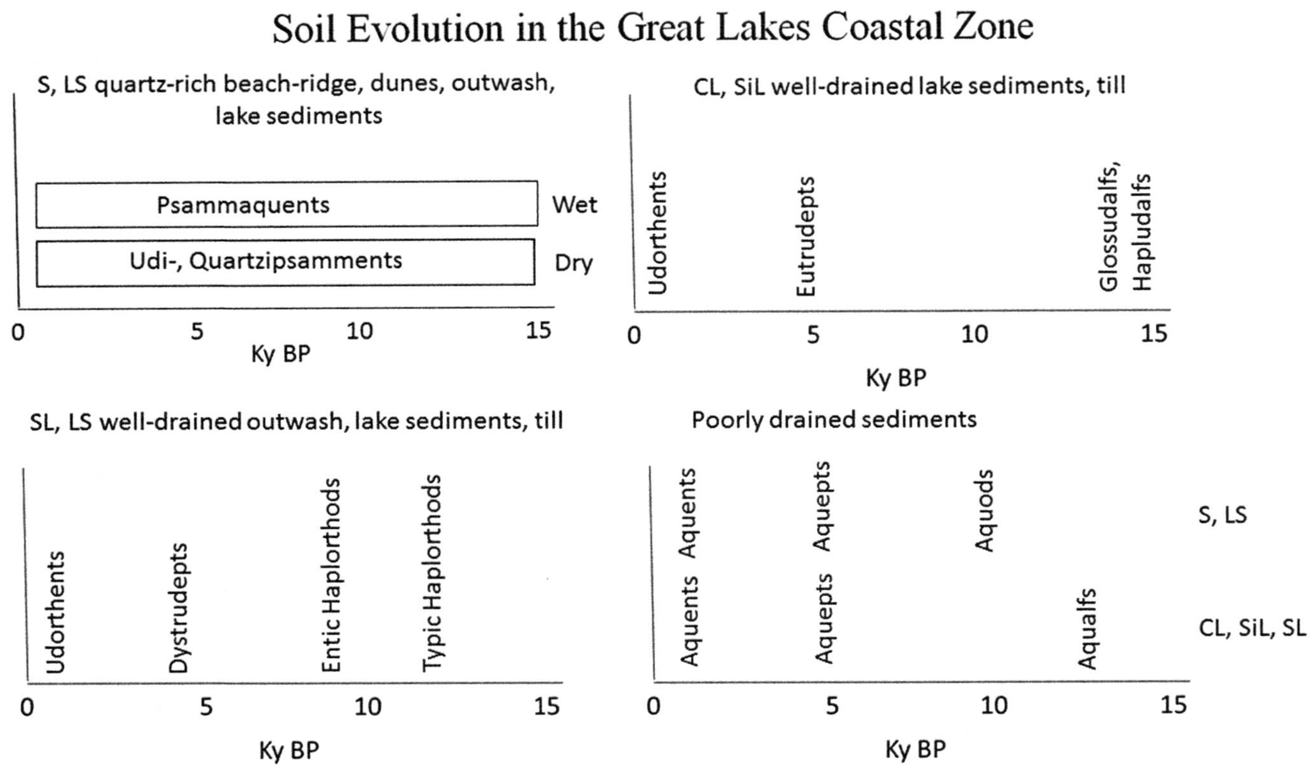

Fig. 4.18 Evolution of soils by parent material and drainage regime in the US Great Lakes Coastal Zone

References

Barrett, L.R. 2001. A strand plain soil development sequence in northern Michigan, USA. *CATENA* 44: 163–186.

Barrett, L.R., and R.J. Schaetzl. 1992. An examination of podzolization near Lake Michigan using chronofunctions. *Canadian Journal of Soil Science* 72: 527–541.

Berg, R.C. 1980. Use of stepwise discriminant analysis to assess soil genesis in a youthful sandy environment. *Soil Science* 129: 353–365.

Bockheim, J.G., and S.A. Schliemann. 2014. Soil richness and endemism across an environmental transition zone in Wisconsin, USA. *CATENA* 113: 86–94.

Franzmeier, D.P., and E.P. Whiteside. 1963a. A chronosquence of Podzols in northern Michigan. I. Ecology and description of pedons. *Michigan State University Agricultural Experiment Station, Quarterly Bulletin* 46: 2–20.

Franzmeier, D.P., and E.P. Whiteside. 1963b. A chronosquence of Podzols in northern Michigan. II. Physical and chemical properties. *Michigan State University Agricultural Experiment Station, Quarterly Bulletin* 46: 21–36.

Hole, F.D. 1976. *Soils of Wisconsin*. Madison: Univ. of Wisconsin Press.

Howard, J.L., C.R. Clawson, and W.L. Daniels. 2012. A comparison of mineralogical techniques and potassium adsorption isotherm analysis for relative dating and correlation of late Quaternary soil chronosequences. *Geoderma* 179–180: 81–95.

Lichter, J. 1998. Rates of weathering and chemical depletion in soils across a chronosequence of Lake Michigan sand dunes. *Geoderma* 85: 255–282.

Olson, J.S. 1958. Rates of succession and soil changes on southern Lake Michigan sand dunes. *Botanical Gazette* 119: 125–170.

VandenBygaart, A.J., and R. Protz. 1995. Soil genesis on a chronosequence, Pinery Provincial Park, Ontario. *Canadian Journal of Soil Science* 75: 63–72.

5 Soils of the Great Lakes Coastal Zone

5.1 Introduction

The previous chapter outlined the taxonomic structure of soils and soil-forming factors in the Great Lakes Coastal Zone, recognizing that Alfisols are the dominant soil order (28% of all soil series), followed by Spodosols (27%), Inceptisols (19%), Entisols (12%), Mollisols (11%), and Histosols (4%) (Table 4.1). This chapter examines the dominant soil great groups, soil climate, vegetation, parent materials, landforms, time required to establish the soil within each order. Chaps. 6 through 10 will examine soils, parent materials, and landforms by segment around each lake.

5.2 Alfisols

Alfisols are the most abundant soil order in the Great Lakes Coastal Zone in terms of number of soil series and area. Alfisols occur mostly along the Lake Michigan and Lake Erie shores, but they also are present on the shores of Lake Huron, Superior, and Ontario. The well-drained Udalfs comprise 72% of the Alfisols and the poorly drained Aqualfs the remaining 28% (see Table 4.1). Hapludalfs (45%) are the predominant great group within the Alfisols, followed by Glossudalfs (26%), Endoaqualfs (16%), and Epiaqualfs (9.3%).

The Collamer, Hegberg, Rimer, Augustana, Bonduel, Boyer, Dunkirk, Emmet, Forbay, Gageville, Hilton, Kewaunee, Kibbie, Manawa, Onaway, Spinks, and Symco soil series are commonly occurring Hapludalfs, especially along Lake Michigan. Two examples of Hapludalfs are the Augustana or Hegberg (upper) and Rimer (lower) series (Fig. 5.1). The Augustana-Hegberg soil series (Fig. 5.1a) is derived from 30 cm of loamy lacustrine sediments over loamy till. The Bt horizons are derived from both sediments and have a reddish brown color with a pronounced subangular blocky structure. The Rimer soil series (Fig. 5.1b) is derived from 28 in. (71 cm) of glaciolacustrine material over clayey till. The profile includes a dark brown Ap horizon from 0 to 10 in. (0–25 cm), a yellowish brown E1 horizon from 10 to 17 in. (25–43 cm), a brown E2 horizon from 17 to 23 in. (43–58 cm), and a Bt horizon from 23 in. (58 cm) to the bottom of the tape with Fe depletions (whitish zones).

The Miskoaki, Big Iron, Flintsteel, Omena, and Sanborg are frequently occurring Glossudalfs, particularly around Lakes Superior and Huron. The Sanborg and Amnicon soils are shown in Fig. 5.2. The Sanborg soil series (Fig. 5.2a) is derived from clayey till. A glossic horizon (E/B and B/E) extends from 5 to 17 in. (12–43 cm), has zones of reddish brown (B) and pink (E colors, and represents the degradation of an argillic (Bt) horizon. The argillic horizon is between 17 and 48 in. (43–122 cm) and has a strong prismatic structure. The Amnicon soil series also is derived from clayey reddish brown till. The profile has a dark brown A horizon from 0 to 2 in. (0–5 cm), a pinkish gray E horizon from 2 to 5 in. (5–12 cm), a glossic horizon (E/B and B/E) from 5 to 16 in. (12–41 cm), and a Bt horizon with prismatic structure to the base of the pit.

The Niagara and Rhinebeck soil series are commonly occurring Endoaqualfs, especially along the Lake Erie and Lake Ontario shores. The Churchville soil series, an Aeric Endoaqualfs, is depicted in Fig. 5.3. The soil is derived from 26 in. (66 cm) (the rock hammer is 11 in. or 27 cm long) of clayey lacustrine materials over gravelly loam till. Gravels are visible in the left center of the photo. The horizons include a very dark grayish brown Ap from 0 to 9 in. (23 cm), a pinkish gray E horizon from 9 to 15 in. (27–38 cm), and a reddish brown Bt horizon from 15 in. (38 cm) to the base of the pit.

The Blount and Eldes soil series are frequently occurring Epiaqualfs. The Hoytville soil series, a Mollic Epiaqualfs, is shown in Fig. 5.4. This soil is derived from clayey till and contains a very dark grayish brown Ap horizon from 0 to 9 in. (23 cm) (base of the shovel handle), a dark gray E horizon from 9 in. to 18 in. (23–46 cm) (end of the visible wood portion of the shovel), a yellowish brown Btg horizon from 18 to 40 in. (46–102 cm) (halfway along the blade of

J. G. Bockheim, *Soils of the Laurentian Great Lakes, USA and Canada*,
https://doi.org/10.1007/978-3-030-52425-8_5

Fig. 5.1 **a** The Augustana or Hegberg soil series, an Aquollic or Mollic Oxyaquic Hapludalfs from St. Louis Co., MN. *Note* the photo shows the upper 2 feet of the profile. Photo courtesy of Larissa Hindman, Soil Scientist, NRCS, Duluth, MN. **b** The Rimer soil series, an Aquic Arenic Hapludalfs. The scale is in feet. Photo from the soil survey of Hancock Co., OH (Robbins and Martin 2006)

the shovel), and a brown C horizon from 40 in. (102 cm) to the base of the pit.

The Cuttre and Londo soil series are commonly occurring Glossaqualfs, especially around Lakes Superior and Huron. The Cuttre soil series is an Aeric Glossaqualfs (Fig. 5.5). The Cuttre soil is derived from clayey till and has a dark reddish brown A horizon from 0 to 3 in. (0–8 cm) (the rule is in inches), a glossic horizon (E/B and B/E) with brown (B) and pinkish gray (E) zones from 3 to 12 in. (8–30 cm), and dark reddish brown Bt horizons from 12 in. (30 cm) to the bottom of the pit. The whitish color between 22 and 44 in. (56–110 cm) are soft masses of calcium carbonate.

More than two-thirds (70%) of the Alfisol soil series have a mesic soil-temperature regime; the remaining 30% are in the frigid soil-temperature class (see Fig. 4.4). Nearly half (49%) of the Alfisol soil series are in loamy particle-size classes and 37% are in clayey (clayey, fine, very-fine) classes. Alfisols in the GLCZ invariably have a mixed mineralogy (84%); 16% contain primarily illite. Nearly two thirds (63%) of the Alfisols are in active or superactive cation-exchange capacity activity classes.

Alfisols in the GLCZ feature mainly temperate broad-leaved forest (58% of soil series) or temperate mixed coniferous and broad-leaved forest (34%). The parent materials are primarily till (43%), lacustrine-glaciolacustrine (40%), and outwash (11%). The landforms featuring Alfisols are primarily till plains (48%) and lake plains (40%). Alfisols occur on landforms that are 11.2 kyr in age or more (see Fig. 4.6).

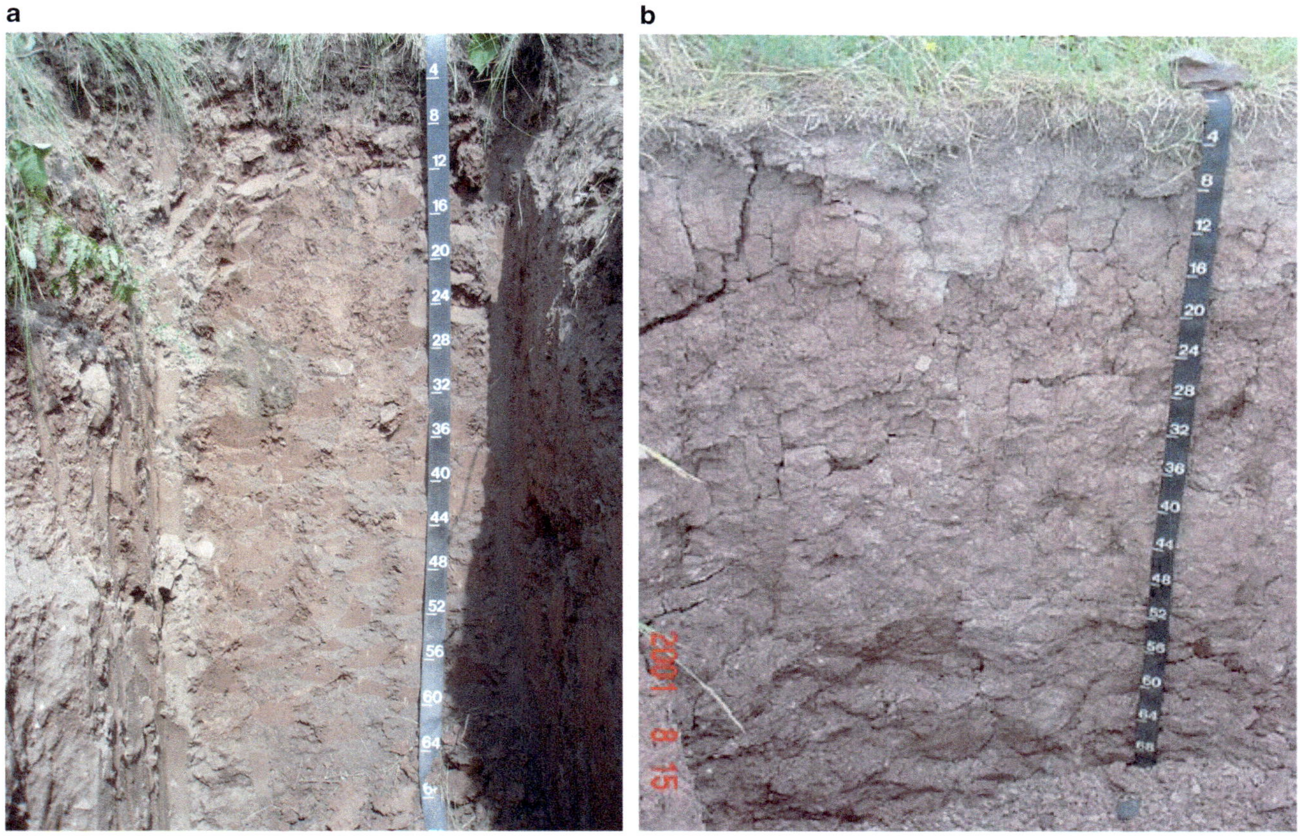

Fig. 5.2 **a** The Sanborg soil series, an Oxyaquic Glossudalfs from Lake County, MN. Photo courtesy of Larissa Hindman, Soil Scientist, NRCS, Duluth, MN. **b** The Amnicon soil series, an Oxyaquic Vertic Glossudalfs from Upper Michigan. *Note* cracking throughout the pedon. Photo courtesy of Dwight Jerome, Resource Soil Scientist, NRCS, Marquette, MI. The scales are in inches

Analytical data for soil series representative of the five dominant great groups of Alfisols in the GLCZ are provided in Table 5.1. The key properties are the presence of an argillic (Bt) horizon enriched in transported clay (shown in bold face) and a base saturation within and below the argillic horizon in excess of 35%. Many Alfisols in the Great Lakes region are alkaline as evidenced by a pH in the C horizon ranging from 7.7 to 8.5. The Miskoaki and Cuttre pedons have argillic horizons enriched in calcium carbonate (Btk horizons).

5.3 Spodosols

Spodosols are the second most dominant soil order in the GLCZ and occur mainly around the shores of Lakes Superior, Huron, and Michigan; they occur to a limited extent around Lake Ontario and are lacking around Lake Erie. The well-drained Orthods comprise 73% of the Spodosols, and the poorly drained Aquods comprise the remaining 27% (see Table 4.1). The Haplorthods (61%) are the predominant great group within the Spodosols, followed by Endoaquods (13%), Fragiorthods (8.4%), and Epiaquods (7.3%).

The most commonly occurring Haplorthods in the GLCZ are the Rubicon, Croswell, Kalkaska, Longrie, Deerton, Manistee, Melita, Waiska, Covert, East Lake, Kellogg, Keweenaw, Leelanau, Michigamme, and Rousseau soil series. Figure 5.6 contains images of the Croswell (Oxyaquic), Kalkaska (Typic), Rubicon (Entic), Manistee (Alfic), and Michigamme (Fragic) soil series. The Croswell soil (Fig. 5.6a) is derived from sandy glaciofluvial materials and has a black A horizon from 0 to 5 cm (the rule is in centimeters), a brown E horizon to 13 cm, and a dark reddish brown spodic (Bs) horizon to 41 cm. The water table is at 70 cm. Derived from sandy outwash, the Kalkaska soil series (Fig. 5.6b) is Michigan's state soil. It contains a thin

Fig. 5.3 The Churchville soil series is an Aeric Endoaqualfs. The rock hammer is 11 in. long. Photo from the soil survey of Akwesasne Territory: St. Regis Mohawk Reservation (Trevail 2006), Franklin Co., NY

(2–5 cm) black A horizon, a thick (to 50 cm) gray albic (E) horizon, a strongly tongued, dark reddish brown Bh horizon, and brown Bs horizons to 90 cm.

The Rubicon soil (Fig. 5.6c) is derived from sandy lacustrine and outwash deposits and has a black A horizon from 0 to 1 in. (0–2.5 cm), a light brownish gray E horizon from 1 to 6 in. (2.5–15 cm), brown and dark yellowish brown Bs (spodic) horizons from 6 to 21 in. (15–53 cm), and a light yellowish brown C horizon from 21 to 48 in. (53–122 cm). The Manistee soil series (Fig. 5.6d) is an Alfic Haplorthods developed in sandy over clayey lacustrine deposits. The profile is bisequal; the upper sequum has an Ap, E, and Bs horizons to 8, 11, and 28 in. (20, 27, and 71 cm), respectively; the lower sequum has an E' and a 2Bt horizon developed in the reddish brown clayey materials. The Michigamme soil series (Fig. 5.6e) is a Fragic Haplorthods derived from loamy sediment over loamy till with metamorphic bedrock at 32 in. (81 cm). The profile is bisequal; the upper sequum contains a thin grayish E horizon or a series of Bhs and Bs horizons to 25 in. (64 cm). The

Fig. 5.4 The Hoytville soil series, a Mollic Epiaqualfs. The shovel is 44 in. long. Photo from the soil survey of Wood County, OH (Robbins and Lantz 2007)

lower sequum has a weakly developed 2E/Bx, with roots concentrated on top of the fragipan.

The Pipestone, Au Gres, Kinross, Wainola, Charlevoix, Ingalls, Iosco, and Zeba soil series are commonly occurring Endoaquods in the GLCZ. The Au Gres (Typic) and Iosco (Argic) are shown in Fig. 5.7. The Au Gres soil series (Fig. 5.7a) is a Typic Endoaquods derived from sandy lacustrine deposits, with a black A horizon to 3 in. (8 cm), a grayish brown E horizon to 14 in. (36 cm), and a strong brown spodic (Bhs, Bs) horizon to 30 in. (76 cm). Prominent yellowish red masses of oxidized Fe appear as blotches in the photo. The Iosco soil series (Fig. 5.7b) is an Agric Endoaquods from sandy over loamy lacustrine deposits. The Iosco is bisequal; the upper sequum contains a dark grayish brown A horizon, a light brownish gray E horizon, and dark reddish brown Bhs and Bs2 to 24 in. (64 cm); the lower sequum has a pale brown E' horizon underlain by a brown argillic (2Bt) horizon.

Fig. 5.5 The Cuttre soil series, an Aeric Glossaqualfs. The scale is in inches. Key features are the blocky structure and the whitish secondary carbonates in the argillic horizon. The photo is from the soil survey of Ontonagon Co, MI (Eversoll and Carey 2010)

The Paavola, Yalmer, Montreal, and Munising soil series are commonly occurring Fragiorthods, primarily around Lake Superior. The Munising and Paavola are Typic Fragiorthods (Fig. 5.8). The Munising soil series (Fig. 5.8a) is a Typic Fragiorthods derived from loamy till. The soil is bisequal, with the upper sequum contain a black A horizon, a pinkish gray E horizon, and a dark reddish brown spodic (Bhs, Bs) horizon to 16 in. (40 cm). The fragipan includes the Bx horizon (16 to 22 in.; 40 to 55 cm) and the (B/E)x (22 to 35 in.; 55 to 90 cm). The latter horizon contains reddish brown material with tongues of pinkish gray material. The Bt horizon has a reddish brown color and extends beyond 55 in. (140 cm). The Paavola soil series (Fig. 5.8b) is an Alfic Oxyaquic Fragiorthods derived from wave-washed till. A bisequal soil, the Paavola has a dark reddish brown A horizon and a series of Bhs and Bs (spodic) horizons to 30 in. (76 cm). The lower sequum has a dark reddish gray 2 (E/B)x horizon from 30 to 38 in. (76–97 cm) and a reddish brown 2Btx (argillic and fragipan) to 59 in. (150 cm). The latter horizon has masses of Fe accumulation.

The Finch and Saugatuck soil series are commonly occurring Duraquods, especially along Lakes Huron and Michigan. The Finch soil series, a Typic Duraquods, is derived from sandy lacustrine sediments (Fig. 5.9). The profile contains a black A horizon, a light gray E horizon, a dark reddish brown Bhsm horizon, and a yellowish red Bsm horizon to about 40 in. (102 cm). A somewhat poorly drained soil, the Finch has standing water at 52 in. (132 cm). The Allendale soil series (Fig. 5.10) is an Alfic Epiaquods derived from sandy over clayey lacustrine deposits. The profile has a very dark gray A horizon, a gray E horizon, a dark brown Bhs horizon, and a yellowish brown Bs horizon to 32 in. (81 cm) in the upper sequum. The lower sequum contains a pale brown E' and a 2Bt (argillic) horizon to 34 in. (86 cm) over brown clayey materials. The Skanee soil series (not shown) is a commonly occurring Epiaquods around Lake Superior.

The Wallace soil series (Fig. 5.11) is a frequently occurring Durorthods around Lakes Michigan, Huron, and Superior. Derived from sandy lacustrine and outwash deposits, the Wallace soil series has a black Oe horizon, underlain by a light brownish gray E, dark reddish brown Bhs, and brown Bhsm and Bsm to a depth of 51 in. (80 cm). The ortstein layer is strongly and exists as a continuous horizon but extends as tongues to 55 in. (140 cm).

The majority (89%) of the Spodosol soil series have a frigid soil-climate regime; the remaining 11% are in the mesic soil-temperature class (see Fig. 4.3). Nearly two-thirds (61%) of the Spodosols in the GLCZ are in the sandy particle-size class and 25% are coarse-loamy. More than three-quarters (82%) of the Spodosol soil series have a mixed mineralogy; 18% are in the isotic soil mineral class. The majority (59%) of the Spodosols in the GLCZ have minimal CEC activity; however, 41% of the Spodosol soil series are in semiactive, active, or superactive classes.

Spodosols in the GLCZ support either temperate mixed coniferous and broad-leaved forests (70%) or temperate broad-leaved forest (23%). It is of interest that about two-thirds of the Spodosols supporting broad-leaved forest have an argillic horizon and/or a fragipan. The dominant parent materials of Spodosols are lacustrine-glaciolacustrine deposits (40%), till (27%), or outwash (24%). Predominant landforms include lake plains-terraces (47%), moraines-till plains (26%), and outwash plains (13%). Although some Spodosols may form in the GLCZ in as little as 5.3 kyr, a more appropriate time frame is 11.2 kyr (see Figs. 4.16 and 4.17).

Analytical data for soil series representative of each of the seven great groups of Spodosols in the GLCZ are provided in Table 5.2. The key properties of Spodosols are the presence of a spodic horizon containing 85% or more spodic

Table 5.1 Primary characterization data for Alfisol (Luvisol) soil series in the Great Lakes Coastal Zone (Natural Resources Conservation Service 2018)

Horizon	Depth	Clay	Silt	Sand	OC	CEC7	Base sat.	pH	Fe_d	Al_d
	(cm)	(%)	(%)	(%)	(%)	(cmol(+)/kg)	(%)	H_2O	(%)	(%)
Collamer series; fine-silty, mixed, smectitic, mesic Glossaquic Hapludalfs; Pedon No. 08N0168; Alleghany Co., NY; glaciolacustrine deposits (fine); lake plain										
Ap	0–30	15.4	57.3	27.3	1.49	9.37	100	7.1	0.9	0.1
E/B	30–46	20.3	67.1	12.6	0.53	7.8	100	6.7	1.3	nd
Bt/E	46–58	24.1	64.5	11.4	0.39	8.5	100	6.9	1.5	0.1
Bt	58–84	25.5	66.1	8.4	0.26	8.1	100	7.2	1.4	0.1
C1	84–114	3.2	51.7	45.1	0.07	1.8	100	7.7	0.5	tr
Emmet series; coarse-loamy, mixed, active, frigid Mollic Hapludalfs; Pedon No. 40A1920; Emmet Co., MI; till (SL); till plain										
Ap	0–18	6.9	16.5	76.6	1.26	6.3	42	7.1	0.2	tr
Bw1	18–26	6.2	22.7	71.1	0.89	5.3	31	7.3	0.2	tr
Bw2	26–46	5.4	15.8	78.8	0.34	3.5	25	7.5	0.1	tr
E	46–56	7.1	19.5	73.4	0.26	4.2	100	7.4	0.1	tr
Bt	56–74	8.2	10.3	81.5	0.33	4.2	100	7.9	0.1	tr
2C1	74–99	10.6	16.5	72.9	0.17	2.4	100	8.4	0.1	tr
Rimer series; loamy, mixed, active, mesic Aquic Arenic Hapludalfs; Pedon No.										
HK-035; Hancock Co., OH; glaciolacustrine sediments/till; till plain (wave-washed)										
Ap	0–30	5.6	18.0	76.4	0.75	7.4	25	4.8	nd	nd
E1	30–36	5.9	18.5	75.6	0.23	5.9	34	5.5	nd	nd
E2	36–56	6.3	10.2	83.5	0.17	5.3	47	5.8	nd	nd
Bt	56–66	18.0	11.2	70.8	0.17	12.4	72	6.3	nd	nd
2Btg	66–86	33.4	31.6	35.0	nd	15.4	81	7.0	nd	nd
2Bt	86–97	30.1	44.3	25.6	nd	nd	nd	7.9	nd	nd
2BC	97–117	34.2	45.0	20.8	nd	nd	nd	7.8	nd	nd
2Cd	117–168	36.4	43.8	19.8	nd	nd	nd	7.8	nd	nd
Kewaunee series; fine, mixed, active, mesic Typic Hapudalfs; Pedon No. 40A1540; Manitowoc Co., WI; loess/till (clayey); till plain										
E	0–5	9.8	60.1	30.1	2.25	9.8	22	4.7	0.4	nd
Bt	5–18	17.3	57.7	25.0	2.47	14	7	4.3	1.1	nd
2Bt1	18–28	32.7	42.2	25.1	0.64	12.5	40	4.4	1.2	nd
2Bt2	28–53	47.0	28.3	24.7	0.46	19.5	100	5.0	1.4	nd
2Bt3	53–76	37.1	37.6	25.3	0.24	12.4	100	7.1	0.8	nd
2C	76–114	36.3	38.8	24.9	0.16	8.9	100	7.9	0.7	nd
Miskoaki series; very-fine, mixed, active, frigid Vertic Glossudalfs; Pedon No. 91P0391; Douglas Co., WI; till (clayey); till plain										
A	0–5	**38.4**	**36.7**	**24.9**	**5.13**	**25.4**	**50**	**4.7**	nd	nd
B/E	5–25	**49.7**	**31.7**	**18.6**	**2.02**	**30.5**	**73**	**4.6**	nd	nd
Bt1	25–46	**71.9**	**18.6**	**9.5**	**0.51**	**32.1**	**100**	**7.4**	nd	nd
Bt2	46–61	**65.3**	**24.5**	**10.2**	**0.21**	**24.3**	**100**	**8.0**	nd	nd
Btk	61–102	**68.7**	**21.4**	**9.9**	**0.20**	**19.9**	**100**	**8.3**	nd	nd
BC	102–157	**64.0**	**25.7**	**10.3**	**0.07**	**17.6**	**100**	**8.5**	nd	nd
Rhinebeck series; fine, illitic, mesic Aeric Endoaqualfs; Pedon No. 83P0126; St. Lawrence Co., NY; glaciolacustrine sediments (fine); lake plain										
Ap	0–18	24.4	56.3	19.3	1.74	14.2	69	7.1	nd	nd
E	18–25	30.3	53.4	16.3	0.36	13.4	72	7.4	nd	nd
Bt1	25–46	36.2	52.1	11.7	0.27	18.9	93	7.5	nd	nd

(continued)

Table 5.1 (continued)

Horizon	Depth (cm)	Clay (%)	Silt (%)	Sand (%)	OC (%)	CEC7 (cmol(+)/kg)	Base sat. (%)	pH H_2O	Fe_d (%)	Al_d (%)
Bt2	46–56	44.7	37.3	18.0	0.25	22.7	100	7.4	nd	nd
Bt3	56–71	36.8	44.0	19.2	0.19	19.5	99	7.5	nd	nd
BC	71–81	20.5	33.8	45.7	0.15	10.2	100	7.8	nd	nd
Cuttre series; very-fine, mixed, active, frigid Aeric Glossaqualfs; Pedon No. 12N7627; Lake Co., MN; till (clayey); till plain										
A	0–8	33.7	41.7	24.6	4.8	22.4	82	5.8	nd	nd
E/B	8–15	28.7	43.3	28.0	1.5	14.4	78	5.6	nd	nd
B/E	15–20	44.8	34.6	20.6	1.8	20.7	77	5.5	nd	nd
Bt	20–58	69.2	20.0	10.8	0.7	31.7	100	6.2	nd	nd
Btk	58–74	72.5	18.0	9.5	0.3	28.7	100	7.9	nd	nd
C1	74–142	80.7	14.0	5.3	0.4	28.7	100	8.1	nd	nd
Blount series; fine, illitic, mesic Aeric Epiaqualfs; Pedon No. ER-031; Eric County, OH; till; till plain (wave-washed)										
Ap	0–20	18.2	50.4	31.4	1.51	15.2	65	6.1	nd	nd
BE	20–25	29.0	46.8	24.2	0.52	15.9	43	4.9	nd	nd
Bt1	25–51	41.1	39.9	19.0	0.41	22.2	46	4.7	nd	nd
Bt2	51–69	35.7	41.0	23.3	0.52	22.8	74	6.4	nd	nd
C1	69–97	29.4	45.6	25.0	0.46	nd	nd	7.8	nd	nd
C2	97–122	21.8	41.1	37.1	nd	nd	nd	7.9	nd	nd

The argillic horizon (Bt) is highlighted in bold face

materials. Spodic materials are evidenced by acidic conditions (pH $\leq$ 5.9); organic C levels exceeding 0.6%; strong, reddish hues, cementation by Al and organic C (optional); 10% or more cracked coatings on sand grains; an optical density of the oxalate extract (ODOE) greater than or equal to 0.25 (and twice that or more of an overlying eluvial horizon); and Al + ½ Fe ratio of 0.50 or more (and twice or more of that of an overlying eluvial horizon). Spodic horizons in Table 5.2 are highlighted in bold face. Most Spodosols have an albic (E) from which Fe, Al, and OC have been leached; and many Spodosols in the Great Lakes region have additional diagnostic subsurface horizons, such as a fragipan (designated by an *x*), ortstein (cemented by Fe and/or Al and OC; designed by an *m*), an argillic horizon (Bt), and a glossic horizon (indicated by E/B, B/E, etc.). For example, the Munising pedon in Table 5.2 contains five diagnostic horizons, including an ochric horizon (A and E), an albic horizon (E), a spodic horizon (Bhs and Bs), a fragipan (Bx1, Ex, Bx2), and an argillic horizon (Bt).

5.4 Inceptisols

Inceptisols are the third most common soil order in the GLCZ. They are most common on Lakes Erie, Ontario, and Superior, but also occur around Lakes Michigan and Huron. The majority (62%) of Inceptisols in the GLCZ are poorly drained Aquepts, with Udepts accounting for the remaining 38%. The dominant great groups include Endoaquepts (29%), Eutrudepts (21%), Epiaquepts (18%), Humaquepts (14%), and Dystrudepts (13%) (Table 4.1).

The most common Endoaquepts re the Canandaigua, Fredon, Toledo, Ensley, Lamson, Angelica, and Sanilac soil series. The Gay (Aeric) and Minocqua (Typic) are depicted in Fig. 5.12. The Gay soil series (Fig. 5.12a) has formed from loamy till and contains 8 in. (20 cm) of black Oe and Oa and dark brown A horizons, a 3 in. (7.5 cm) light gray E horizon, and a brown Bw horizon to 22 in. (56 cm). The Minocqua soil series (Fig. 5.12b) is derived from silt-loam

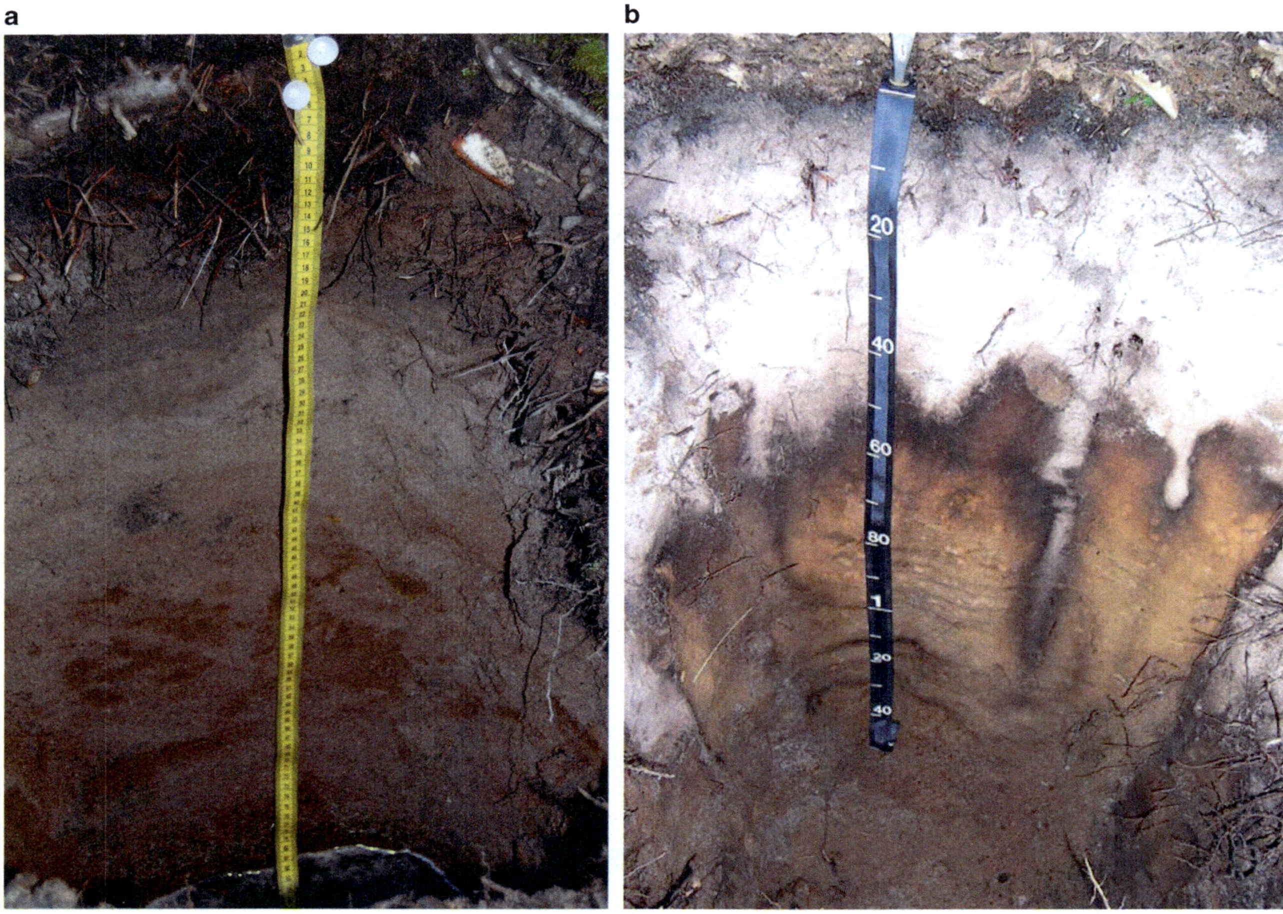

Fig. 5.6 **a** The Croswell series, an Oxyaquic Haplorthods from Marquette Co., MI. The scale is in centimeters. Photo by J. Bockheim. **b** The Kalkaska series is a Typic Haplorthods and the state soil of Michigan. The scale is in decimeters. The albic (E) and upper spodic (Bh) horizons have an irregular lower boundary. Photo from the soil survey of Alger County, MI (Schwenner 2013). **c** A Rubicon soil series, an Entic Haplorthods. The scale is in inches. Photo from soil survey of Luce Co., MI (Whitney and Rodock 2006). **d** The Manistee soil series is an Alfic Haplorthods. The scale is in feet. The photo is from the soil survey of Alcona Co., MI (Williams 1998). **e** The Michigamme soil series, a Fragic Haplorthods. The fragipan underlies the rooted zone. The exposure is about 2 feet in depth. Photo from soil survey of Isle Royale National Park, MI (Carey 2012)

alluvium over stratified outwash and has a black A horizon to 6 in. (15 cm) that overlies a thin grayish brown E horizon and grayish brown Bg (gleyed) horizons.

The most common Eutrudepts are the Minoa, Greysolon, Mesaba, Barto, Detour, Rockland, Summerville, Ensign, and Normanna soil series. They are especially prominent along the northwest shore of Lake Superior. The Barto and Mesaba soil series (Eutrudepts) are shown in Fig. 5.13. The Barto soil series (Fig. 5.13a) is a Lithic Eutrudepts and originates from 33 cm of loamy till over unweathered bedrock and has a black A horizon over a brown Bw (cambic) horizon. The Mesaba soil series (Fig. 5.13b) is a Dystric Eutrudepts and also has developed from loamy till over bedrock. The horizonation is similar to that of the Barto except that the cambic horizon is thicker, i.e., extends to 29 in. (74 cm).

The Lenawee, Parkhill, Painesville, and Conneaut soil series are the most frequently occurring Epiaquepts in the GLCZ. The Nevens soil series is a Typic Epiaquepts (Fig. 5.14). The Nevens soil series is derived from loamy material overlying dense till. The profile is composed of a very dark brown A horizon to 6 in., followed by a cambic horizon (Bw) to 37 in. (94 cm).

The Ira, Sodus, and Williamson soil series are common Fragiudepts and occur primarily along the southeast shore of Lake Ontario. The Chenango soil series is a Dystrudepts that is common on Lake Erie. The Tatches soil series, a Lamellic Dystrudepts, occurs along Lake Michigan (Fig. 5.15). The Tatches soil series is derived from loamy over sandy outwash and has a thin very dark gray A horizon over strong brown Bw horizon to a depth of 32 in. (81 cm). The lamellae

c d

Fig. 5.6 (continued)

e

Fig. 5.6 (continued)

occur in the underlying sand to a depth of more than 80 in. (203 cm).

The Sabattis soil series (Fig. 5.16) is a Histic Humaquepts that is common on till plains of Lake Superior. Formed from loamy materials over sandy loam till, the Sabattis soil series has a black Oa, a dark grayish brown AB, and a Bg horizon to a depth of 21 in. (53 cm).

Nearly two thirds (61%) of the Inceptisols have a mesic soil-temperature regime; the others (39%) have a frigid soil-temperature class. Nearly two-thirds (65%) of the Inceptisols in the GLCZ are in loamy (coarse-loamy, fine-loamy, loamy, loamy-skeletal) particle-size classes. More than three-quarters (81%) of the Great Lakes Inceptisols are in the mixed soil mineral class; and 75% are in semiactive, active, or superactive CEC activity classes.

The vegetation on Inceptisols is variable, including temperate mixed forest (38%), lowland mixed forest (24%), temperate broad-leaved forest (21%), lowland broad-leaved forest (16%), and nonforested wetlands (2%). The predominant parent materials for Great Lakes Inceptisols include till (39%), lacustrine-glaciolacustrine sediments (38%), alluvium (10%), and outwash (5.6%). Predominant landforms include lake terrace-plains (39%), moraine-till plains (30%), glacial lake benches (10%), alluvial fans-floodplains (9%), and outwash plains (9%). Many Inceptisols have a cambic horizon that requires $\leq$ 5.3 kyr to form (see Figs. 4.16 and 4.17).

Analytical data for soil series representative of six of the seven great groups of Inceptisols in the GLCZ are provided in Table 5.3. The key properties are the presence of a cambic horizon, or in the case of Sodus soil series a fragipan (Bx) and the Newton soil series an umbric (Ap, A1, and A2) horizon. The analytical properties of Inceptisols are quite variable, since these soils represent early stages in soil development.

5.5 Entisols

Nearly two-thirds (62%) of the Entisols in the Great Lakes region are well-drained Psamments, Orthents, or Fluvents; the remaining 38% are Aquents. Entisols are most common on Lakes Michigan, Superior, and Huron and least common on Lakes Ontario and Erie. The dominant great groups are Udipsamments (36%), Psammaquents (22%), and Udorthents (18%).

The Plainfield, Oakville, Deer Park, Tyner, Colonie, Elnora, Wurtsmith, and Eastport soil series are common Udipsamments, occurring on all of the Great Lakes. The Deer Park (Spodic), Plainfield (Typic), and Coloma (Lamellic) soil series are shown in Fig. 5.17. The Deer Park soil series (Fig. 5.17a) is derived from eolian materials (sand dunes) or beach ridges and is composed of a thick, pale brown E horizon over a brown spodic-like Bs horizon to a depth of 38 in. (97 cm). The Plainfield soil series (Fig. 5.17b) is formed from sandy outwash and contains a dark brown A horizon, a brown Bw horizon (to 22 in.; 56 cm), a brown BC (transitional) horizon to 36 in. (91 cm), and a yellowish brown C horizon. The Coloma soil series (Fig. 5.17c) is derived from sandy till and features a thin, very dark brown A horizon over brown Bw horizons to 20

Fig. 5.7 The Au Gres soil series (left) is a Typic Endoaquods. Photo from the soil survey of Alcona County, MI (Williams 1998). The Iosco soil series (right) is an Argic Endoaquods. Photo from the soil survey of Alpena Co., MI (Williams 2007). The scales are in feet

in. (50 cm) and lamellae (E and Bt) to more than 79 in. (200 cm).

The Roscommon, Deford, and Tobico soil series are common Psammaquents on all of the Great Lakes, except Lake Ontario. The Jeske soil series, a Typic Psammaquents is shown in Fig. 5.18. Derived from glaciofluvial sediments over sandstone bedrock, the Jeske soil series has a thin (3 in.; 8 cm) organic layer over pink sandy material (C horizon), with weathered sandstone bedrock occurring at 28 in. (70 cm).

The Quetico and Cosad soil series are common Udorthents in the GLCZ. The Quetico soil series (Fig. 5.19) is derived from 7 in. (18 cm) of loamy, noncalcareous drift over granite bedrock. The profile contains a thin organic layer over dark brown Bw horizon. The underlying granite bedrock is striated from the passing of the glacier.

Other common soils series common along the Great Lakes are in the Quartzipsamments (Nordhouse) and Endoaquents (Brevort) great groups. The Levasseur soil series (Fig. 5.20) is an Aeric Endoaquents that is common on Isle Royale and the Keweenaw Peninsula on Lake Superior. It has formed from glaciofluvial deposits over the Jacobsville sandstone. The profile is compressed into a thin organic layer (O horizon) over a dark reddish gray E horizon (upper two-third of shovel blade), and a dark yellowish brown Bw horizon.

Entisols are roughly split in the mesic and frigid soil-temperature classes. Entisols tend to be coarse-textured

Fig. 5.8 The Munising soil series (left) is a Typic Fragiorthods. The scale is in decimeters. Photo from Soil survey of Keweenaw Co., MI (Tardy 2006). The Paavola soil series (right) is a Typic Fragiorthods. Photo from soil survey of Alger Co., MI (Schwenner 2013). The scale is in inches

with 88% being in sandy or sandy-skeletal particle-size classes; 87% in the mixed soil mineral class; and only 23% in a CEC activity class.

Entisols are covered by temperate mixed forest (39%), temperate broad-leaved forest (21%), lowland mixed forest (16%), and low broad-leaved (16%). The parent materials for Great Lakes Entisols include lacustrine-glaciolacustrine sediments (48%), outwash (16%), alluvium (12%), eolian (10%), and till (8%). The landforms include lake terraces-plains (44%), glacial lake benches (14%), beach ridges (10%), outwash plains (10%), floodplains (9%), till plains (4%), and dunes (4%). Some Entisols may be among the oldest in the Great Lakes region. An example is the Plainfield series, which has remained a Udipsamments for 13 kyr or more, due to a lack of pedogenic intertia, i.e., lack of weatherable minerals.

Analytical data for soil series representative of five of the eight great groups of Entisols in the GLCZ are provided in Table 5.4. The Entisols are the most poorly developed soils, often containing only an ochric horizon and occasionally having an albic subsurface horizon (Nordhouse soil series).

5.6 Mollisols

There are 73 Mollisol soil series in the Great Lakes region. They are most evident on the Lake Michigan shore, occur to a lesser extent along the Huron, Erie, and Superior shores, and are absent along Lake Ontario. Nearly two-thirds (63%) of the Mollisols in the GLCZ are Aquolls with restricted drainage; the remaining 37% are Udolls or Rendolls with good drainage. The dominant great groups are Endoaquolls

Fig. 5.9 The Finch soil series is a Typic Duraquods. Because of restricted drainage, the water table is at 52 in. (132 cm). The scale is in inches. The photo is from the soil survey of Luce Co., MI (Whitney and Rodock 2006)

Fig. 5.10 The Allendale soil series is an Alfic Epiaquods. The scale is in inches. The profile is from the soil survey of Marquette Co., MI. The profile contains sandy materials to 32 in. (81 cm) that are underlain by clayey lacustrine deposits or till. The argillic horizon occurs in the clayey materials. The photo is courtesy of Dwight Jerome, Resource Soil Scientist, NRCS, Marquette, MI

Fig. 5.11 The Wallace soil series is a Typic Durorthods. The duripan is from 20 to 28 in. (50–70 cm). The scale is in decimeters. The photo was taken by Randy Schaetzl (Earth Science Picture of the Day)

(41%), Hapludolls (18%), Argiudolls (14%), Argiaquolls (12%), Epiaquolls (9.5%), and Haprendolls (5.5%).

Common Mollisol soil series in the GLCZ include the Solona (Argiudolls), Ruse (Endoaquolls), Tappan (Epiaquolls), Alpena and Namur (Hapludolls), and Shelter (Haprendolls). The Shag (Endoaquolls), Namur (Hapludolls), and Pewamo (Argiaquolls) soil series are shown in Fig. 5.21. The Shag soil series (Fig. 5.21a) is a Typic Endoaquolls derived from silty glaciolacustrine sediments and has a black A horizon to 10 in. (25 cm), a thin gray E horizon, and a brown cambic (Bw) horizon to 25 in. (64 cm), underlain by a massive, brown C horizon with Fe accumulation. The Namur soil series (Fig. 5.21b) is a Lithic Hapludolls from loamy till over bedrock. The profile contains a very dark brown mollic (A) horizon to 12 in. (30 cm) directly overlying dolostone bedrock. The Pewamo soil series (Fig. 5.21c) is a Typic Argiaquolls derived from till subject to wave-washing during the Highest Lake Maumee approximately 13.8 kyr ago. The Pewamo soil profile contains a very dark brown mollic (Ap, A horizons) epipedon to 13 in. (33 cm) and is underlain by dark gray and gray gleyed argillic horizons (Btg) to 37 in. (94 cm).

Mollisols in the GLCZ tend to be in the mesic soil-temperature class (64%), with the remaining soils in the frigid class (36%). The soils tend to be loamy (coarse loamy, fine-loamy, loamy, loamy-skeletal) (49%), followed by clayey (20%), silty (19%), and sandy (12%). Most (84%) of the Mollisols in the GLCZ have a mixed soil mineralogy; 76% of the Mollisols are in superactive, active, and semiactive CEC activity classes.

The predominant vegetation covering Mollisols in the GLCZ is lowland broad-leaved forest (30%), followed by temperature mixed forest (18%), prairie grasses (14%), and nonforested wetlands (10%). Major parent materials include lacustrine-glaciolacustrine sediments (40%), till (27%), outwash (11%), and alluvium (11%). Landforms include lake plains (51%), till plains (16%), and glacial lake benches (12%). Mollisols in the Great Lakes region may form in as little as 5.3 kyr, but are most common on surfaces > 11.2 kyr in age (see Figs. 4.16 and 4.17).

Analytical data for soil series representative of each of the six great groups of Mollisols in the GLCZ are provided in Table 5.5. The key property is the presence of a mollic epipedon (highlighted in boldface). The mollic horizons of the six soil series shown in Table 5.5 range from 18 to 28 cm in thickness and have a dark color (not shown but

Table 5.2 Primary characterization data for Spodosols (Podzols) soil series in the Great Lakes Coastal Zone (Natural Resourses Conservation Service 2018)

Horizon	Depth (cm)	Clay (%)	Silt (%)	Sand (%)	OC (%)	CEC7 (cmol(+)/kg)	Base sat. (%)	pH H_2O	ODOE	Fe_o (%)	Al_o (%)
Saugatuck series; sandy, mixed, mesic, shallow, ortstein Typic Duraquods; Pedon No. 99P0348; Van Buren Co., MI; glaciofluvial deposits; lake plain											
Ap	0–18	1.5	7.2	91.3	1.9	6.0	30	4.7	0.15	0.04	0.11
E	18–31	0.2	4.8	95.0	0.3	1.4	29	4.6	0.09	tr	0.04
Bhsm1	31–41	1.8	2.3	95.9	1.5	11.1	4	4.5	0.79	0.01	0.48
Bhsm2	41–56	0	2.2	97.8	1.2	7.0	6	4.6	0.38	0.04	0.46
Bs1	56–71	0	1.5	98.5	0.8	4.1	7	4.7	0.22	0.05	0.31
Bs2	71–84	0	1.3	98.7	0.4	2.5	12	4.6	0.23	0.01	0.17
C1	84–135	0	1.9	98.1	0.4	1.8	33	4.8	0.13	tr	0.11
Tula series; coarse-loamy, mixed, superactive, frigid Argic Fragiaquods; Pedon No. 05N0178; Washtenaw Co., MI; till; till plain (wave-washed)											
A	3–13	10.5	48.7	40.8	6.88	23.4	53	4.5	0.25	0.33	0.21
Bs1	18–28	7.4	36.7	55.9	1.31	8.4	31	5.0	0.22	0.40	0.27
Bs2	28–51	4.8	34.2	61.0	0.64	5.3	21	5.3	0.12	0.25	0.31
2E/Bx	51–79	17.0	36.8	46.2	0.05	8.0	100	6.1	0.07	0.27	0.12
2B/Ex	79–130	16.3	48.8	34.9	0.08	8.5	85	5.8	0.08	0.27	0.14
Pipestone series; sandy, mixed, mesic Typic Endoaquods; Pedon No. 95P0529; Benzie Co., MI; outwash; till plain (wave-washed)											
A	4–7	2.0	8.8	89.2	3.30	9.3	11	4.5	0.04	0.03	0.05
E	7–27	0.8	5.0	94.2	0.21	1.2	100	3.9	0.02	0.01	0.01
Bhs	27–35	1.8	4.2	94.0	2.20	10.8	1	4.5	0.55	0.05	0.64
Bs	35–54	1.3	4.0	94.7	1.26	5.2	17	4.7	0.19	0.04	0.50
BC	54–76	0.4	2.9	96.7	0.56	2.6	1	4.8	0.08	0.02	0.25
C	76	0.2	1.3	98.5	0.12	0.8	13	4.9	0.03	tr	0.05
Wallace series: sandy, mixed, frigid, shallow, ortstein Typic Durorthods; Pedon No. 40A0302;											
St. Lawrence Co., NY; lacustrine deposits (sandy); lake plain										Fe_d	
E	0–30	0.4	2.3	97.3	0.17	1.1	27	4.1	nd	tr	nd
Bh	30–36	2.8	3.3	93.9	1.73	31.4	4	4.4	nd	0.80	nd
Bsm	36–44	1.2	4.7	94.1	1.33	27.2	2	4.3	nd	0.80	nd
Bsm	44–61	1.2	1.8	97	1.22	5.8	2	5.2	nd	0.40	nd
BC	61–102	0.5	1.2	98.3	0.14	2.8	4	5.8	nd	0.40	nd
C1	102–147	0.7	1.1	98.2	0.08	1.4	14	5.8	nd	0.10	nd
Munising series; coarse-loamy, mixed, active, frigid Alfic Oxyaquic Fragiorthods; Pedon No. 40A1969; Baraga Co., MI; outwash; lake plain											
A	0–3	5.8	22.6	71.6	6.60	26.5	32	4.8	0.16	0.04	0.02
E	3–23	3.8	21.7	74.5	1.21	7.5	23	4.5	0.07	0.04	0.02
Bhs	23–33	10.6	23.0	66.4	2.55	25.9	6	4.7	0.87	0.62	0.60
Bs	33–53	7.8	22.5	69.7	1.74	14.8	5	4.8	0.34	0.16	0.27
Bx1	53–73	4.4	11.3	84.3	0.23	3.5	11	5.0	0.04	0.05	0.06
Ex	73–100	4.7	17.2	78.1	0.16	2.1	33	5.2	0.02	0.06	0.02
Bx2	100–120	12.4	15.4	72.2	0.03	5.2	60	5.0	0.02	0.12	0.03
Bt	120–155	13.1	15.5	71.4	0.03	5.8	62	4.9	0.02	0.20	0.03
C1	155–205	10.1	16.5	73.4	0.02	4.4	82	5.7	0.01	0.03	0.01
Kalkaska series; sandy, isotic, frigid Typic Haplorthods; Pedon No. 40A1971; Emmet Co., MI;											
outwash; outwash plains						CEC8					

(continued)

Table 5.2 (continued)

Horizon	Depth	Clay	Silt	Sand	OC	CEC7	Base sat.	pH		Fe_o	Al_o
	(cm)	(%)	(%)	(%)	(%)	(cmol(+)/kg)	(%)	H_2O	ODOE	(%)	(%)
E	0–23	1.0	3.2	95.8	0.08	0.7	3	4.7	0.01	nd	nd
Bh	23–28	3.1	3.3	93.6	0.90	10.3	2	4.3	0.33	0.15	0.09
Bhs	28–38	1.8	2.1	96.1	0.73	9.2	1	4.8	0.22	0.12	0.24
Bs	38–58	nd	nd	nd	0.28	nd	nd	5.0	0.08	0.02	0.17
BC	58–95	0.9	0.7	98.4	0.18	1.9	1	5.0	0.03	0.05	0.09
C1	95–188	0.2	2.1	97.7	0.06	1.0	2	5.3	0.01	0.02	0.05
Rubicon series; sandy, mixed, frigid Entic Haplorthods; Pedon No. 40A1975; Alger Co., MI;											
lacustrine deposits (sandy); lake plain						CEC8					
A	0–4	4.9	4.8	90.3	2.88	11.4	20	4.9	0.04	0.19	0.20
E	4–10	2.2	5.1	92.7	0.47	3.2	16	4.6	0.01	0.18	0.18
Bs1	10–23	3.7	5.8	90.5	0.88	10.0	3	5.4	0.2	0.40	0.63
Bs2	23–35	2.1	2.1	95.8	0.22	3.2	1	5.7	0.03	0.26	0.54
BC1	35–53	1.1	0.6	98.3	0.08	1.3	2	5.8	tr	0.19	0.29
BC2	53–69	0.9	1.1	98.0	0.06	1.0	2	5.6	tr	0.19	0.26
C1	69–101	0.6	0.3	99.1	0.04	0.7	3	5.4	tr	0.19	0.23
Annalake series; coarse-loamy, mixed, superactive, frigid Alfic Oxyaquic Haplorthods; Pedon No. 01N0430; Ontonagon Co., MI; glaciofluvial deposits; lake terrace											
Oe	0–8	24.6	54.6	20.8	46.8	nd	nd	4.1	0.17	0.16	0.16
E	8–25	2.3	8.4	89.3	0.27	1.4	29	4.2	0.01		0.01
Bhs	25–30	7.9	2.6	89.5	1.49	11	12	4.3	0.54	0.57	0.36
Bs	30–64	1.8	2.6	95.6	1.06	6.2	19	4.8	0.24	0.33	0.52
B/E	64–107	1.9	16.6	81.5	0.25	2.4	58	5.4	0.05	0.16	0.14
C	107–152	1.9	15.1	83.0	0.03	nd	100	5.8	0.01	0.10	0.03
Alcona series; coarse-loamy, mixed, active, frigid Alfic Haplorthods; Pedon No. 86P0133; Cheboygan Co., MI; glaciolacustrine deposits (fine); lake plain											
A	0–5	5.9	33.7	60.4	3.96	12.8	100	5.8	0.11	0.18	0.09
E	5–18	5.0	31.9	63.1	1.60	6.7	75	5.2	0.09	0.21	0.13
Bs1	18–23	2.7	29.7	67.6	1.03	7.3	82	6.3	0.21	0.57	0.34
Bs2	23–28	0.9	31.0	68.1	0.84	5.2	83	6.4	0.15	0.32	0.34
B/E	28–41	3.0	42.6	54.4	0.30	2.47	100	7.1	0.04	0.07	0.14
2Bt	41–48	23.3	32.6	44.1	0.36	11.1	100	7.2	0.04	0.19	0.22
2C	48–152	3.7	51.3	45.0	0.17	1.8	100	8.0	0.01	0.05	0.02
Sedgwick series; coarse-loamy over clayey, mixed active, frigid Alfic Epiaquods; Pedon No. 00P1214; Bayfield Co., WI; alluvium/till (clayey); till plain											
E	0–13	4.7	11.7	83.6	0.73	4.1	66	5	0.07	0.17	0.06
Bs	13–20	8.0	14.1	77.9	0.58	5.2	49	5.1	0.09	0.35	0.13
2B/E	20–41	40.5	30.9	28.6	0.33	14.0	44	5.2	0.06	0.62	0.17
2Bt1	41–48	49.7	28.4	21.9	0.24	20.7	74	5.5	0.07	0.48	0.18
2Bt2	48–61	48.4	28.6	23.0	nd	22.8	92	6.3	nd	nd	nd
2Bt3	61–86	49.1	28.8	22.1	nd	23.7	100	7.2	nd	nd	nd
2Btk	86–135	42.2	44.4	13.4	nd	16.2	100	8.1	nd	nd	nd

The spodic horizon is highlighted in bold face

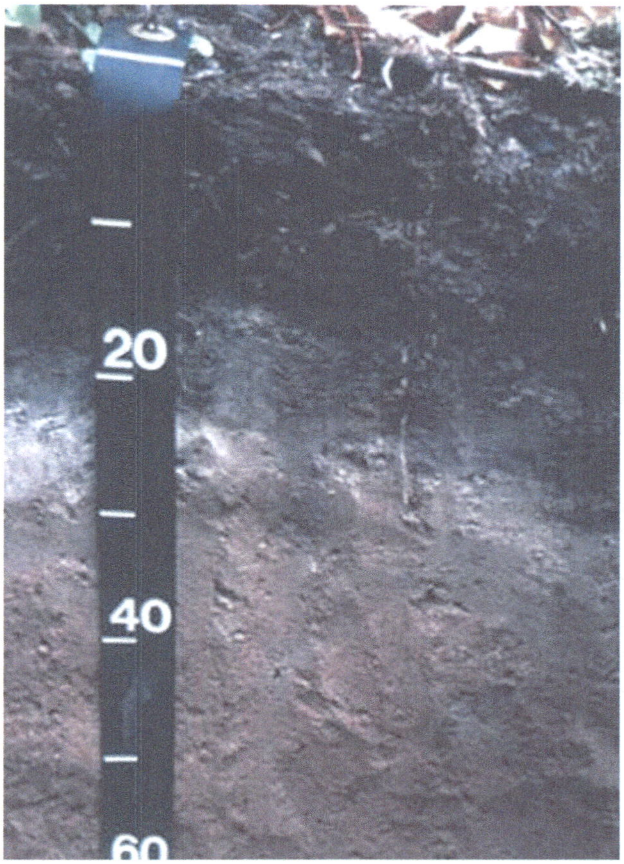

Fig. 5.12 The Gay soil series (left) is an Aeric Endoaquepts. The scale is in decimeters. The photo is from the soil survey of Keweenaw County, MI (Tardy 2006). The Minocqua soil series (right) is a Typic Endoaquepts. The pit is about 20 in. (50 cm) deep. The photo is from the soil survey of Isle Royale National Park, MI (Carey 2012)

visible in Fig. 5.21), an OC $\geq$ 0.6%, and a base saturation $\geq$ 50%.

5.7 Histosols

Histosols are the least abundant of the six soil orders present in the GLCZ. These soils occur most commonly on Lakes Superior, Michigan, and Huron and least commonly on Lakes Erie and Ontario. All of the 26 soil series except the Minong series are in wetlands. The great groups are distributed Haplosaprists (81%), Haplohemists (15%), and Udifolists (4%).

Extensive Haplosaprists include the Cathro, Tawas, Lupton, Carbondale, Markey, Dawson, Houghton, and Carlisle soil series. The Carbondale and Dawson mucks are shown in Fig. 5.22. The Carbondale soil series (Fig. 5.22a) is a Hemic Haplosaprists with a sapric horizon (Oa) to 24 in. (60 cm) (bottom of tape). The Dawson soil series (Fig. 5.22b) is a Terric Haplosaprists with a dark reddish brown Oi horizon to 8 in. (20 cm), a black Oa to 38 in. (97 cm), and an A horizon at the bottom of the pit.

The Rifle soil series is the most common Haplohemists. The Greenwood peat also is a Typic Haplohemist that occurs around Lakes Superior, Michigan, and Michigan (Fig. 5.23). The Greenwood peat contains 6 in. (15 cm) of fibric material

Fig. 5.13 The Barto soil series (top) is a Lithic Eutrudepts from Lake County, MN. The Mesaba soil series (bottom) is a Dystric Eutrudepts from Cook Co., MN. The scales are in centimeters. Both photos were provided by Larissa Hindman, Soil Scientist, NRCS, Duluth, MN

Fig. 5.14 The Nevens soil series is a Typic Epiaquepts. The scale is in inches. The photo is from the soil survey of Isle Royale National Park, MI (Carey 2012)

Fig. 5.15 The Tatches soil series is a Lamellic Dystrudepts that occurs along Lake Michigan. The scale is in inches. The photo is from the soil survey of Benzie-Manistee Counties (Kroell 2008)

Fig. 5.16 The Sabattis soil series is a Histic Humaquepts that is common on till plains of Lake Superior. The scale is in decimeters. The photo is from the soil survey of Isle Royale National Park, MI (Carey 2012)

Table 5.3 Primary characterization data for Inceptisols (some Brunisols) soil series in the Great Lakes Coastal Zone (Natural Resourses Conservation Service 2018)

Horizon	Depth	Clay	Silt	Sand	OC	CEC7	Base sat.	pH	Fe_d	Al_d
	(cm)	(%)	(%)	(%)	(%)	(cmol(+)/kg)	(%)	H_2O	(%)	(%)
Toledo series; fine, illitic, nonacid, mesic Mollic Endoaquepts; Pedon No. SN-015; Sandusky Co., OH; glaciolacustrine deposits (fine); lake plain										
Ap	0–18	52.7	41.3	6.0	2.45	35.6	88	7.0	nd	nd
Bg1	18–30	53.8	41.0	5.2	1.21	36.9	86	6.7	nd	nd
Bg2	30–56	58.3	38.1	3.6	0.97	37.6	87	6.7	nd	nd
Bg3	56–84	62.3	35.3	2.4	0.40	39.1	92	7.0	nd	nd
Bg4	84–112	63.3	34.2	2.5	0.39	40.2	95	7.5	nd	nd
Cg	112–140	63.3	34.3	2.4	nd	nd	nd	7.9	nd	nd
Ensley series; coarse-loamy, mixed, active, nonacid, frigid Aeric Endoaquepts; Pedon No. 40A1941; Delta Co., MI; till; till plain (wave-cut)										
A	0–7	11.7	32.9	55.4	1.51	11.0	55	5.3	nd	nd
Bg	7–15	9.0	43.8	47.2	0.21	6.2	83	5.8	nd	nd
Bw1	15–25	8.7	38.4	52.9	0.11	4.9	100	6.5	nd	nd
Bw2	25–46	7.4	22.9	69.7	0.08	2.8	100	7.4	nd	nd
C	46–74	7.6	25.2	67.2	0.06	2.5	100	7.7	nd	nd
Cg	74–94	8.4	25.0	66.6	0.06	2.1	100	8.0	nd	nd
Normanna series; coarse-loamy, isotic, frigid Oxyaquic Eutrudepts; Pedon No. 14N0995; St. Louis Co., MN; lacustrine/till; till plain										
A	0–16	10.3	47.8	41.9	3.7	14.6	76	5.6	nd	nd
Bw1	16–29	7.9	52.9	39.2	0.7	7.2	50	5.7	nd	nd
Bw2	29–56	5.9	51.7	42.4	0.2	7.1	83	5.8	nd	nd
Bw3	56–104	7.3	31.3	61.4	0.1	9.9	100	6.2	nd	nd
2Cd	104–200	3.6	41.3	55.1	tr	8.3	100	6.8	nd	nd
Minoa series; coarse-loamy, mixed, active, mesic Aquic Dystric Eutrudepts; Pedon No. 40A0379; Erie Co., PA; lacustrine deposits (fine); lake plain										
A	0–10	6.8	31.4	61.8	5.32	33.3	19	3.9	1.1	nd
E1	10–18	11.2	31.1	57.7	3.00	13.0	3	4.2	1.6	nd
E2	18–28	11.2	30.6	58.2	1.89	9.6	5	4.3	1.7	nd
Bw1	28–36	9.6	28.9	61.5	0.82	5.8	3	4.4	1.0	nd
Bw2	36–56	6.8	19.2	74.0	0.55	4.0	38	4.4	1.1	nd
BC	56–66	3.8	10.7	85.5	0.39	2.6	12	4.6	0.8	nd
C1	66–89	5.3	12.1	82.6	0.18	2.9	69	5.4	0.6	nd
Painesville series; coarse-loamy, mixed, active, nonacid, mesic Aeric Epiaquepts; Pedon No. 40A0381; Erie Co., PA; glaciolacustrine deposits (fine); lake plain										
Ap1	0–8	9.7	21.6	68.7	3.45	13.5	36	4.9	1.3	nd
Ap2	8–23	10.2	23.2	66.6	2.00	9.1	30	5.1	1.5	nd
Ag	23–38	5.2	16.4	78.4	0.23	2.4	71	5.6	1.1	nd
Bg1	38–48	7.8	14.4	77.8	0.12	3.0	83	6.1	1.8	nd
Bg2	48–56	8.2	11.9	79.9	0.11	3.5	94	6.4	1.7	nd
Bg3	56–81	9.4	16.5	74.1	0.16	4.5	100	6.8	2.0	nd
C	81–97	8.2	23.2	68.6	0.26	3.6	nd	7.4	1.8	nd
2C	97–168	19.5	66.7	13.8	1.08	3.0	nd	7.2	0.2	nd
Lenawee series; fine, mixed, semiactive, nonacid, mesic Mollic Epiaquepts; Pedon No. SN-014; Sandusky Co., OH; lacustrine deposits (fine); lake plain										
Ap	0–23	29.8	49.4	20.8	2.47	26.1	86	7.1	nd	nd
Bg1	23–43	29.4	47.1	23.5	1.05	22.3	89	7.3	nd	nd

(continued)

Table 5.3 (continued)

Horizon	Depth (cm)	Clay (%)	Silt (%)	Sand (%)	OC (%)	CEC7 (cmol(+)/kg)	Base sat. (%)	pH H_2O	Fe_d (%)	Al_d (%)
Bg2	43–61	29.9	44.0	26.1	0.48	19.3	91	7.5	nd	nd
Bg3	61–79	27.5	40.9	31.6	0.33	15.2	91	7.6	nd	nd
Bg4	79–99	28.5	41.4	30.1	0.24	14.1	90	7.6	nd	nd
Bg5	99–114	47.2	41.9	10.9	0.28	18.6	92	7.6	nd	nd
BCg	114–127	26.5	49.1	24.4	0.36	13.3	89	7.6	nd	nd
Cg	127–152	24.8	65.0	10.2	0.46	12.6	87	7.7	nd	nd
Sodus series; coarse-loamy, mixed, active, mesic Typic Fragiudepts; Pedon No. 40A0286; Cayuga Co., NY; till; till plain										
Ap	0–18	9.4	38.0	52.6	2.77	15	39	5.7	1.0	nd
Bw	18–41	6.8	38.9	54.3	0.82	8.7	21	5.4	0.9	nd
E	41–51	5.4	37.0	57.6	0.25	4.1	27	5.2	0.4	nd
Bx1	51–76	11.5	36.6	51.9	0.1	5.9	59	5.8	1.0	nd
Bx2	76–104	9.8	37.2	53.0	0.06	5.2	71	6.3	0.7	nd
Bx3	104–135	9.3	35.4	55.3	0.08	4.6	72	6.2	0.7	nd
C1	135–185	10.0	36.7	53.3	0.06	4.4	82	6.3	1.0	nd
C2	185–216	8.4	37.5	54.1	0.06	3.6	36	8.1	0.6	nd
Chenango series; loamy-skeletal, mixed, superactive, mesic Typic Dystrudepts; Pedon No. 65PA039008; Crawford Co., PA; outwash; outwash plain										
Ap	0–25	11.2	54.8	34.0	2.98	21.5	50	6.7	1.6	nd
AB	25–43	8.2	51.6	40.2	1.17	13.2	29	6.0	1.1	nd
2Bw	43–64	7.9	39.3	52.8	0.57	8.4	27	5.5	1.4	nd
3BC1	64–86	3.4	23.7	72.9	0.42	7.2	18	5.5	1.6	nd
3BC2	86–127	6.3	9.1	84.6	0.43	7.8	21	5.4	1.7	nd
3C	127–160	10.0	7.4	82.6	0.18	7.7	42	5.3	1.9	nd
Newton series; sandy, mixed, mesic Typic Humaquepts; Pedon No. 40A1951; Ottawa Co., MI; lacustrine deposits (sandy); lake plain										
Ap	0–28	4.2	6.4	89.4	3.59	6.1	92	5.6	0.2	0.1
A1	28–46	2.6	3.2	94.2	1.97	2.0	60	5.3	tr	0.1
A2	46–66	1.6	1.2	97.2	0.90	0.9	44	5.1	tr	tr
E	66–91	1.5	1.0	97.5	0.37	0.6	67	5.1	tr	tr
Bg1	91–99	2.3	0.5	97.2	0.36	1.1	55	5.1	tr	tr
Bg2	99–112	1.7	1.4	96.9	0.29	1.0	50	5.1	tr	tr
Cg	112–155	0.8	0.4	98.8	0.19	0.5	40	5.0	tr	tr

The cambic horizon (Bg, Bw) and fragipan (Bx) are highlighted in bold face

Fig. 5.17 The Deer Park soil series (top left) is a Spodic Udipsamments. The photo is from the soil survey of Luce Co., MI (Whitney and Rodock 2006). The Plainfield soil series (top right) is a Typic Udipsamments. The photo is from the soil survey of Benzie-Manistee Counties (Kroell 2008). The scales of the two upper photos are in inches. The Coloma soil series (bottom) is a Lamellic Udipsamments. The scale of the lower photo is in centimeters. The photo is from J. Bockheim

Fig. 5.18 The Jeske soil series is a Typic Psammaquents. The scale is in decimeters. The photo is from the soil survey of Alger Co., MI (Schwenner 2013)

(Oi horizon) over hemic materials (intermediate stage of decomposition) that extend beyond 24 in. (60 cm).

The Minong soil series (Fig. 5.24) is the only Folist identified in the Great Lakes region. Classified as a Lithic Udifolists. The Minong soil series occurs only on Isle Royale and is derived from organic materials that have accumulated since the Nipissing lake stage on wave-cut platforms. The Minong profile has 12 in. (30 cm) of brown organic materials over igneous bedrock.

Histosols in the GLCZ most commonly (69%) have a frigid soil-temperature regime; the remaining soil series (31%) occur in the mesic soil-temperature class. More than three-quarters (77%) of the Histosols are in the euic soil reaction class, meaning that they have a pH of 4.5 or more; the remaining Histosols are dysic, i.e., they have a lower pH.

The most common vegetation on Great Lakes Histosols is lowland mixed forest (50%), followed by roughly equal amounts of lowland conifer forest, lowland broad-leaved forests, and nonforested wetlands. The parent materials are organic materials that range from 40 to over 200 cm in thickness. The most common landforms are lake terraces-plains (77%) and glacial lake benches (15%). Histosols require a minimum of 5.3 kyr to form but are most common on surfaces that are either 11.2 or 12.4 kyr in age (see Figs. 4.16 and 4.17).

Analytical data for soil series representative of the three great groups of Histosols in the GLCZ are provided in Table 5.6. The key property is the presence of organic materials that are at least 40 cm thick. Haplosaprists contain highly decomposed organic remains (sapric materials), and Haplohemists contain a predominance of materials intermediate in decomposition stage (hemic materials). Histosols in Terric subgroups have a mineral layer ≥ 30 cm thick within the control section (normally the upper 130 to 160 cm). All of the Histosol pedons in Table 5.6 are euic, meaning that they have a pH of ≥ 4.5 (dysic is less than 4.5).

5.8 Conclusions

In this chapter, the beauty and diversity of soils in the Great Lakes Coastal Zone is richly illustrated by digital images of soil profiles representative of Alfisols, Spodosols, Inceptisols, Entisols, Mollisols, and Histosols. The soil horizons and taxa are readily evidenced by differences in color, structure, redoximorphic features, glossic features, and secondary carbonates.

Fig. 5.19 The Quetico soil series is a Lithic Udorthents that is composed of till underlain by granite. The scale is in centimeters (left) and inches (right). The photo is from the soil survey of Isle Royale National Park, MN (Carey 2012)

Fig. 5.20 The The Levasseur soil series is an Aeric Endoaquents. The pit is about 20 in. deep. The photo is from the soil survey of Alger Co., MI (Schwenner 2013)

Table 5.4 Primary characterization data for Entisol (Regosol) soil series in the Great Lakes Coastal Zone (Natural Resourses Conservation Service 2018)

Horizon	Depth	Clay	Silt	Sand	OC	CEC7	Base sat.	pH	Fe_d	Al_d
	(cm)	(%)	(%)	(%)	(%)	(cmol(+)/kg)	(%)	H_2O	(%)	(%)
Elnora series; mixed, mesic Aquic Udipsamments; Pedon No. 03N0672; Lake Co., OH; lacustrine deposits (sandy); former longshore bar										
Ap1	0–15	5.1	19.6	75.3	0.85	2.9	100	7.5	0.5	0.1
Ap2	15–26	5.8	18.7	75.5	0.91	3.1	100	8	0.5	0.1
Ap3	26–43	4.3	10.4	85.3	0.66	2.0	100	8.2	0.5	tr
Bw1	43–66	7.0	30.2	62.8	0.15	1.8	100	7.7	0.4	0.2
Bw2	66–104	7.4	32.0	60.6	0.19	2.0	75	6.7	0.4	0.2
Bw3	104–132	7.3	33.3	59.4	0.27	1.9	42	5.7	0.8	0.2
C1	132–179	0.8	3.6	95.6	0.06	1.2	100	6.5	0.7	0.1
Oakville series; mixed, mesic Typic Udipsamments; Pedon No. 10N0852; Muskegon Co., MI; eolian deposits; dune										
A	0–6	2.5	6.3	91.2	2.45	6.2	56	5.1	0.1	tr
AE	6–11	3.0	4.9	92.1	1.47	3.3	39	4.7	tr	tr
Bw1	11–24	2.7	3.9	93.4	1.12	2.0	25	4.6	0.1	0.1
Bw2	24–56	1.3	2.8	95.9	0.24	1.0	10	5.3	0.1	0.1
C	56–203	0.1	1.0	98.9	0.03	0.2	50	5.7	nd	nd
Quetico series; loamy, isotic, acid, frigid Lithic Udorthents; Pedon No. 02N0811; St. Louis Co., MN; till; till/bedrock; wave-cut bench										
A	0–5	20.0	43.2	36.8	8.4	36.3	44	5.6	2.9	0.7
Bw	5–18	12.0	41.7	46.3	6.1	31.1	9	5.7	3.2	1.8
R	18+									
Roscommon series; mixed, frigid Mollic Psammaquents; Pedon No. 02N0811; St. Louis Co., MN; lacustrine deposits (sandy); lake plain										
A	0–28	nd	nd	nd	nd	nd	nd	nd	nd	nd
Bg1	28–51	4.9	14.0	81.1	0.62	5.3	15	4.7	0.5	0.2
Bg2	51–71	1.9	3.4	94.7	0.10	1.4	100	5.4	0.4	tr
BCg	71–112	2.4	2.2	95.4	0.09	2.1	100	5.7	0.5	tr
Nordhouse series; mesic, uncoated Spodic Quartzipsamments; Pedon No. 99P0369; Manistee, MI; eolian deposits; dune										
A	0–8	nd	nd	nd	nd	nd	nd	nd	nd	nd
E	8–28	0	2.1	97.9	0.26	0.9	44	4.8	0.1	tr
Bs	28–102	0	0.4	99.6	0.21	0.8	63	5.2	0.1	tr
C1	102–152	0	0.4	99.6	nd	0.3	100	5.4	tr	tr
Brevort series; sandy over loamy, mixed, active, nonacid, frigid Mollic Endoaquents; Pedon No. 92P0374; Montmorency Co., MI; lacustrine deposits (sandy); lake plain										
A	0–8	7.9	34.7	57.4	32.1	138	32	6.5	nd	nd
Eg	8–53	1.3	8.2	90.5	0.11	1.7	100	6.8	nd	nd
Bw	53–107	8.0	14.9	77.1	0.13	4.0	100	7.9	nd	nd
C	107–203	10.6	20.6	68.8	0.05	2.1	100	8.4	nd	nd

Fig. 5.21 The Shag soil series (top left) is a Typic Endoaquolls. The photo is from the soil survey of Isle Royale National Park, MI (Carey 2012). The Namur soil series (top right) is a Lithic Hapludolls. The photo is from the soil survey of Alpena Co., MI (Williams 2007). The Pewamo soil series is a Typic Argiaquolls. The photo is from the soil survey of Hancock Co., OH (Robbins and Feusner 2006). The scale of the photo in the top right is in inches. The scale of the other two images are in feet

Table 5.5 Primary characterization data for Mollisol (some Brunisols) soil series in the Great Lakes Coastal Zone (Natural Resourses Conservation Service 2018)

Horizon	Depth	Clay	Silt	Sand	OC	CEC7	Base sat.	pH	Fe_d	Al_d
	(cm)	(%)	(%)	(%)	(%)	(cmol(+)/kg)	(%)	H_2O	(%)	(%)
Granby series; sandy, mixed, mesic Typic Endoaquolls; Pedon No. 40A1950; Ottawa Co., MI; glaciolacustrine deposits (sandy); lake plain										
Ap	0–20	9.0	17.7	73.3	5.84	16.4	57	4.8	0.3	0.2
A	20–28	10.9	20.0	69.1	3.22	12.4	90	5.2	0.2	0.1
Bg	28–38	2.0	6.0	92.0	0.98	3.2	97	5.4	0.1	tr
Cg1	38–53	0.7	1.2	98.1	0.08	0.6	100	5.4	0.1	tr
Cg2	53–76	0.6	1.0	98.4	0.05	0.6	100	6.6	0.1	tr
Tappan series; fine-loamy, mixed, active, calcareous, mesic Typic Epiaquolls; Pedon No. 40A1956; Huron Co., MI; till; till plain (wave-cut)										
Ap	0–28	21.0	33.2	45.8	1.8	13.3	61	7.6	0.4	0.1
A	28–32	19.7	33.8	46.5	2.6	17.6	55	7.6	0.3	0.1
Bg11	32–37	15.4	32.7	51.9	0.46	7.9	72	7.7	0.3	tr
Bg12	37–54	24.2	42.3	33.5	0.40	8.2	100	7.9	1.3	0.1
Bg2	54–78	22.6	40.1	37.3	0.45	6.7	100	7.9	0.6	tr
C1	78–123	22.1	37.0	40.9	0.41	5.4	100	7.8	0.7	tr
Castalia series; loamy-skeletal, carbonatic, mesic Inceptic Haprendolls; Pedon No. WD-132; Wood Co., OH; residuum (limestone); glacial lake bench										
Ap	0–18	17.6	33.9	48.5	4.61	nd	nd	7.7	nd	nd
Bw	18–41	3.1	20.5	76.4	0.87	nd	nd	7.8	nd	nd
C	41–53	2.0	16.6	81.4	nd	nd	nd	7.9	nd	nd
R	53									
Solona series; coarse-loamy, mixed, superactive, frigid Aquic Argiudolls; Pedon No. 78P0505; Shawano Co., WI; till; till plain										
Ap	0–28	15.9	39.3	44.8	1.68	18.4	100	7.7	nd	nd
Bt1	28–53	13.0	34.1	52.9	0.23	9.8	100	8.0	nd	nd
Bt2	53–61	11.0	33.5	55.5	0.14	7.0	100	8.3	nd	nd
C1	61–96	10.5	32.5	57.0	0.09	4.5	100	8.4	nd	nd
Lachine series; loamy, mixed, superactive, frigid Lithic Hapludolls; Pedon No. 97P0444; Alpena Co., MI; till; glacial lake bench										
Ap	0–23	14.3	41.7	44.0	2.7	15.6	100	7.4	nd	nd
Bw	23–33	15.8	38.5	45.7	1.17	12	100	7.8	nd	nd
C	33–41	6.7	33.7	59.6	0.35	2.4	100	8.3	nd	nd
R	41									
Pewamo series; fine, mixed, superactive, mesic Typic Argiaquolls; Pedon No. 68IL043006; DuPage Co., IL; till; till plain (wave-washed)										
Ap	0–23	27.0	68.0	4.0	3.58	26.7	72	5.6	nd	nd
A1	23–38	35.0	61.0	3.0	1.66	27.0	82	5.6	nd	nd
Bg1	38–53	42.0	53.0	4.0	0.57	29.3	91	6.0	nd	nd
Bg2	53–69	40.0	56.0	3.0	0.41	24.2	98	6.6	nd	nd
Bg3	69–86	42.0	55.0	1.0	0.38	23.2	100	7.0	nd	nd
BCg1	86–117	40.0	57.0	1.0	0.30	20.4	100	6.3	nd	nd
BCg2	117–145	32.0	65.0	2.0	0.30	12.2	nd	7.9	nd	nd
C1	145–178	32.0	66.0	1.0	0.38	10.0	nd	8.2	nd	nd

The mollic epipedon (Ap, A) is highlighted in bold face

Fig. 5.22 The Carbondale soil series (top) is a Hemic Haplosaprists (photo by J. Bockheim). The scale is in decimeters. The Dawson soil series (bottom) is a Terric Haplosaprists. The mineral layer is readily visible at the bottom of the pit at a depth of 38 in. (97 cm). The photo is from the soil survey of Ontonagon Co., MI (Eversoll and Carey 2010)

Fig. 5.23 The Greenwood peat soil series is a Typic Haplohemists. The scale is in decimeters. The photo was provided by Dwight Jerome, Resource Soil Scientist, NRCS, Marquette, MI

Fig. 5.24 The Minong soil series is the only Folist (Typic Udifolists) identified in the Great Lakes region. The soil occurs on wave-cut platforms and is composed of 23–38 cm of organic sediments over residuum derived from igneous rocks. The photo is from the soil survey of Isle Royale National Park, MI (Carey 2012)

Table 5.6 Primary characterization data for Histosol (Organic) soil series in the Great Lakes Coastal Zone (Natural Resourses Conservation Service 2018)

Horizon	Depth (cm)	Clay (%)	Silt (%)	Sand (%)	OC (%)	CEC7 (cmol(+)/kg)	Base sat. (%)	pH H_2O	Fiber, un-rubbed (%)	Fiber, rub-bed (%)
Lupton series; euic, frigid, Typic Haplosaprists; Pedon No. 40A1648; Oconto Co., WI; organic/lacustrine deposits (sandy); lake plain										
Oa1	0–20				36.1	107.0	83	6.5	37	8
Oa2	20–38				34.8	116.0	79	6.5	28	4
Oe1	38–46				46.6	164.0	72	6.1	57	11
Oa3	46–70				45.5	199.0	69	6.0	16	3
Oa4	70–117				nd	nd	nd	nd	nd	nd
Oa5	117–150				nd	nd	nd	nd	nd	nd
Oe2	150–215				nd	nd	nd	nd	nd	nd
C2	215–358				nd	nd	nd	nd	nd	nd
Cathro series; loamy, mixed, euic Terric Haplosaprists; Pedon No. 40A5522; Door Co., WI; organic/till; till plain										
Oa1	0–18				31.5	113		6.3	nd	nd
Oa2	18–84				42.1	178		6.4	nd	nd
2Cg1	84–114	21.3	62.2	16.5	0.82	5.7	100	7.3	nd	nd
2Cg2	114–152	25.9	69.1	5.0	0.5	7.1	100	7.7	nd	nd
Dawson series; loamy, mixed, euic, frigid Terric Haplosaprists; Pedon No. 02N0717; Ontonagon Co., MI; organic/outwash; outwash plain										
Oe	10–23				48.0	115	9	nd	54	40
Oa	23–86				52.3	103	4	4.0	60	36
2E	86–91	1.3	16.4	82.3	0.72	2.2	41	4.3	nd	nd
2Bhs	91–127	0.9	12.0	87.1	2.14	9.2	11	4.3	nd	nd
2C	127–151	0.6	7.8	91.6	0.44	1.9	53	4.7	nd	nd
Rifle series; euic, frigid Typic Haplohemists; Pedon No. 40A1740; St. Louis Co., MN;										
organic/till; till plain								$CaCl_2$		
Oap	0–25							5.1	53	3
Oe1	25–60							4.8	65	20
Oe2	60–70							4.8	44	17
Oe3	70–130							5.0	72	21
Oa2	130–165							5.3	32	7
Oa3	165–185							5.3	10	3
2AB	185–192							nd	nd	nd
2Cg	192–202							nd	nd	nd
Minong series; euic, frigid Lithic Udifolists; Pedon No. 07N0167; Keweenaw Co., MI; organic/bedrock; wave-cut platform										
Oi1	0–8				49.9	91.6	30	4.6	76	46
Oi2	8–18				46.3	113	24	4.0	88	48
Oa	18–38	10.8	41.3	47.9	19.1	92.2	10	5.1	20	12
R	38+									

The histic epipedon is highlighted in bold face

References

Carey, L.M. 2012. Soil survey of Isle Royale National Park, Michigan. *Natural Resources Conservation Service.*

Eversoll, J.S., and L.M. Carey. 2010. Soil survey of Ontonagon County, Michigan. *Natural Resources Conservation Service.*

Kroell, M.L., III. 2008. Soil survey of Benzie and Manistee Counties, Michigan. *Natural Resources Conservation Service.*

Natural Resources Conservation. Service. 2018. Soil characterization database. https://ncsslabdatamart.sc.egov.usda.gov. Accessed 03 June 2018.

Robbins, R.A., and N.H. Martin. 2006. Soil survey of Erie County, Ohio. *Natural Resources Conservation Service.*

Robbins, R.A., and M.M. Feusner. 2006. Soil survey of Hancock County, Ohio. *Natural Resources Conservation Service.*

Robbins, R.A., and A.M. Lantz. 2007. Soil survey of Wood County, Ohio. *Natural Resources Conservation Service.*

Schwenner, C.F. 2013. Soil survey of Alger County, Michigan. *Natural Resources Conservation Service.*

Tardy, S.W. 2006. Soil survey of Keweenaw County area, Michigan. *Natural Resources Conservation Service.*

Trevail, T.D. 2006. Soil survey of Akwesasne territory: St. Regis Mohawk reservation. *Natural Resources Conservation Service* 252.

Whitney, G., and S. Rodock. 2006. Soil survey of Luce County, Michigan. *Natural Resources Conservation Service.*

Williams, T.E. 1998. Soil survey of Alcona County, Michigan. *Natural Resources Conservation Service.*

Williams, T.E. 2007. Soil survey of Alpena County, Michigan. *Natural Resources Conservation Service.*

6 Soils of the Lake Superior Coastal Zone

6.1 Introduction

With a surface area of 82,100 km^2, Lake Superior is the largest freshwater lake in the world. Lake Superior is the highest (183 m), western-most (92° W), and northern-most (49° N) of the great Lakes. The Lake Superior basin has lon, cold, and snowy winters and short, cool summers. Three-quarters (75%) of the Lake Superior shore contains Precambrian igneous and metamorphic rocks, with sandstone being present in the other 25% on the south shore. Boreal forest occurs alon the north shore of Lake Superior; the native vegetation of the remaining shore is temperate mixed coniferous-broadleaved forest. Lake Superior was the last great Lake to be located, with the glaciers retreating 9.5 kyr ao.

6.2 Soils by Natural Resource Sement

The US portion of the Lake Superior Coastal Zone contains dominantly Haplorthods and Haplosaprists, with lesser amounts of lossudalfs, Eutrudepts, Hapludalfs, and Endoaquods (Table 6.1, Fig. 6.1). Lake Superior can be divided into six segments based on Major Land Resource Areas (US only), parent materials, and dominant soil series. From Duluth, Minnesota to Thunder Bay, Ontario–the first segment–the dominant parent material is loamy till that forms a moraine parallelin Lake Superior. This moraine has been wave-washed by ancestral lakes. Key soils on the moraine are Eutrudepts (Barto, reysolon, Mesaba, and Normanna soil series) and lossaqualfs (Cuttre). lossudalfs (Miskoaki soil series) occur on wave-cut till plains, and Hapludalfs (Auustana and Heber soil series) occur in areas of discontinuous lacustrine sediments that frine the coastline and overlie till in places. Segment 1 is part of MLRA 93A Superior Stony and Rocky Loamy Plains and Hills, Western Part. Some of the key soil series were depicted in Chap. 5, including the Barto (Fig. 5.13a), the Mesaba (Fig. 5.13b), the Cuttre (Fig. 5.5), and the Auustana-Heber (Fig. 5.1).

Discontinuous sandy and sandy loam till deposits over bedrock extend along the second segment from Thunder Bay to Red Rock, Ontario. Provisional mapping by the Great Lakes Forestry Centre suggests that the soils are primarily Dystric Brunisols, with some Luvisols (Hapludalfs), Mesisols (Haplohemists), and rock land.

The north and eastern shore of Lake Superior from Red Rock to Sault Ste. Marie, Ontario–the third segment–is composed of discontinuous sandy till over bedrock with a discontinuous, narrow Fringe of sandy lacustrine materials containing Orthic Humo-Ferric Podzols (Haplorthods) and Precambrian rock land.

The southeast shore of Lake Superior from Sault Ste. Marie to Munising, Michigan (segment 4) is composed of a wide band of sandy lacustrine material backed by a prominent end moraine, the Munising Moraine (Fig. 6.2). The Kingston sandy outwash plains occur near Grand Marais; a large outwash plain also exists south of Munising. The Rubicon (Fig. 5.6c), Croswell (Fig. 5.6a), and Kalkaska soil series (Fig. 5.6b) (Haplorthods) are common on these sandy materials. Beach ridges feature Udipsamments (Deer Park soil series) and Endoaquods (Au Gres and Kinross soil series); the swales contain Haplosaprists (Carbondale, Tawas, and Lupton soil series) and Psammaquents (Deford soil series). The Grand Sable Dunes, which occur west of Grand Marais, are the largest dune system on Lake Superior. This segment is part of MLRA 94B, Michian Eastern Upper Peninsula Sandy Drift. Several of these soil series were shown in Chap. 5, including the Deer Park (Fig. 5.17a), Au Gres (Fig. 5.7a), and Carbondale (Fig. 5.22a).

The fifth segment along Lake Superior extends from Munising to Marquette, Michigan. This segment features sandy till and lacustrine sediments over sandstone bedrock (Fig. 6.3). This segment contains primarily Haplorthods (Croswell and Kalkaska soil series) and Haplosaprists (Carbondale and Tawas soil series), along with Fragiorthods (Munising and Yalmer) on the Marquette end moraine and Udipsamments (Deer Park) on beach rides. This segment is included in MLRA 93B, the Superior Stony and Rocky

J. G. Bockheim, *Soils of the Laurentian Great Lakes, USA and Canada*,
https://doi.org/10.1007/978-3-030-52425-8_6

Table 6.1 Segments of the Great Lakes shorelines delineated from parent materials, soils, and Major Land Resource Areas

Lake	Segment no.	Segment	Length (km)	MLRA[1]	Dominant shoreline parent materials[2]	Lake stages[3]	Dominant great groups	Dominant soil series
Superior	1	Duluth, MN to Thunder Bay, ON	922	93A	C till	D, Mi, N	Eutrudepts, Hapludalfs	Barto, Cuttre, Greysolon, Mesaba, Miskoaki, Augustana, Hegberg, Quetico, Normanna,
	2	Thunder Bay, ON to Red Rock, ON	739	–	discontin. S, SL till	D, A, N	Dystrudepts, Haplohemists, Hapludalfs	[insufficient mapping]
	3	Marathon, ON, to Sault Ste. Marie, ON	860	–	Precambrian bedrock	D, A, S, N	Haplorthods, Haplohemists	[insufficient mapping]
	4	Sault Ste. Marie, MI to Munising, MI	443	94B	lacustrine S	A, Mi, N	Haplorthods, Haplosaprists, Endoaquods	Croswell, Carbondale, Deer Park, Rubicon, Kinross, Au Gres, Deford, Kalkaska, Tawas, Lupton
	5	Munising, MI to Marquette, MI	186	93B	lacustrine S; S till	A, N	Haplorthods, Haplosaprists	Croswell, Kalkaska, Munising, Deer Park, Carbondale, Tawas
	6	Marquette, MI to Superior, WI Subtotal	1645 4796	92	C, SL till; lacustrine Si, C	D, N	Haplorthods, Glossudalfs, Haplosaprists	Deer Park, Rubicon, Munising, Yalmer, Croswell,
								Big Iron, Flintsteel, Kellogg, Cuttre, Miskoaki, Kinross, Au Gres, Deford, Kalkaska, Tawas, Lupton
Michigan	1	Chicago, IL to Milwaukee, WI	232	110	C till	G, C	Hapludalfs, Haplosaprists	Kewaunee, Manawa, Boyer, Houghton, urban land
	2	Milwaukee, WI to Escanaba, MI (includes Door Peninsula)	980	95A	C till; lacustrine Si, C	G, C, A, N	Haplosaprists, Hapludalfs, Hapludolls	Kewaunee, Manawa, Onaway, Deford, Oakville, Lupton, Cathro, Markey, Tawas, Carbondale, Ensley, Summerville, Alpena, Namur, Wainola, Longrie, Rousseau
	3	Escanaba, MI to Mackinaw City, MI	368	94B	lacustrine S	A, N	Haplosaprists, Endoaquods	Lupton, Markey, Tawas, Deford, Carbondale, Au Gres, Wainola
	4	Mackinaw City, MI to Manistee, MI	416	96	SL till; dune S	G, C, A, N, Al	Haplorthods, Hapludalfs, Udipsamments	Spinks, Emmet, Nordhouse, Eastport, Deer Park, Pipestone, Leelanau, East Lake, Kalkaska
	5	Manistee, MI to Muskegon, MI	210	98	dune S; lacustrine S; C till	G, C	Endoaquods, Udipsamments	Nordhouse, Plainfield, Houghton, Saugatuck, Pipestone
	6	Muskegon, MI to Chicago, IL Subtotal	468 2673	97	lacustrine S; dune S; C till	G, C	Udipsamments, Endoaquods	Blount, Rimer, Plainfield, Oakville, Houghton, Saugatuck, Pipestone

(continued)

Table 6.1 (continued)

Lake	Segment no.	Segment	Length (km)	MLRA[1]	Dominant shoreline parent materials[2]	Lake stages[3]	Dominant great groups	Dominant soil series
Huron	1	Port Huron, MI to Au Gres, MI	583	99	wave-washed L till; lacustrine Si, C	WA, A, N	Epiaquepts, Endoaquolls	Londo, Tobico, Eastport, Plainfield, Parkhill, Tappan, Iosco, Covert
	2	Au Gres, MI to Alpena, MI	312	94A	lacustrine S, G	WA, A, N	Haplorthods, Endoaquods	Deford, Au Gres, Croswell
	3	Alpena, MI to Mackinaw City, MI	558	94C	lacustrine S, G	A, N	Haplosaprists, Endoaquods, Haplorthods	Roscommon, Deer Park, Lupton, Cathro, Tawas, Ruse, Alpena, Iosco, Au Gres, East Lake, Croswell
	4	Sault Ste. Marie, ON to Blind River, ON	325	–	lacustrine; SL till	A, N, Al	Haplorthods, Endoaquods	Monteagle, Kenabeek, Mallard, Baldwin, [marsh]
	5	St. Ignace to Cockburn Is.; Cockburn Is. to Sault Ste. Marie; Manitoulin Is.; Bruce Penin.	2279	–	dolo. ls bedrock; intermitt. glaciolacustrine Si, C; SL till	A, S, N	Eutrudepts, Endoaquolls	Shelter, Farmington, Wendigo, Mallard, Wolsey, Breypen, Ferndale, Wauseon, Harkaway, Sargent, [marsh]
	6	Blind River, ON to Port Severn, ON	890	–	Precambrian bedrock	A, N	Haplorthods, Haplosaprists	Monteagle, Chartrand, Wendigo, Atherley
	7	Port Severn, ON to Port Elgin, ON (excludes Bruce Peninsula)	435	–	lacustrine S	A, N	Eutrudepts, Hapludalfs, Hapludolls	Brady, Brighton, Waterloo, Kemble, Brookston, Sargent, Dunedin, Vincent, Lily, Wiarton, Breypen, Elderslie, Osprey, Brisbane, Tioga, Vasey, Gwillimbury, Farmington, Wyevale, Eastport, Parkhill
	8	Port Elgin, ON to Sarnia, ON Subtotal	448 5869	–	Si, SiC till	WA, A, N	Endo-, Epiaqualfs; Hapludalfs	Brady, Brighton, Waterloo, Kemble, Brookston, Sargent, [muck], Brisbane, Elderslie, Perth, Listowel, Sullivan, Fox, Plainfield, Eastport, [marsh], Toledo, Perth
Erie	1	Detroit, MI to Sandusky, OH	406	99	lacustrine Si, C	WA	Epiaquepts, Epiaqualfs	Blount, Lamson, Toledo, Lenawee, urban land
	2	Sandusky, OH to Buffalo, NY	686	101	lacustrine Si, C, S, G	W, WA	Endoaquepts, Epiaquepts, Udipsamments	Niagara, Platea, Elnora, Colonie,Canandaigua, Lamson, Conneaut, Painesville, Chenango, Minoa
	3	Buffalo, NY to Ridgetown, ON	639	–	lacustrine S, Si, C; GCL till	WA	Hapludalfs, Endo-, Epiaqualfs	Muriel, Gobles, Kelvin, Fox, Brady, Granby, Watiford, Normandale, St. Williams, Lowbanks, Brantford, Beverly,

(continued)

Table 6.1 (continued)

Lake	Segment no.	Segment	Length (km)	MLRA[1]	Dominant shoreline parent materials[2]	Lake stages[3]	Dominant great groups	Dominant soil series
								Toledo, Smithville, Haldimand, Lincoln, Ontario, Niagara, Welland
	4	Ridgetown, ON to Windsor, ON Subtotal	290 2020	–	lacustrine S, Si, C; SiCL till	WA	Endo-, Epiaqualfs, Hapludalfs	Brookston, Perth, Harrow, Fox, Berrien, Caistor, Eastport, [marsh], Plainfield, Haldimand, Granby, Beverly
Ontario	2	Youngstown, NY to Cape Vincent, NY	547	101	lacustrine S; L till	I, Cs	Hapludalfs, Fragiudepts, Endoaqualfs	Niagara, Rhinebeck, Collamer, Dunkirk, Hilton, Colonie, Canandaigua, Ira, Sodus, Williamson
	2	Kingston, ON to Niagara-on-the-Lake, ON Subtotal	781 1328	–	lacustrine S, Si, C; L till	I, Cs	Endo-, Epiaqualfs, Hapludalfs	Grimsby, Tavistock, Vineland, Chinguacousy, Lansdowne, Otonabee, Lindsay, Solmesville, Farmington, Napanee, Hillier, Eastport, Ameliasburg, Athol, Brighton, Gerow, [marsh], Tecumseth, Smithfield, Bondhead, Colborne, Guerin, Dundonald, Bookton, Edenvale, Gilford, Fox, Newcastle, [ravines]
		Total	16,686					

[1]NRCS (2006)

[2]From *Quaternary Atlas of the USA*; *Quaternary Geology of Ontario*; and *Quaternary Geology of the Unites States and Canada* (dates 1983 to 1993)

[3]Modified from Karrow (1984) and Larsen and Schaetzl (2001): A = Algonquin; Al = Algoma; C = Calumet; Cs = Champlain Sea; D = Duluth; G = Glenwood; I = Iroquois; Mi = Minong; N = Nipissing; S = Stanley (Wyebridge, Penetang, Cedar Point, Payette, Sheguiandah, Korah); W = Whittlesey; WA = Warren

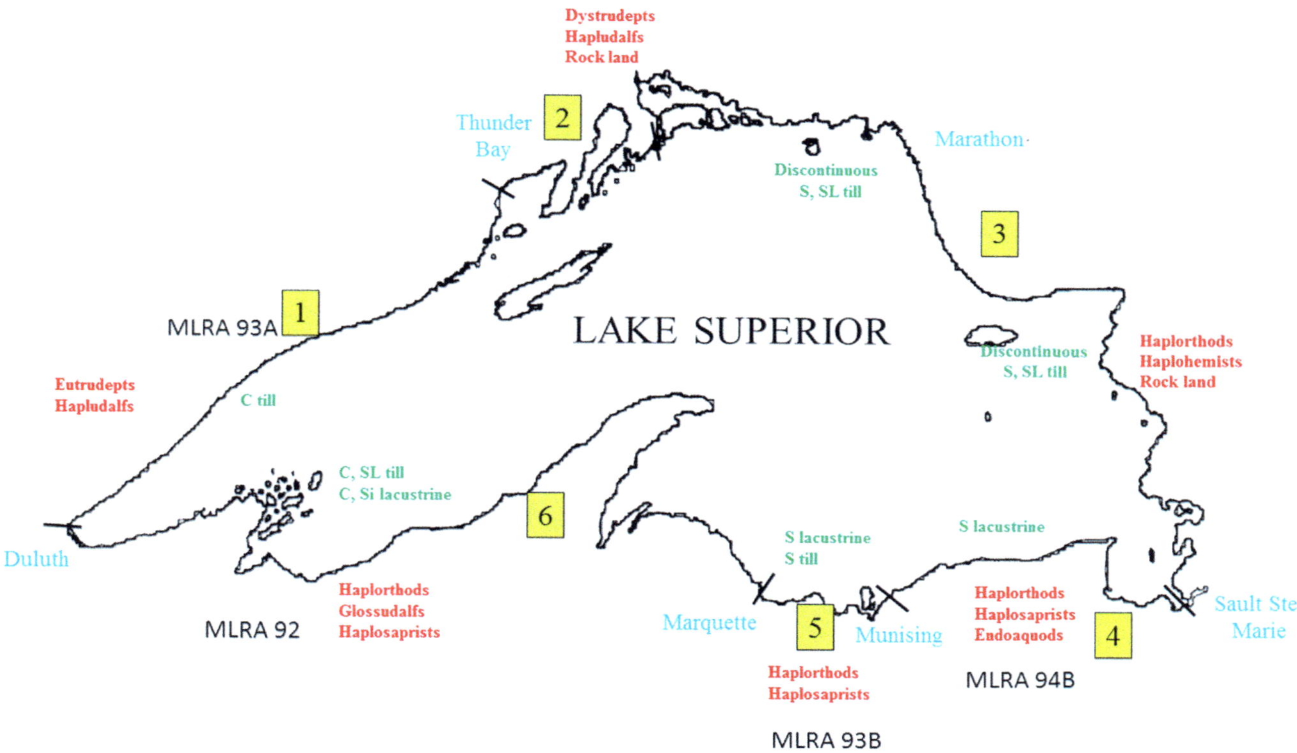

Fig. 6.1 Key segments of the Lake Superior Coastal Zone

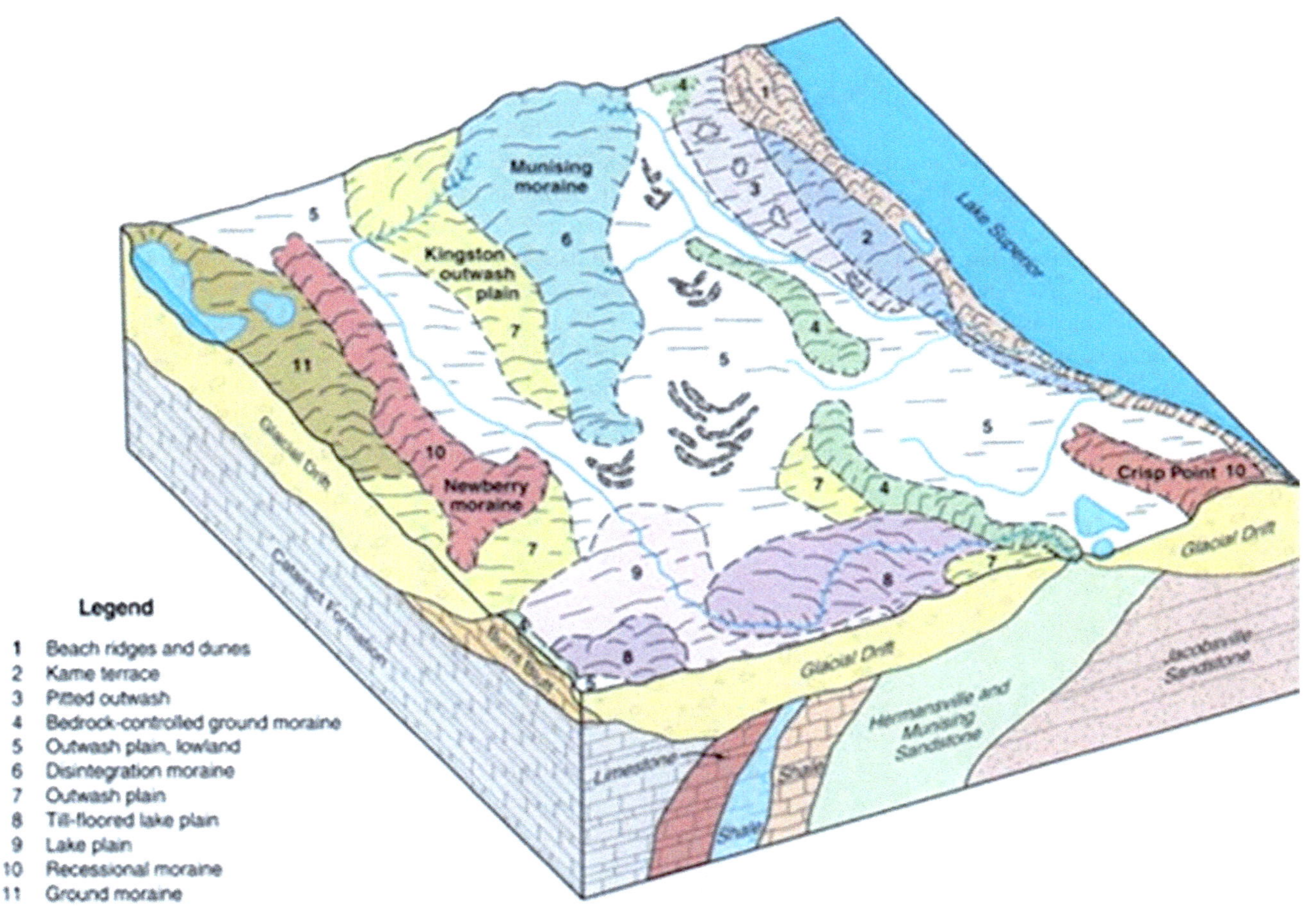

Fig. 6.2 Landforms alone segment 4 of Lake Superior (MLRA 98). *Source* Luce County Soil Survey (Rodock et al. 2006)

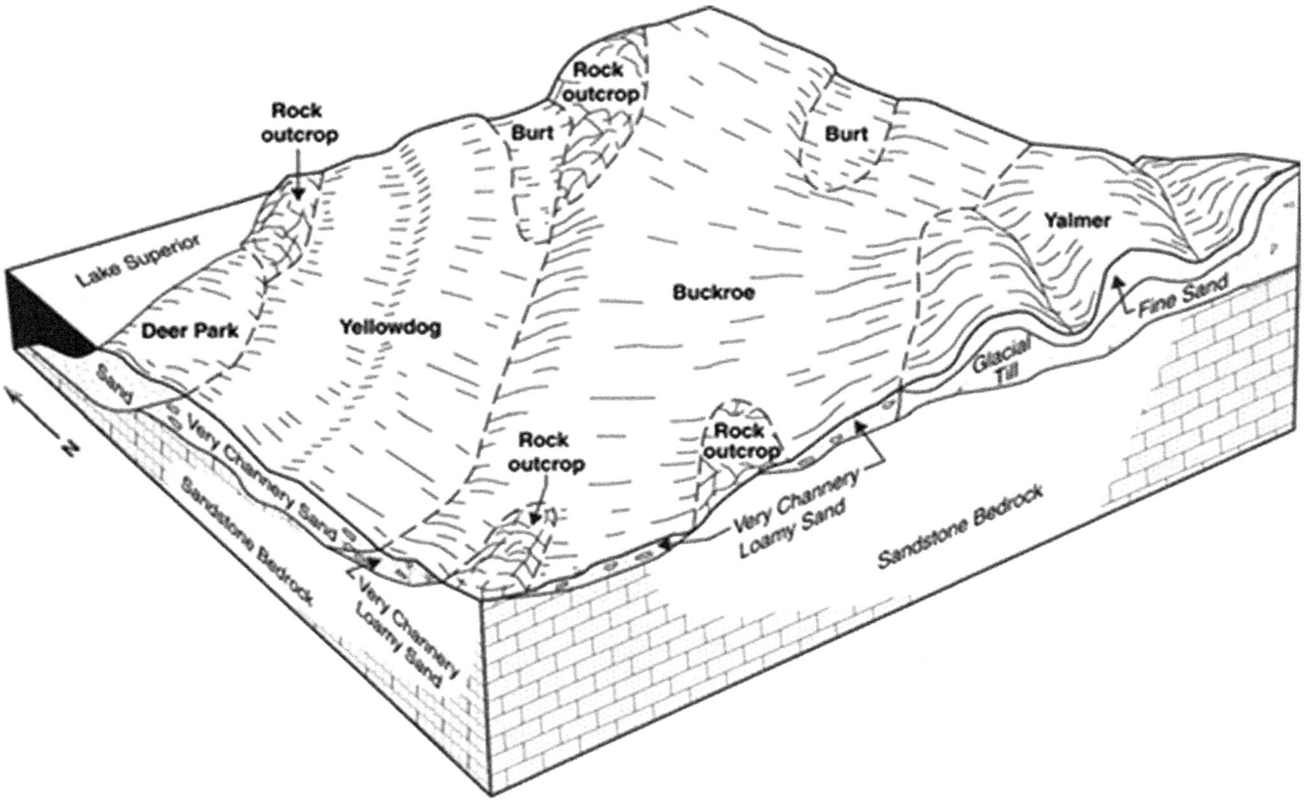

Fig. 6.3 Parent materials, landforms, and soil series in segment 5 of Lake Superior (MLRA 93B). The soils are on a wave-washed sandstone bench from the Nipissing period. *Source* Soil survey of Marquette County, Michigan (Schwenner 2007)

Loamy Plains and Hills, Eastern Part. The Munising soil series is shown in Fig. 5.8a.

The sixth segment along Lake Superior is in MLRA 92, the Lake Superior Lake Plain, and extends from Marquette around the Keweenaw Peninsula to Superior, Wisconsin. This segment is composed largely of wave-washed clayey and sandy loam till and sandy dune and beach deposits. Clay-rich till and lacustrine deposits contain Glossudalfs (Big Iron, Flintsteel, and Miskoaki soil series), Glossaqualfs (Cuttre), and Fragiorthods (Munising, and Yalmer soil series). The Yalmer soil series is a bisequal soil derived from outwash over till (Fig. 6.4). The upper sequum contains a black A horizon (cut off from top of photo), a reddish ray E horizon to 12 in. (30 cm), a dark reddish brown Bhs horizon to 24 in. (60 cm), and yellowish red Bs horizons to 30 in. (76 cm). The lower sequum contains a fragipan and glossic horizon (2(E/B)x and 2(B/Ex) with a mixture of reddish gray and dark reddish brown colors that is underlain by a reddish brown argillic horizon (2Bt).

Areas of wave-washed sandy till in Segment 6 contain Haplorthods (Keweenaw soil series) and Fragiorthods (Gogebic soil series) (Fig. 6.5). Sandy sediments feature Haplorthods (Kello, Rubicon, Kalkaska, and Croswell soil series), Udipsamments (Deer Park) and Psammaquents

Fig. 6.4 The Yalmer soil series is an Alfic Oxyaquic Fragiorthods. The soil is bisequal with an upper Spodosol (A, E, Bhs, Bs1, Bs2) in sandy outwash and a lower Alfisol (2E/B)x, 2(B/E)x, 2Bt in loamy till. The scale is in inches. The photo was provided by Dwight Jerome, Resource Soil Scientist, NRCS, Marquette, MI

(Deford) on beach rides, and Endoaqods (Au Gres and Kinross) in wet, sandy depressions (Fig. 6.6). Haplosaprists (Tawas and Lupton soil series) are common in organic-rich depressions. The Kello soil series is an Alfic Oxyaquic Haplorthods derived from sandy over clayey lacustrine materials (Fig. 6.7). The Kello soil profile contains an Oe to 2 in. (5 cm), a pinkish gray E horizon to 8 in. (20 cm), brown to strong brown Bs (spodic) horizons to 18 in. (45 cm), a 2B/E lossic horizon to 29 in. (74 cm), and a 2Bt argillic horizon to 40 in. (102 cm).

6.3 Conclusions

The Lake Superior Coastal Zone can be divided into six sements. Haplorthods are the dominant soil great-group, followed by Haplosaprists, Hapludalfs, and Haplohemists. These soils are derived from discontinuous sandy and sandy-loam till over bedrock, lacustrine sands, and clayey till. Soils of the Lake Superior CZ are used primarily for wilderness appreciation and forestry.

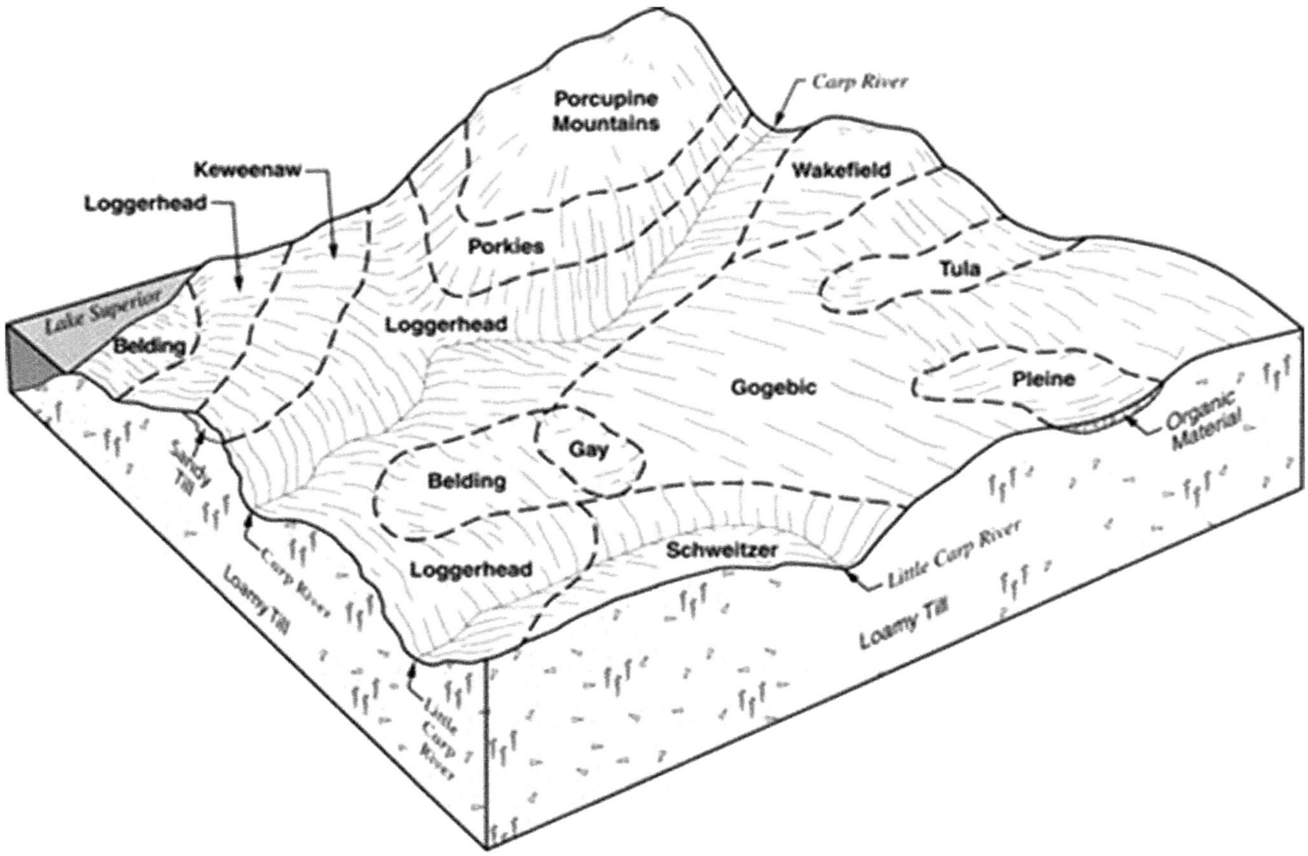

Fig. 6.5 Soil series on sandy and loamy till in segment 6 of Lake Superior (MLRA 92). *Source* Soil survey of Ontonagon County, Michigan (Eversoll and Carey 2010)

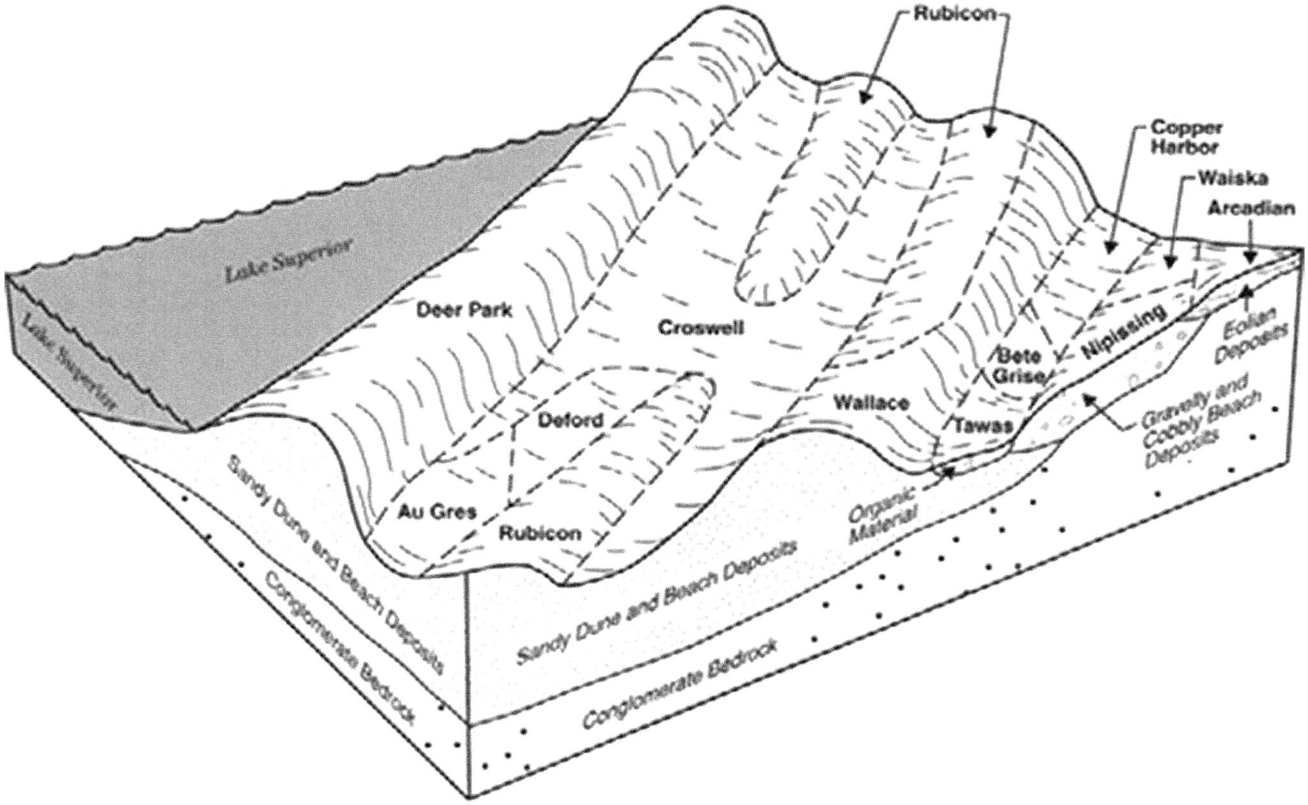

Fig. 6.6 Soil series on sandy dunes and beach deposits in segment 6 of Lake Superior (MLRA 92). *Source* Soil survey of Keweenaw County, Michigan (Tardy 2006)

Fig. 6.7 The Kello soil series is an Alfic Oxyaquic Haplorthods. The scale is in feet. The photo is from the soil survey of Otsego County, MI (Perkis 2004)

References

Eversoll, J.S., and Carey, L.M. 2010. Soil survey of Ontonagon County, Michigan. *Natural Resources Conservation Service.*

Tardy, S.W. 2006. Soil survey of Keweenaw County area, Michigan. *Natural Resources Conservation Service.*

Perkis, W.E. 2004. Soil survey of Otsego County, Michigan. *Natural Resources Conservation Service.*

Natural Resources Conservation Service. 2006. *Land resource regions and major land resource areas of the United States, the Caribbean, and the Pacific Basin.* U.S. Dep. Agric. Handbook 296.

Larson, G., and R. Schaetzl. 2001. Origin and evolution of the Great Lakes. *Journal of Great Lakes Research* 27: 518–546.

Soils of the Lake Michigan Coastal Zone

7

7.1 Introduction

Lake Michigan is the third-largest Great Lake by area and the second-largest by volume. It is the only Great Lake with its axis oriented roughly north-to-south. Lake Michigan is joins Lake Huron via the Straits of Mackinac, leading some to consider the two lakes as one large lake, with an elevation of 176 m. However, because of ice blocking the Straits of Mackinac during glacial cycles, the two lakes responded differently to post-glacial lake processes. The Lake Michigan shore has the most complex bedrock geology of the Great Lakes; it features the Niagara Escarpment on the north, west, and south shores and shale and sandstone on the east shore. The floristic tension zone cuts across Lake Michigan between Sheboygan, WI and Muskegon, MI, separating the temperate mixed coniferous-broadleaved forest to the north and the temperate broadleaved forest to the south. The easternmost extension of prairie reaches the southern tip of Lake Michigan. Some of the oldest geomorphic surfaces in the Great Lakes Coastal Zone are found along the southern Lake Michigan shore, including soils derived from the 13.8 kyr Glenwood lake stage.

7.2 Soils by Natural Resource Segment

Lake Michigan is entirely in the US, and the coastal zone contains primarily Hapludalfs, Haplosaprists, and Endoquods, with lesser amounts of Haplorthods, Hapludolls, and Udipsamments (Table 6.1; Fig. 7.1). Lake Michigan can be divided into six segments based on MLRAs, parent materials and dominant soil series. The first segment is a lake-floored till plain extending from Chicago, Illinois to Milwaukee, Wisconsin. Urban land composes nearly 70 miles (120 km) of shoreline between Gary, Indiana and Milwaukee, Wisconsin. The primary native soils are Hapludalfs (Manawa, Kewaunee, and Boyer soil series) and Haplosaprists (Houghton soil series). This segment is part of MLRA 110, the Northern Illinois and Indiana Heavy Till Plain. A common soil in this segment the Kewaunee series is derived from a thin mantle of loess over clayey till (Fig. 7.2). The Kewaunee profile contains a brownish Ap horizon to 10 in. (25 cm), a grayish brown E horizon to 14 in. (36 cm), and a reddish brown argillic (Bt1, 2Bt2-4) to 29 in. (74 cm).

The second segment, which extends from Milwaukee to Escanaba, Michigan, includes the Door Peninsula, which projects into Lake Michigan. Part of MLRA 95A, the Northeastern Wisconsin Drift Plain, this segment contains clayey till and lacustrine sand and gravel. The till contains Hapludalfs (Kewaunee, Manawa, and Onaway soil series) (Fig. 7.3). The Ensley soil series (Endoaquepts) occupy lake terraces created by wave action in the till. Glacial lake benches mantled with till support Haplorthods (Longrie soil series), Hapludolls (Alpena and Namur), and Eutrudepts (Summerville). Beach ridges and stable dunes contain Udipsamments (Oakville) and Haplorthods (Rousseau). Haplosaprists (Lupton, Cathro, Markey, Tawas, and Carbondale) occupy swales of beach ridges and wetlands. Poorly drained glaciofluvial sediments in lake plains contain Psammaquents (Deford) and Endoaquods (Wainola). The Rousseau soil series is an Entic Haplorthods derived from sandy eolian deposits on dunes and lake plains (Fig. 7.4).

J. G. Bockheim, *Soils of the Laurentian Great Lakes, USA and Canada*,
https://doi.org/10.1007/978-3-030-52425-8_7

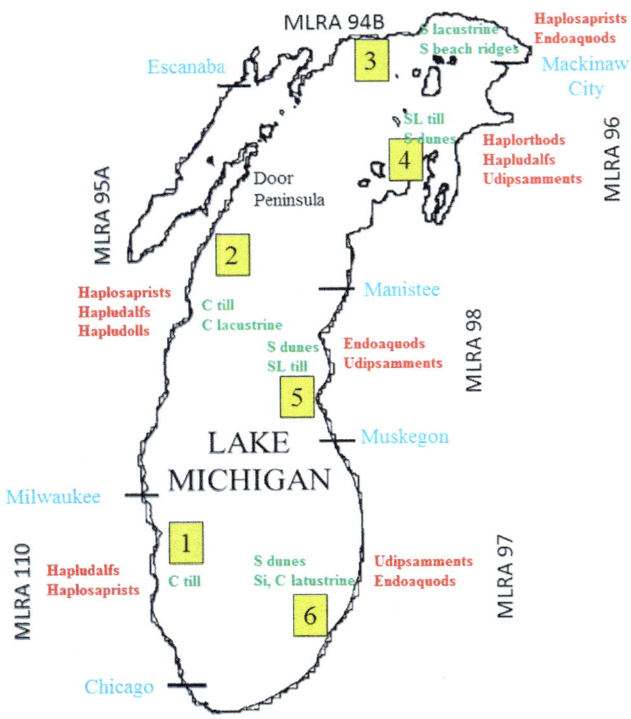

Fig. 7.1 Key segments of the Lake Michigan Coastal Zone

The Rousseau has a black A horizon to 1 in., a pinkish gray E horizon to 7 in., and a dark reddish brown to yellowish red spodic horizon to 25.

The Markey soil series is a Terric Haplosaprists derived from organic sediments in depression on lake plains (Fig. 7.5). The Markey soil has 32 in. of organic materials over gleyed C. The Namur and Carbondale soil series are shown in Chap. 5 (Fig. 5.21B and 5.22, respectively). The Emmet soil series is an Inceptic Hapludalfs derived from sandy loam till on drumlins and till plains (Fig. 7.6). The Emmet soils is bisequal and the profile shown in Fig. 7.6 has a very thin, grayish brown Ap horizon over a brown Bw, a grayish brown E′, and a brown Bt. The Cunard soil series is a Typic Hapludalfs originating from loamy till over limestone bedrock on a glacial bench (Fig. 7.7). The Shawano soil series is a Typic Udipsamments derived from eolian or outwash sand on a lake plain (Fig. 7.8).

Fig. 7.2 The Kewaunee series is a Typic Hapludalfs. *Photo* from the Marbut Memorial Slide Collection (Soil Society of America)

The third segment, part of MLRA 94B, Michigan Eastern Upper Peninsula Sandy Drift region, extends from Escanaba across the north shore of Lake Michigan to Mackinaw City. This segment contains primarily sandy lacustrine deposits with extensive beach ridges. Many of the soils have restricted drainage. The sandy lacustrine deposits contain mainly Endoaquods (Au Gres and Wainola soil series) and Psammaquents (Deford). Haplosaprists (Carbondale,

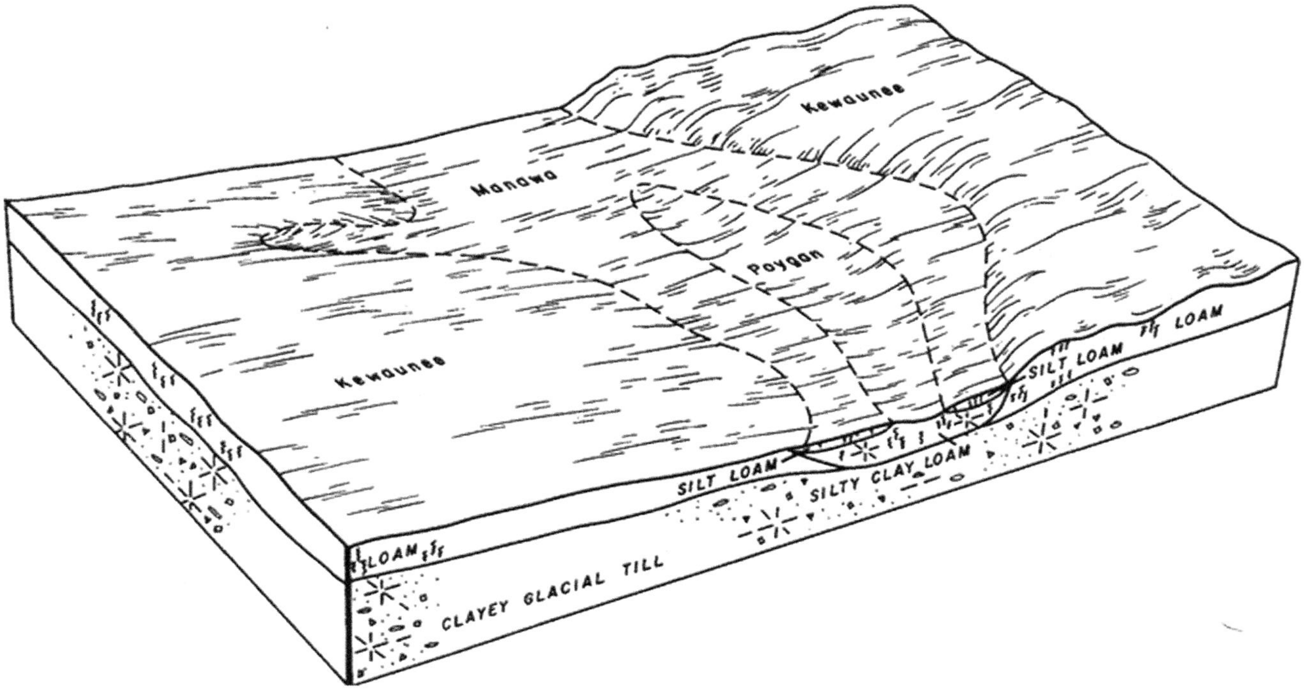

Fig. 7.3 Landforms and soil series on clayey till along segment 2 of Lake Michigan (MLRA 110). *Source* Soil Survey of Calumet and Manitowoc Counties, Wisconsin (Otter 1980)

Lupton, Tawas, and Markey) are common in swales of the beach ridges and in sandy depressions (Fig. 7.9). The Wainola soil series is a Typic Endoaquods derived from fine sandy glaciofluvial deposits on lake plains (Fig. 7.10). The Wainola soil profile contains a black Oa to 2 in. (5 cm), a pinkish gray E horizon to 10 in. (25 cm), a reddish brown Bhs to 12 in. (30 cm), and strong brown and reddish brown Bs horizons to 32 in. (81 cm). The material below 32 in. (81 cm) has massed of oxidized Fe throughout. The Au Gres is shown in Fig. 5.7A.

The fourth segment along Lake Michigan is part of MLRA 96, the Western Michigan Fruit Belt and extends from Mackinaw City down the eastern shore to Manistee, Michigan. This segment is a highly dissected sandy loam till plain with abundant drumlins that is fringed in places by beach ridges and dunes (Fig. 7.11). The till plain contains mainly Haplorthods (East Lake, Kalkaska, and Leelanau soil series) and Hapludalfs (Emmet). The Pipestone soil series (Endoaquods) has developed in water-worked till plains and contains a very dark brown A horizon to 8 in. (20 cm), a grayish brown E horizon to 10 in. (25 cm), a dark reddish brown Bhs horizon to 14 in. (35 cm), and a yellowish brown Bs horizon to 31 in. (80 cm) (Fig. 7.12; note: the scale is upside down). Stable dunes and beach ridges feature Udipsamments (Deer Park and Eastport soil series), Quartzipsamments (Nordhouse), and Hapludalfs (Spinks). The Spinks soil series is a Lamellic Hapludalfs formed in sandy outwash (Fig. 7.13). The Spinks profile contains a brownish A horizon to the top of the shovel handle, a yellowish brown Bw horizon to the first lamella, and a series of lamellae (E and Bt horizon) to more than 2 m (below the base of the photo). The Kalkaska and Deer Park soil series were shown in Figs. 5.6B and 5.17 upper, left), respectively.

The fifth segment extends along the eastern shore of Lake Michigan from Manistee to Muskegon, Michigan. This segment contains primarily sandy loam till and lacustrine

Fig. 7.4 The Rousseau soil series is an Entic Haplorthods. The scale is in feet. *The photo* is from the soil survey of Alpena Co., MI (Williams 2007)

Fig. 7.5 The Markey soil series is a Terric Haplosaprists. The scale is in feet. *This photo* is from the soil survey of Forest Co., WI (Boelter and Barnes 2005)

Fig. 7.6 The Emmet soil series, an Inceptic Hapludalfs, is derived from sandy loam till on drumlins. The horizon sequence is Ap (0–6 in.), A (6–10 in.), Bw (10–18 in.), E/Bt (18–30 in.), and Bt (30 in. to base of pit). The scale is in feet. *Source* Soil survey of Marinette County, Wisconsin (Lorenz 1991)

Fig. 7.7 The Cunard soil series, a Typic Hapludalfs, contains loamy till over limestone bedrock on glacial benches. The horizon sequence is: Oe and A (0–6 in.), E/B (6–13 in.), Bt (13–27 in.), and 2R (27 in.). The scale is in feet. *Source* Soil survey of Marinette County, Wisconsin (Lorenz 1991)

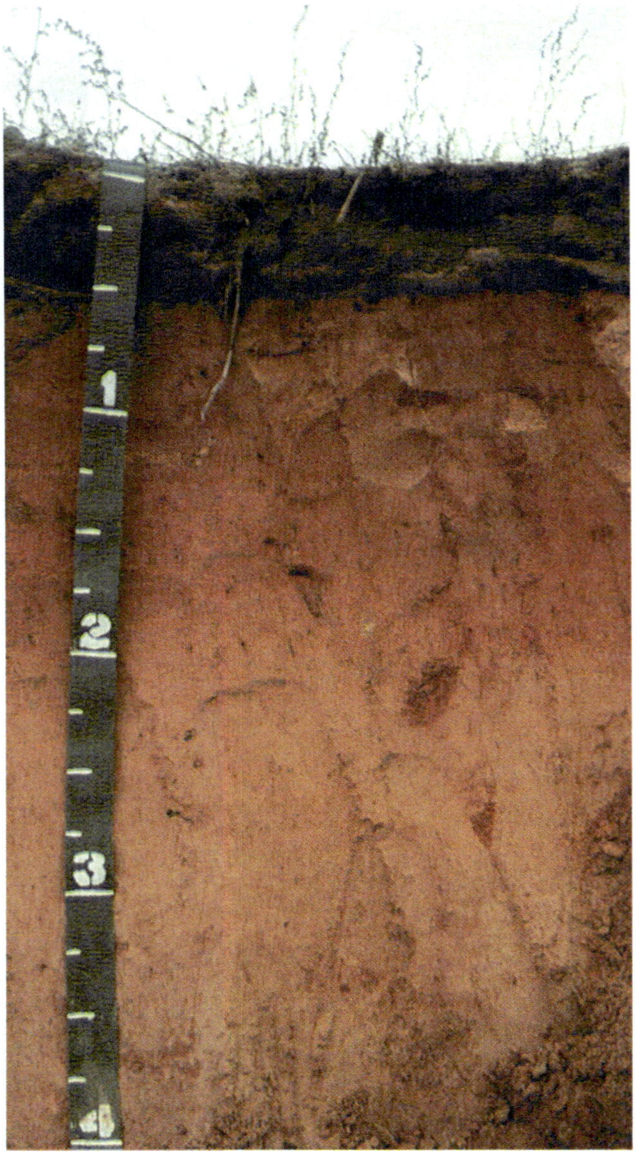

Fig. 7.8 The Shawano soil series, a Typic Udipsamments, is derived from eolian or outwash sand in a lake plain. The horizon sequence is: A (0–2 in.), BA (2–4 in.), Bw (4–26 in.), and C (26 in.). The scale is in feet. Soil survey of Marinette County, Wisconsin (Lorenz 1991)

sand and gravel that has been covered with spectacular dunes in places. The dominant soils on till are Duraquods (Saugatuck). The major dune systems from north to south are the Sleeping Bear Dunes, Acadia Dunes, Ludington and Hamlin Lake Dunes, Silver Lake Dunes, and Nordhouse Dunes. Stabilized dunes and beach ridges feature Udipsamments (Plainfield) and Quartzipsamments (Nordhouse). The Pipestone soil series (Endoaquods) has developed in water-worked till plains. The Houghton soil series (Haplosaprists) occupies wetlands and swales of beach ridges. The Plainfield soil series is depicted in the upper, right panel of Fig. 5.17.

The sixth segment extends from Muskegon down around the base of the lake to Chicago and is part of MLRA 97, the Southwestern Michigan Fruit and Truck Crop Belt. This segment contains mainly water-worked loamy till and lacustrine silts and clays overlain in places by spectacular dunes (Figs. 7.14 and 7.15). Soil series on water-worked till include Hapludalfs (Rimer), Epiaqualfs (Blount), and Endoaquods (Pipestone). The major dune systems from north to south include Saugatuck Dunes, Warren Dunes, and the Indiana Dunes. Stabilized dunes feature Udipsamments (Oakville and Plainfield soil series). Wetlands contain Haplosaprists, dominantly the Houghton series. The Saugatuck series (Duraquods) occupies sandy glaciofluvial deposits on lake plains and till plains.

7.3 Conclusions

The Lake Michigan Coastal Zone can be divided into six segments. Consistent with its diversity of climate and vegetation, Lake Michigan has a high pedodiversity. Dominant soil great-groups along the Lake Michigan Coastal Zone include Haplosaprists, Hapludalfs, Endoaquods, and Udipsamments. These soils are derived from dune sand, clayey till, and silty and clayey lacustrine deposits. Soils of the Lake Michigan CZ are heavily urbanized, but are also used for recreation, agriculture, and forestry.

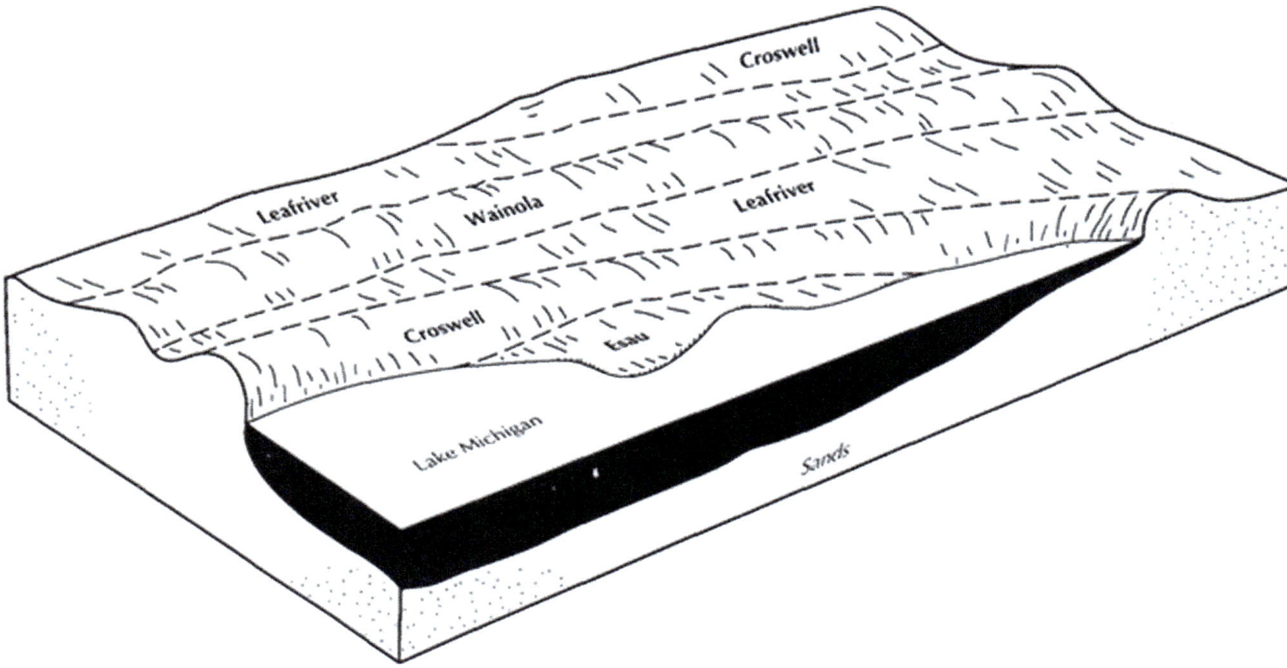

Fig. 7.9 Soil series on beach ridges on segment 3 of Lake Michigan. *Source* Soil survey of Mackinac County, Michigan (Whitney 1997)

Fig. 7.10 The Wainola soil series is a Typic Endoaquods. Redoximorphic features from a seasonally high water table are shown below 35 in. The scale is inches. *The photo* is from the soil survey of Ontonagon Co, MI (Eversoll and Carey 2010)

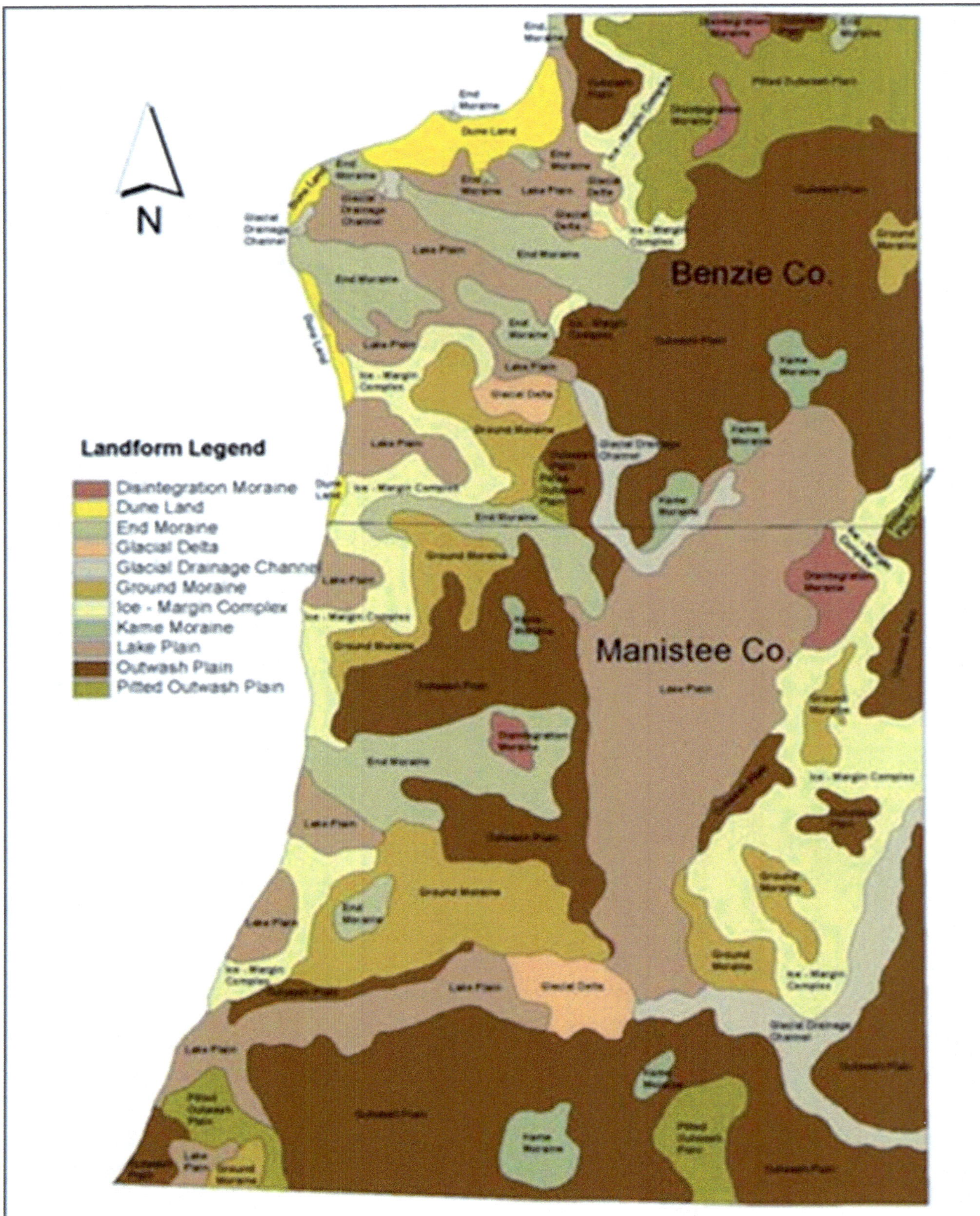

Fig. 7.11 Dominant landforms along segment 4 of Lake Michigan. *Source* Soil survey of Benzie and Manistee Counties, Michigan (Kroell 2008)

Fig. 7.12 The Pipestone soil series is a Typic Endoaquods. The water table 10 cm from the bottom of the pit, i.e., at a depth of 90 cm. The scale is in decimeters. *Photo* from NRCS Site Classification (https://esis.sc.egov.usda.gov/ESDReport/fsReport.aspx?id=F097XA006MI&rptLevel=all&approved=yes&repType=regular&scrns=&comm=)

Fig. 7.13 The Spinks soil series, a Lamellic Hapludalfs, is derived from sandy eolian materials over fine-sandy outwash. The lamellae are readily visible in the subsoil. *The photo* is from the soil survey of Erie County, OH (Robbins and Martin 2006)

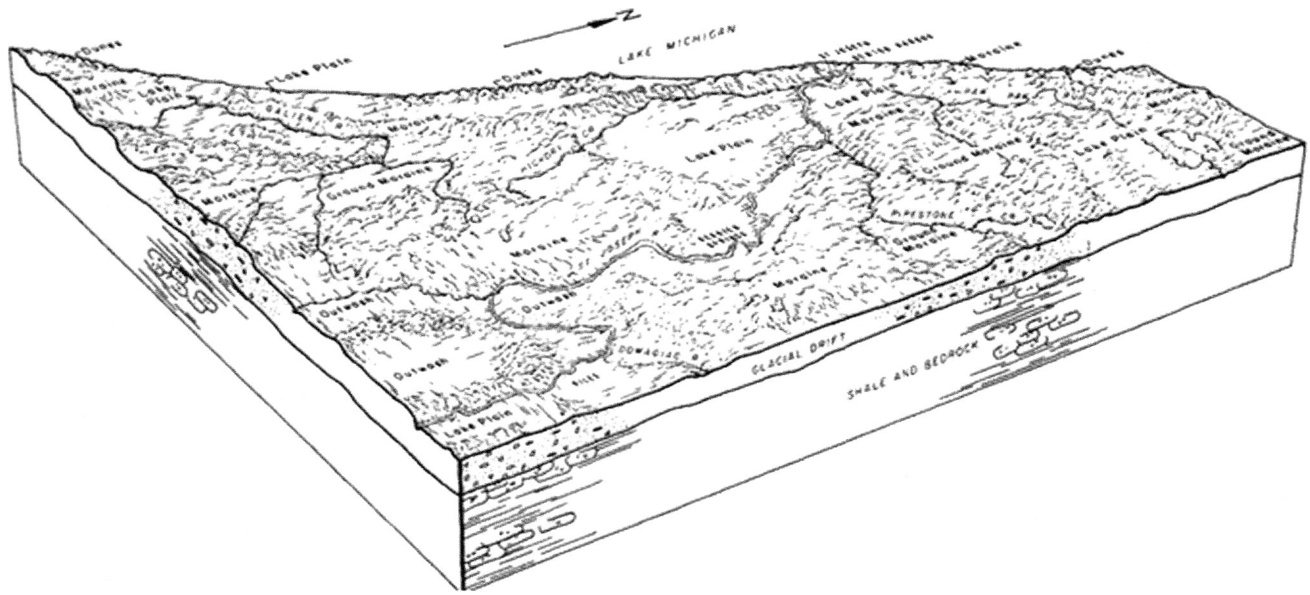

Fig. 7.14 Landforms and soil series derived from glacial drift over shale bedrock on the southeast shore of Lake Michigan. *Source* Soil survey of Berrien County, Michigan (Larson 1980)

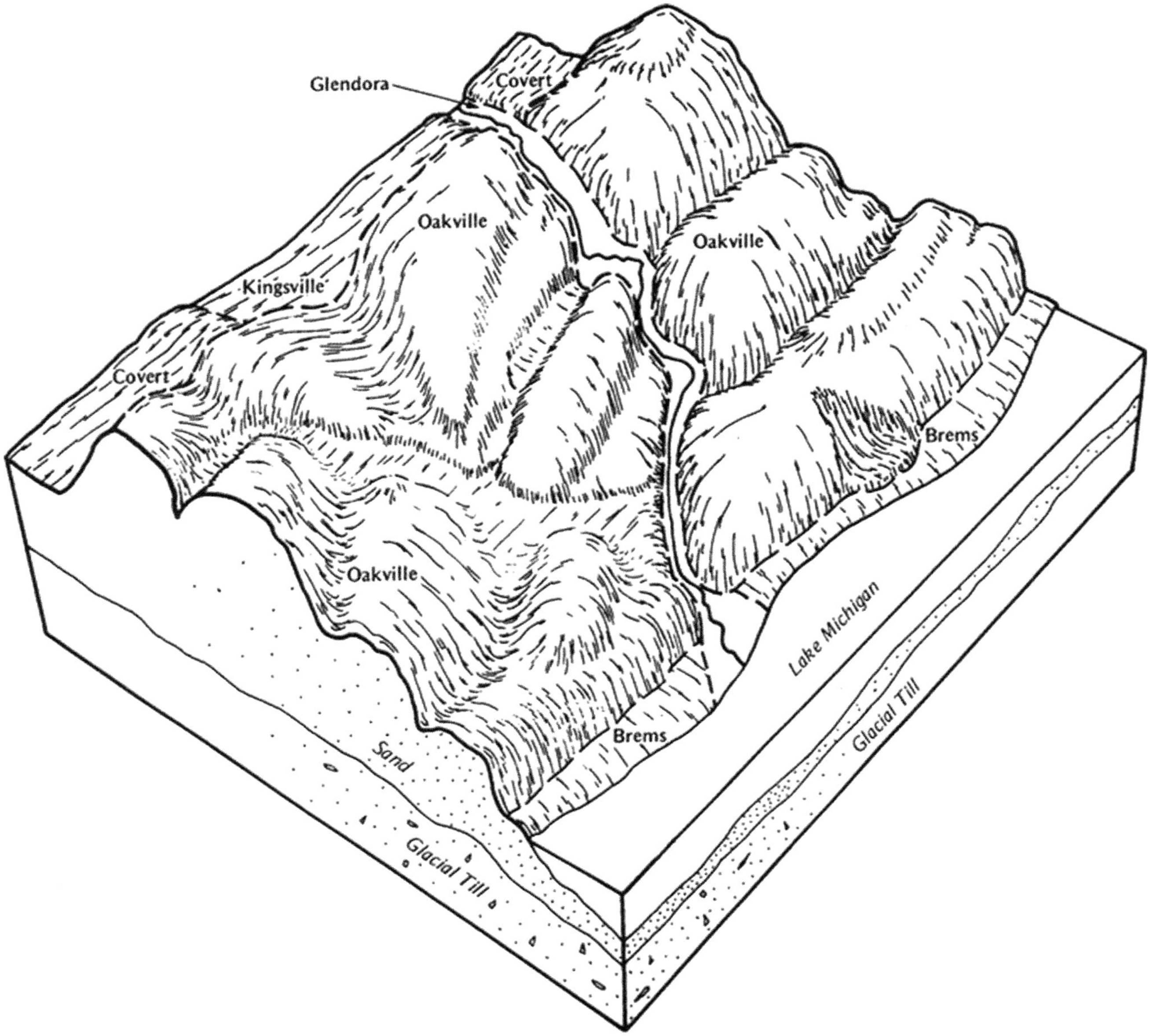

Fig. 7.15 Landforms and soil series on perched dunes and beach ridges along the south shore of Lake Michigan *Source* Soil survey of Van Buren Co., MI (Bowman 1986)

References

Boelter, J.M., A.M. Elg, and J.R. Barnes. 2005. Soil survey of Forest County, Wisconsin. *Natural Resources Conservation Service.*

Bowman, W.L. 1986. Soil survey of Van Buren county, Michigan. *US Department of Agriculture, Soil Conservation Service*

Eversoll, J.S., and L.M. Carey. 2010. Soil survey of Ontonagon County, Michigan. *Natural Resources Conservation Service.*

Kroell III, M.L. 2008. Soil survey of Benzie and Manistee Counties, Michigan. *Natural Resources Conservation Service.*

Larson, J.D. 1980. Soil survey of Berrien county, Michigan. *US Department of Agriculture, Soil Conservation Service*

Lorenz, H.E. 1991. Soil survey of Marinette county, Wisconsin. *US Department of Agriculture, Soil Conservation Service*

Otter, A.J. 1980. Soil survey of Calumet and Manitowoc counties. *US Department of Agriculture, Soil Conservation Service*

Robbins, R.A., and N.H. Martin. 2006. Soil survey of Erie County, Ohio. *Natural Resources Conservation Service.*

Whitney, G. 1997. Soil survey of Mackinac county, Michigan. *US Department of Agriculture, Natural Resources Conservation Service* and *US Forest Service*

Williams, T.E. 2007. Soil survey of Alpena County, Michigan. *Natural Resources Conservation Service.*

Soils of the Lake Huron Coastal Zone

8

8.1 Introduction

Lake Huron is the second largest Great Lake by area and third largest by volume. It is conjoined with Lake Michigan and has the same elevation of 176 m. The Lake Huron shore is composed of Precambrian igneous and metamorphic rocks in the north along Georgian Bay; the Niagara Escarpment links the Bruce Peninsula with Manitoulin, Cockburn, and Drummond Islands; sandstone and shale are present elsewhere. The floristic tension zone extends across Lake Huron from Saginaw, MI to Grand Bend, ON, separating the temperate mixed coniferous-broadleaved forest from the temperate broadleaved forest. Soils along the Lake Huron shore occur on geomorphic surfaces from the Algonquin, Warren, and Nipissing lake stages.

8.2 Soils by Natural Resource Segment

The US Lake Huron coastal zone is mantled dominantly with Haplorthods and Endoaquods, along with some Endoaquolls, Haplosaprists, and Eutrudepts (Table 6.1; Fig. 8.1). Lake Huron can be divided into eight segments based on MLRAs, parent materials, and dominant soil series. The first segment extends from Port Huron to Au Gres, Michigan. Part of MLRA 99, the Erie-Huron Lake Plain. This segment contains wave-washed till of the Port Huron Moraine with discontinuous lacustrine silt and clay (Fig. 8.2). The wave-washed till contains Glossaqualfs (Londo), Epiaquepts (Parkhill), and Epiaquolls (Tappan). Glacial lake benches overlain by sandy lacustrine sediments support Psammaquents (Tobico). Sandy lacustrine sediments support Endoaquods (Iosco), and Haplorthods (Covert). Stabilized dunes and beach ridges have Udipsamments (Eastport and Plainfield soil series). The Iosco soil series is shown in Fig. 5.7B.

The second segment along the Lake Huron coastal zone extends from Au Gres to Alpena, Michigan. This segment is part of MLRA 94A, Northern Michigan and Wisconsin [*sic*] Sandy Drift. This segment contains mainly sandy lacustrine sediments from the Algoma, Nipissing, and Algonquin lake stages that is draped over the Port Huron moraine further inland (Fig. 8.3). Beach ridges are common along the shore and have Haplorthods (Croswell), Endoaquods (Au Gres), and Psammaquents (Deford) (Figs. 8.4 and 8.5). The Croswell and Au Gres soil series are shown in Fig. 5.6A and the left panel of Fig. 5.7.

The third segment extends from Alpena to Mackinaw City (MLRA 94C, Michigan Northern Lower Peninsula Sandy Drift). The sandy lacustrine materials feature (Haplorthods (Croswell and East Lake soil series), Endoaquods (Iosco and Au Gres), and Psammaquents (Roscommon). Beach ridges and stabilized dunes contain Udispamments (Deer Park). Glacial lake benches feature Hapludolls (Alpena) and Endoaquolls (Ruse). Haplosaprists (Cathro, Lupton, and Tawas soil series) occupy depressions of lake plains. The Deer Park soil series is shown in Fig. 5.17.

The fourth segment in the Lake Huron coastal zone extends from Sault Ste. Marie, Ontario, along the north shore of the Georgian Bay to Blind River. This segment contains discontinuous sandy loam lacustrine deposits and drift underlain by Precambrian granite and gneiss rocks of the Canadian Shield. The dominant great groups are Haplorthods (Monteagle soil series), Hapludalfs-Glossudalfs (Baldwin), Endo-, Epiaquods (Mallard), and Endo-, Epiaquepts (Kenabeek soil series). Beach ridges and dunes are uncommon along this segment. Areas mapped as marshes are likely Haplosaprists, Haplohemists, and possibly wet mineral soils.

J. G. Bockheim, *Soils of the Laurentian Great Lakes, USA and Canada*,
https://doi.org/10.1007/978-3-030-52425-8_8

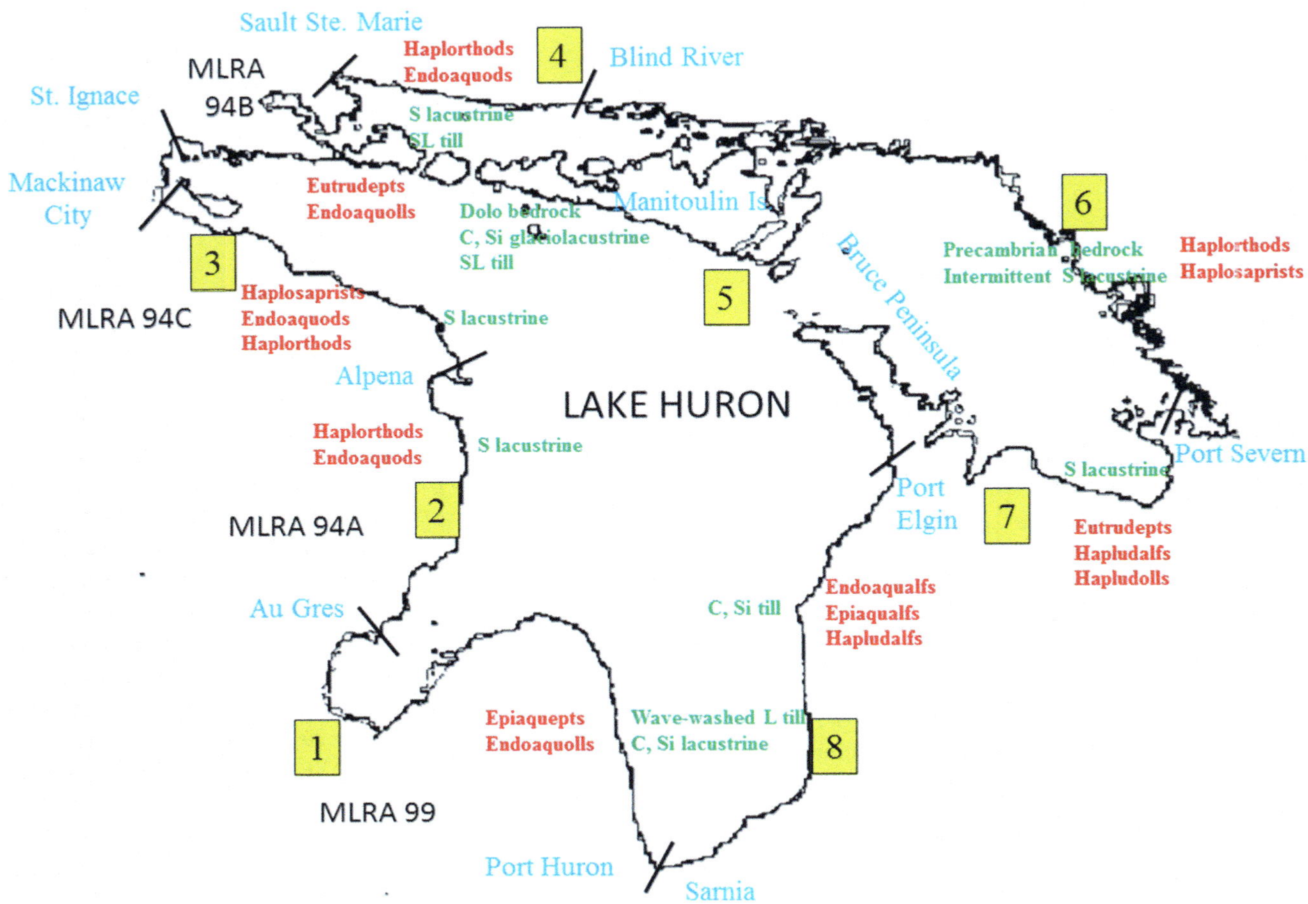

Fig. 8.1 Key segments in the Lake Huron coastal zone

The fifth segment includes land that separates Georgian Bay from the main body of Lake Huron, including the eastern portion of Michigan's Upper Peninsula, Manitoulin, Cockburn, and Drummond Islands and the Bruce Peninsula. The first part of this segment is part of MLRA 94B, Michigan Eastern Upper Peninsula Sandy Drift. This segment is underlain by dolostone of the Niagara Escarpment; Haprendolls (Shelter) occupy glacial lake benches cut into dolostone, which is mined throughout the area. The surficial deposits on Manitoulin Island and the Bruce Peninsula include dominantly exposed limestone bedrock, with lesser areas of fine-textured (c, sic, and sil lacustrine deposits, loamy till, and outwash (Figs. 8.6 and 8.7). The main soils on Manitoulin Island and the Bruce Peninsula are Eutrudepts (Harkaway, Farmington, and Breypen soil series), Endo-, Epiaquolls (Ferndale, Wauseon, and Wolsey), and Hapludolls (Farmington and Sargent soil series). Haplorthods (Wendigo) and Endo-, Epiaquods (Mallard soil series) occupy fluvial and glaciofluvial materials.

The sixth segment extends from Blind River to Port Severn, Ontario, which is composed largely of rock land from Precambrian granites and gneisses covered discontinuously by sandy loam till and lacustrine silts and clays. Dominant soils are Haplorthods (Monteagle and Wendigo soil series), Endo-, Epiaqualfs (Chartrand), and Endo-, Epiaquepts, Endo-, Epiaquolls (Atherly soil series).

The seventh segment along the Lake Huron coastal zone extends from Port Severn to Port Elgin (excludes Bruce Peninsula). Near Port Severn, the bedrock changes to from the Canadian Shield to Paleozoic sedimentary rocks, and the coastal zone contains sandy loam and finer tills (Fig. 8.8). The major soils are Eutrudepts (Farmington, Dunedin,

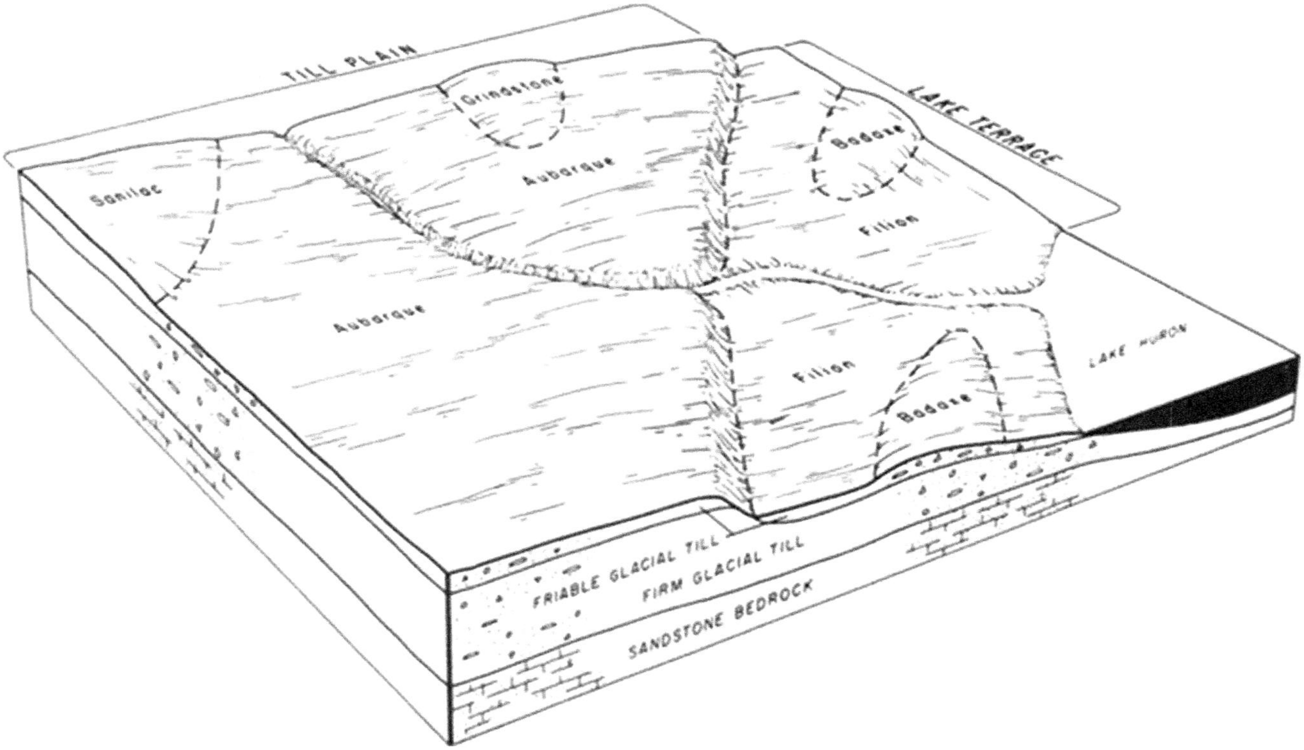

Fig. 8.2 Landforms and soil series on lake terraces and wave-cut till plains along the southwestern Lake Huron shore. *Source* soil survey of Huron County, Michigan (Linsemeier 1980)

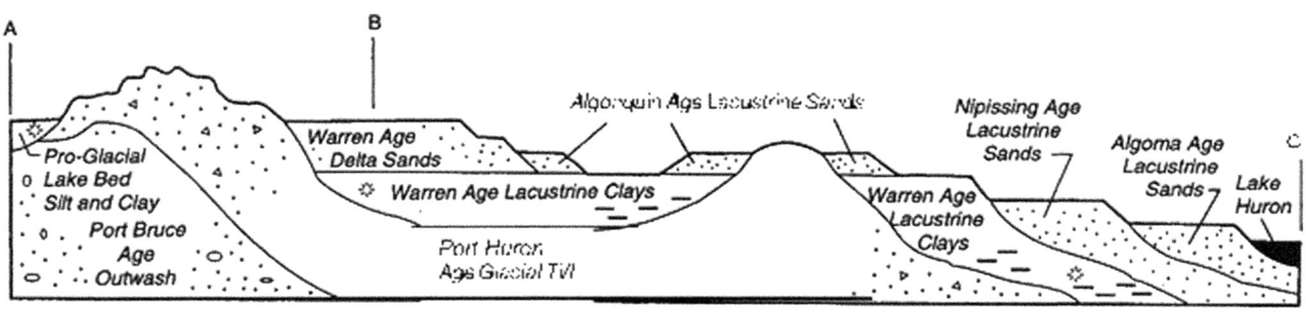

Fig. 8.3 Cross-section of Alcona County, Michigan, along the northwest shore of Lake Huron showing key landforms (Williams 1998)

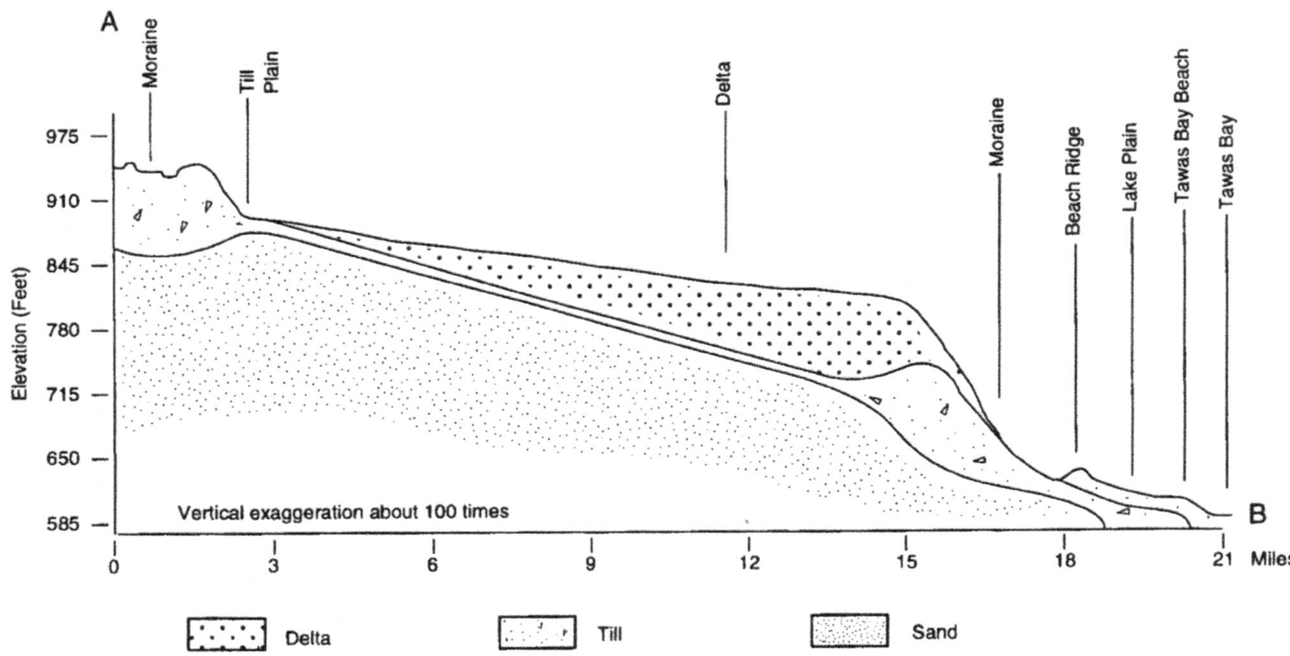

Fig. 8.4 Landforms and parent materials in Iosco County, Michigan, along the northwest shore of Lake Huron (Johnson 2002)

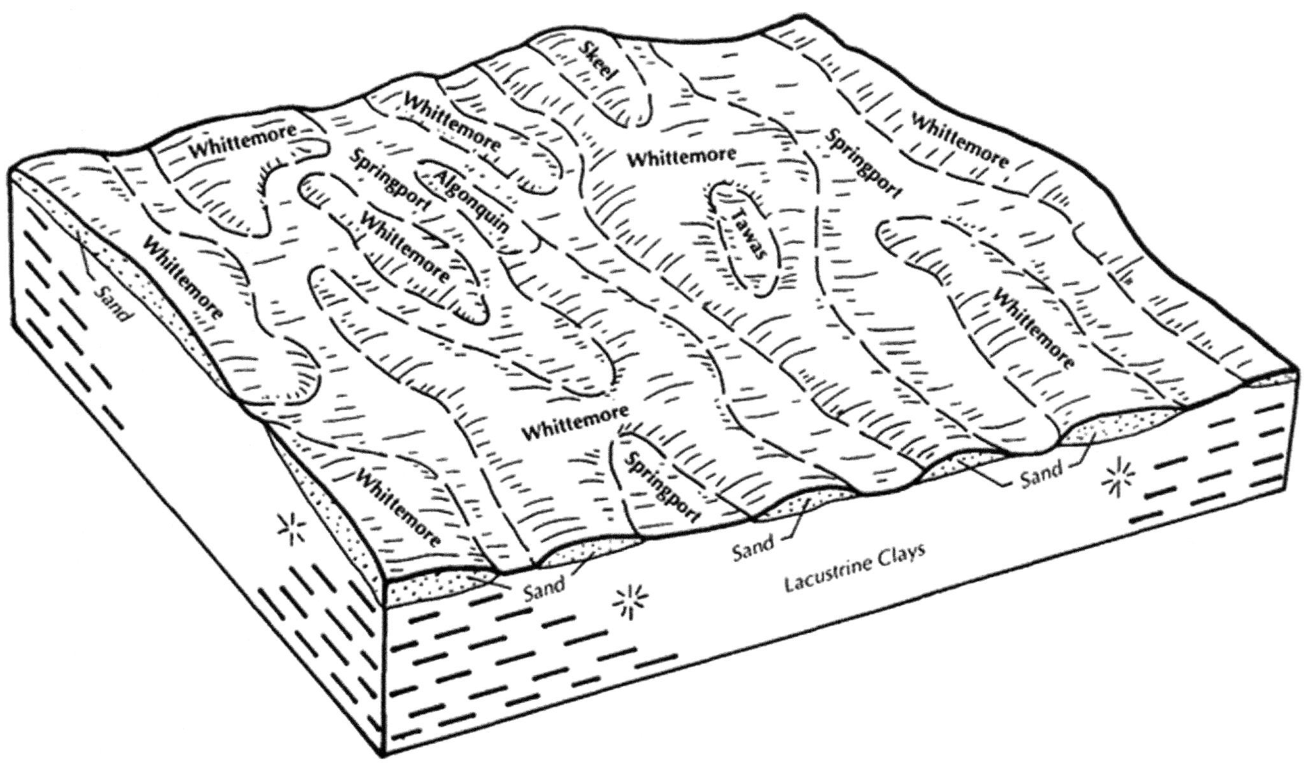

Fig. 8.5 Soil series on beach ridges derived from sandy material over lacustrine clays on an Algonquin shoreline, northwest Lake Huron. *Source* soil survey of Iosco County, Michigan (Johnson 2002)

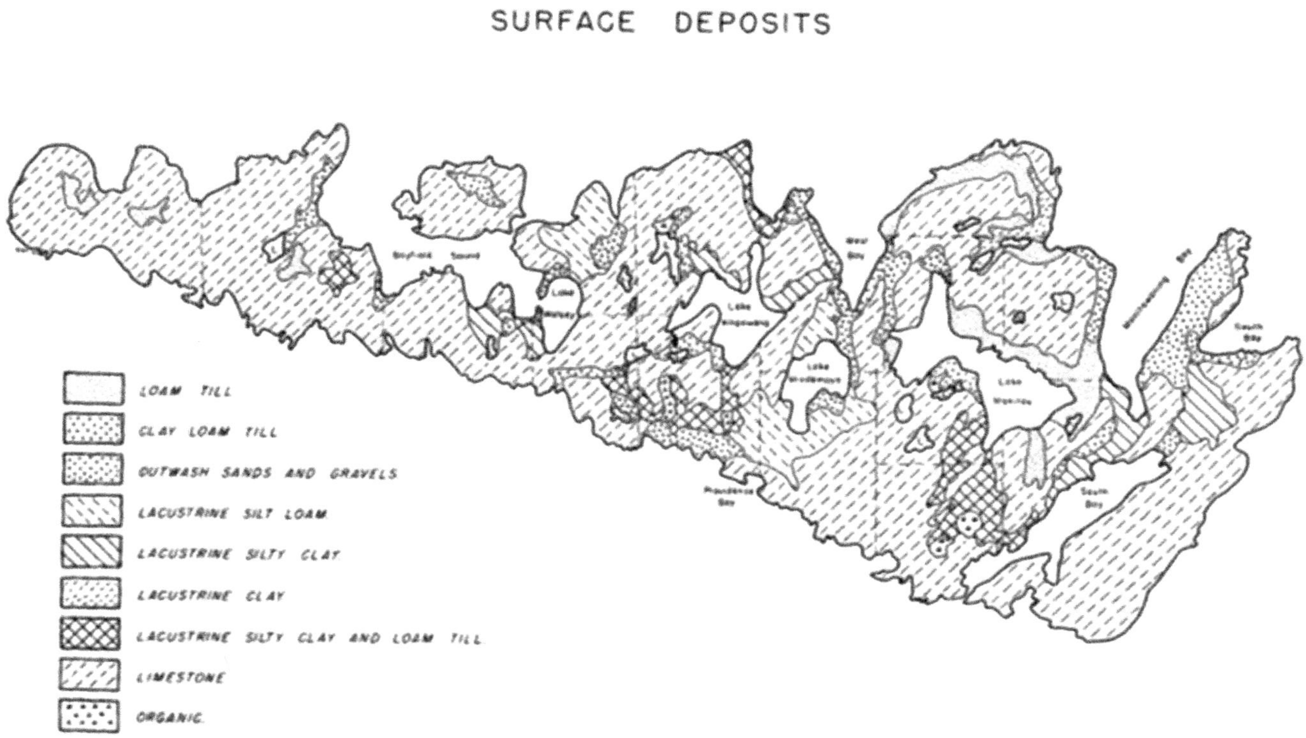

Fig. 8.6 Surficial deposits on Manitoulin Island, northern shore (segment 5) of Georgian Bay, Lake Huron (Hoffman et al. 1959)

Breypen, and Osprey soil series), Hapudalfs (Vincent, Vasey, and Tioga), Endo-, Epiaquepts (Lily and Gwillbury), Endo-, Epiaquolls (Wiarton and Parkhill), Humaquepts (Elderslie), Endo-, Epiaqualfs (Brisbane), and Dystrudepts (Wyevale). Stabilized dunes supporting Udipsamments (Eastport soil series) occur along Nottawasaga Bay.

The last segment, which extends from Port Elgin to Sarnia, contains silty and loamy till, glaciolacustrine sediments, and some sandy stabilized dunes (Fig. 8.9). The dominant soils great groups are Endo-, Epiaqualfs (Brady, Brisbane, Listowel, and Perth soil series), Hapludalfs (Fox, Plainfield, and Waterloo), Hapludolls (Sargent and Sullivan), Endo-, Epiaquolls (Kemble and Toledo), along with Eutrudepts (Brighton), Humaquepts (Elderslie), and Endo-, Epiaquepts (Brookston). Udipsamments (Eastport) occupy beach ridges and stabilized sand dunes. Major sand-dune areas occur at Pinery Provincial Park and Lambton Shores.

8.3 Conclusions

The Lake Huron Coastal Zone can be divided into eight natural resource segments. Haplorthods are the dominant soil great group, followed by Haplosaprists, Endoaquods, Endoaquolls, and Eutrudepts. These soils occur primarily on lacustrine deposits with textures ranging from sandy to clayey and varying thicknesses of sandy loam till. Bedrock commonly occurs near the surface in soils around Georgian Bay. The soils are used for forestry and agriculture.

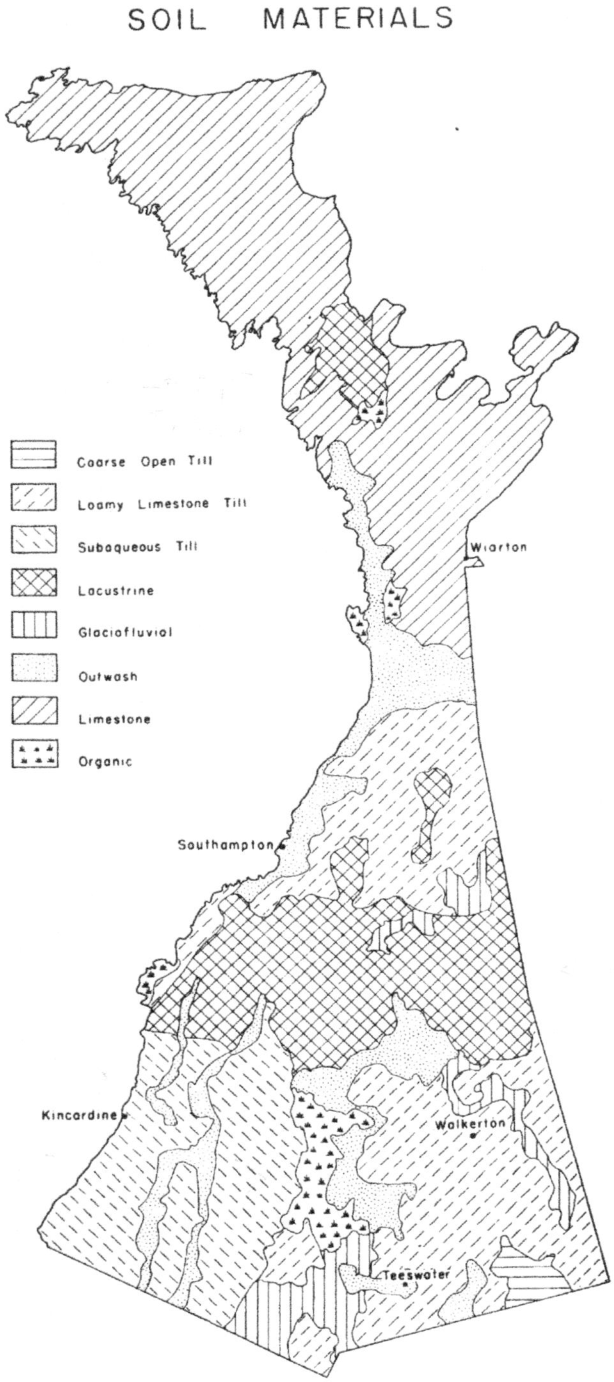

Fig. 8.7 Surficial deposits on the Bruce Peninsula, southeastern shore of Lake Huron. *Source* soil survey of Bruce County, Ontario (Hoffman and Richards 1954)

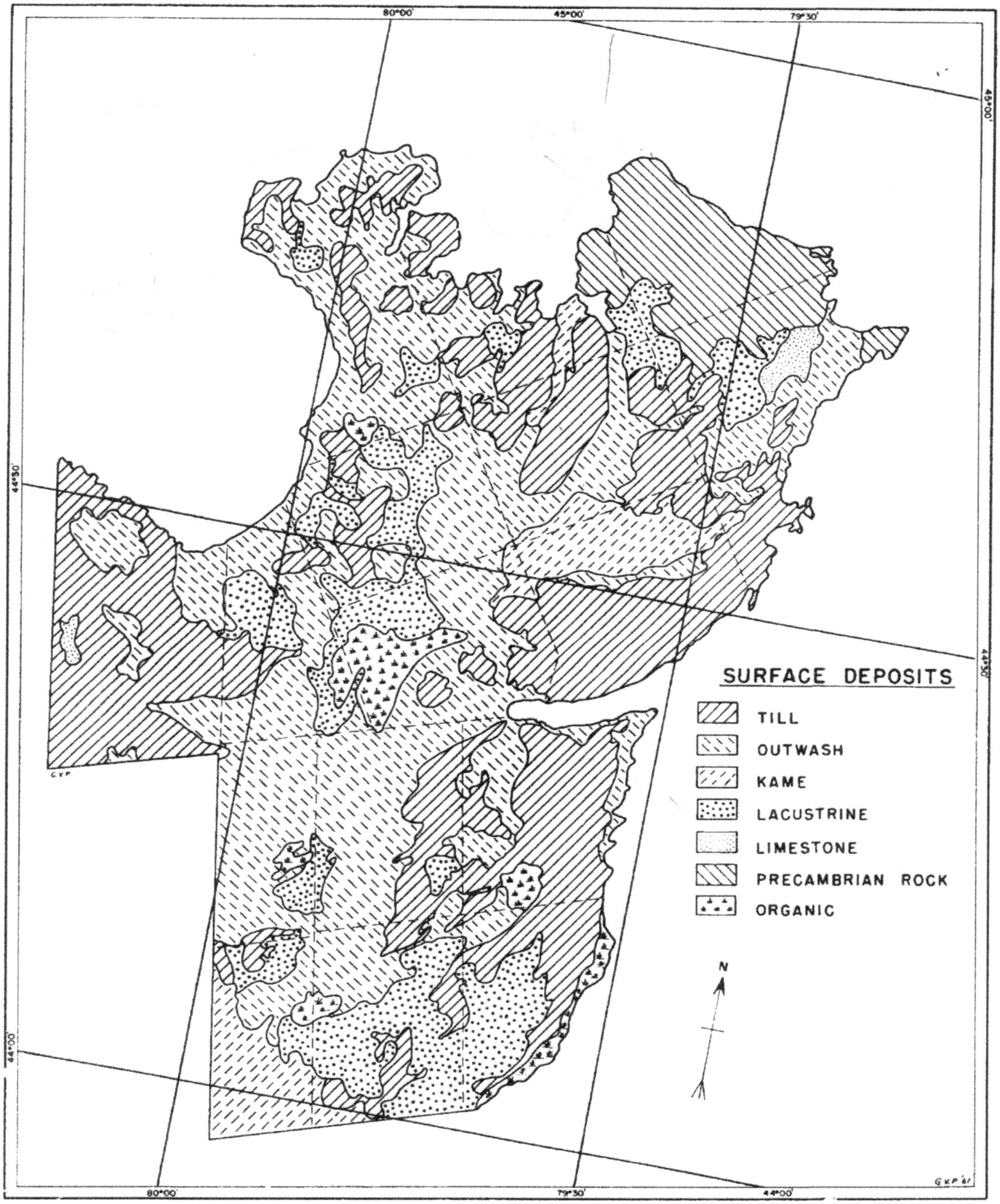

Fig. 8.8 Surficial deposits in Simcoe County, segment 7, southeastern Lake Huron (Hoffman et al. 1962)

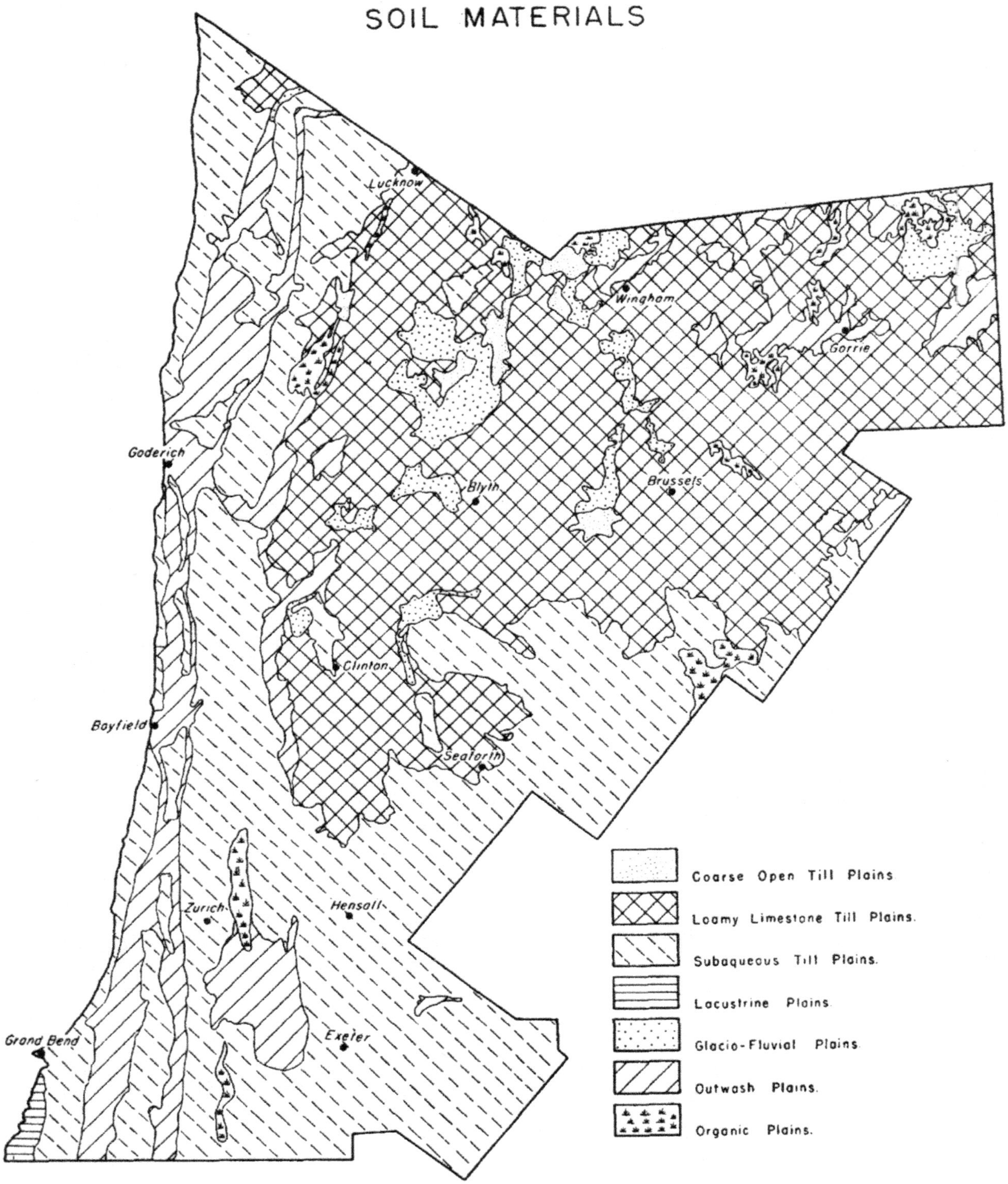

Fig. 8.9 Surficial deposits of segment 8, the southern shore of Lake Huron. *Source* soil survey of Huron County, Ontario (Hoffman et al. 1952)

References

Hoffman, D.W., N.R. Richards, and F.F. Morwick. 1952. *Soil survey of Huron county, Ontario*. Ontario Soil Survey, Rep. No. 13.
Hoffman, D.W., and N.R. Richards. 1954. *Soil survey of Bruce county, Ontario*. Ontario Soil Survey, Rep. No. 16.
Hoffman, D.W., R.E. Wicklund, and N.R. Richards. 1959. *Soil survey of Manitoulin Island, Ontario*. Ontario Soil Survey, Rep. No. 26.
Hoffman, D.W., R.E. Wicklund, and N.R. Richards. 1962. *Soil survey of Simcoe county, Ontario*. Ontario Soil Survey, Rep. No. 29.
Johnson, E.P. 2002. *Soil survey of Iosco county, Michigan*. US Department of Agriculture, Natural Resources Conservation Service and US Forest Service.
Linsemeier, L.H. 1980. *Soil survey of Huron county, Michigan*. US Department of Agriculture, Soil Conservation Service.
Williams, T.E. 1998. *Soil survey of Alcona County, Michigan*. Natural Resources Conservation Service.

9 Soils of the Lake Erie Coastal Zone

9.1 Introduction

Lake Erie is the fourth largest (25,667 km^2), shallowest (19 m), and southernmost (41°26′N) of the Great Lakes. The Lake Erie shore is comprised entirely dolomitic limestone rocks and supports temperate broadleaved forest. Soils along the Lake Erie shore occur on geomorphic surfaces from the Warren, Whittlesey, and Maumee lake stages and are, therefore, are among the oldest (13.8 kyr) in the Great Lakes basin.

9.2 Soils by Natural Resource Segment

The Lake Erie coastal zone is composed dominantly of Endoaquepts-Epiaquepts, Endoaqualfs-Epiaqualfs, Hapludalfs, and Endoaquolls-Epiaquolls (Table 6.1; Fig. 9.1). Urban land is extensive around Detroit, Toledo, Cleveland, Erie, and Buffalo. Lake Erie can be divided into four segments based on MLRAs, parent materials, and dominant soil series. The first segment extends from Detroit, Michigan to Sandusky, Ohio, a part of MLRA 99, the Erie-Huron Lake Plain. This segment contains the Lake Shore Moraine System fringed by clayey and silty lacustrine deposits (Fig. 9.2). Soils on lacustrine deposits include Epiaquepts (Lenawee soil series) and Endoaquepts (Lamson and Toledo soil series) (Fig. 9.3). Epiaqualfs (Blount soil series) occur on the wave-washed till plain. The Hoytville soil series, a Mollic Epiaqualfs, is a common soil on the Lake Erie shores (see Fig. 5.4).

The second segment extends from Sandusky to Buffalo, NY and is part of MLRA 101, the Ontario-Erie Plain and Finger Lakes Region. The parent materials are dominantly fine lacustrine deposits and fine till (Fig. 9.4). The dominant great groups are Endoaquepts (Canandaigua and Lamson soil series), Epiaquepts (Conneaut and Painesville), Endoaqualfs (Niagara), along with Fragiaqualfs (Platea), Dystrudepts (Chenango), and Eutrudepts (Minoa soil series). Beach ridges and stabilized dunes feature Udipsamments (Elnora and Colonie soil series). The Gay soil series, an Aeric Endoaquepts, is common on the Lake Erie shores (see Fig. 5.12). The Grayling and Zimmerman soil series are Typic and Lamellic Udispamments, respectively, and are common in the Lake Erie CZ (Fig. 9.5).

The third segment extends from Buffalo to Ridgetown, Ontario and features moraines and till plains and deep-water (clayey) and shallow-water (sandy) lacustrine deposits underlain by Bertie dolostone (Figs. 9.6 and 9.7). The soils are dominantly Endo-, Epiaqualfs (Beverly, Brady, Gobles, Haldimand, Niagara, Normandale, and Welland soil series), Hapludalfs (Brantford, Fox, Muriel, Ontario, Smithville, Wattford), and Endo-, Epiaquolls (Granby, Kelvin, Lowbanks, St. Williams, and Toledo), with some Endo-, Epiaquepts (Lincoln). Sand dunes occur at Port Maitland and from Long Point to Port Stanley.

The fourth segment, which extends from Ridgetown to Windsor, Ontario, contains lacustrine clays, silts, and sands and silty clay loam till, with intermittent eolian deposits. The dominant great groups are Endo-, Epiaqualfs (Berrien, Beverly, Caistor, Haldimand, and Perth soil series), Hapludalfs (Fox, Harrow, and Plainfield), with some Haplosaprists (marsh), Endo-, Epiaquepts (Brookston), Endo-, Epiaquolls (Granby), and Udipsamments (Eastport soil series). Sand dunes occur at Port Glasgow and Point Pelee Provincial Park.

J. G. Bockheim, *Soils of the Laurentian Great Lakes, USA and Canada*,
https://doi.org/10.1007/978-3-030-52425-8_9

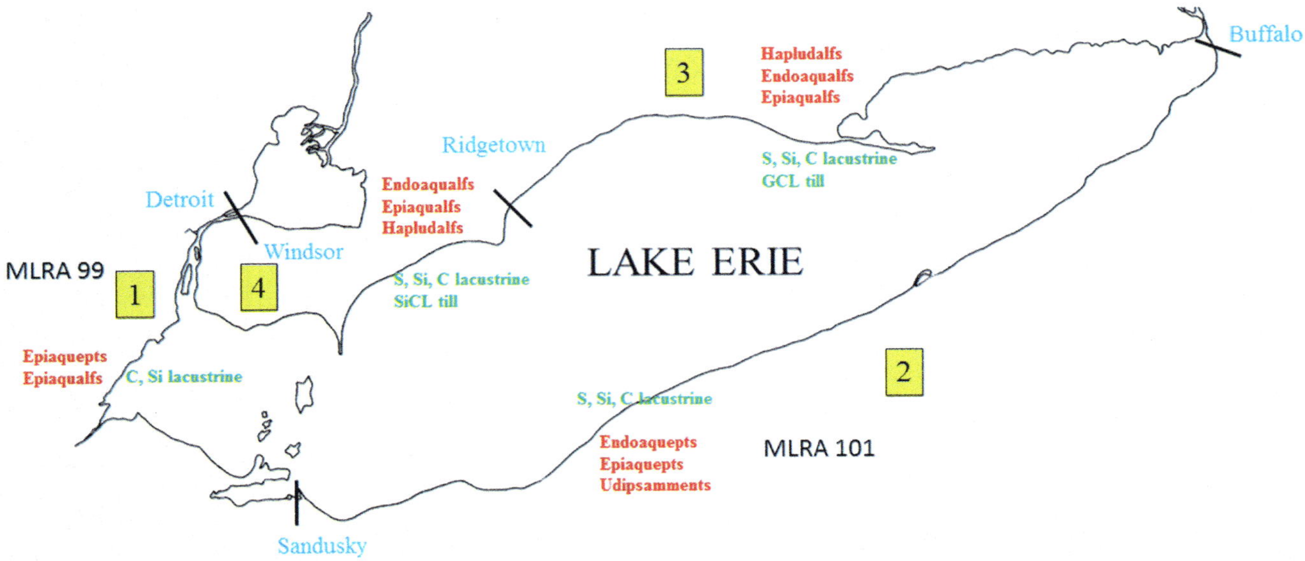

Fig. 9.1 Key segments of the Lake Erie Coastal Zone

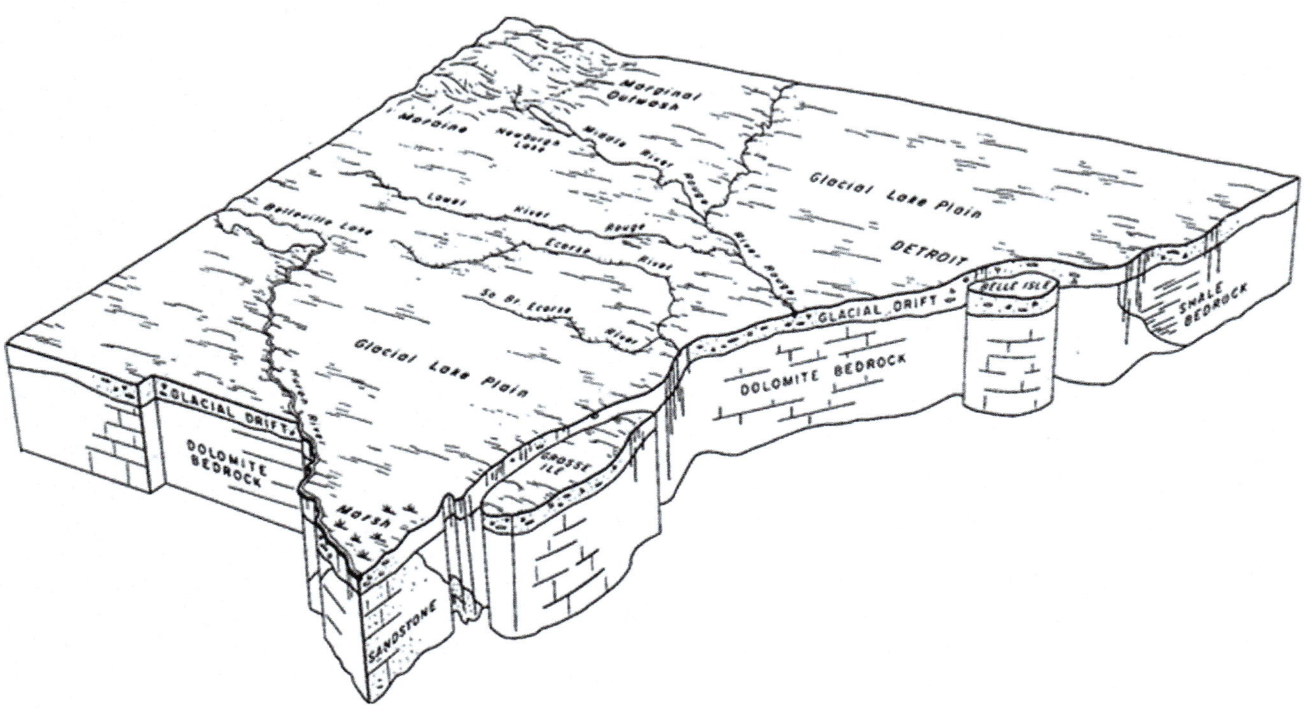

Fig. 9.2 Surficial geology along segment 1, the northwestern shore of Lake Erie. *Source* soil survey of Wayne County, Michigan (Larson 1977)

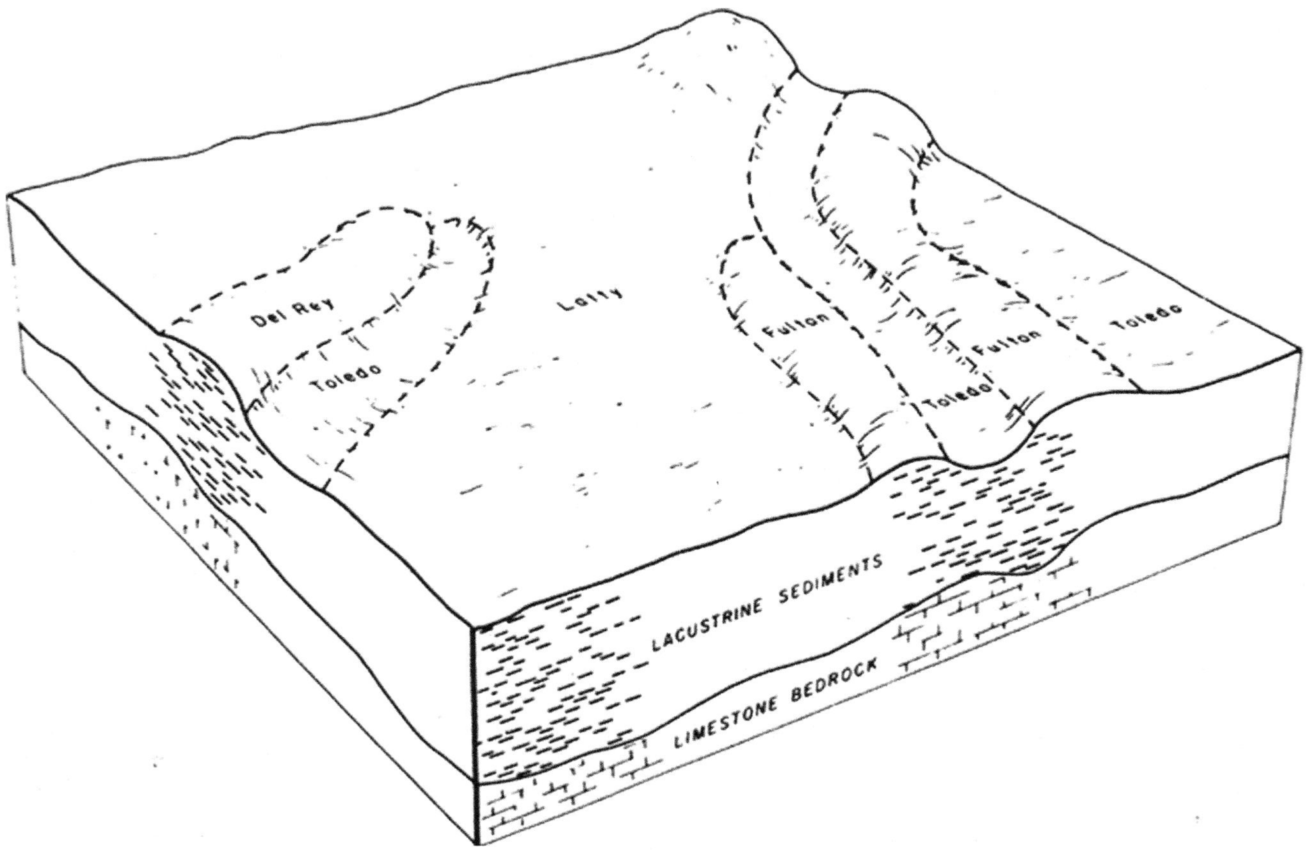

Fig. 9.3 Landforms, parent materials, and soil series of segment 1, the southwestern shore of Lake Erie. *Source* soil survey of Lucas County, Ohio (Stone et al. 1980)

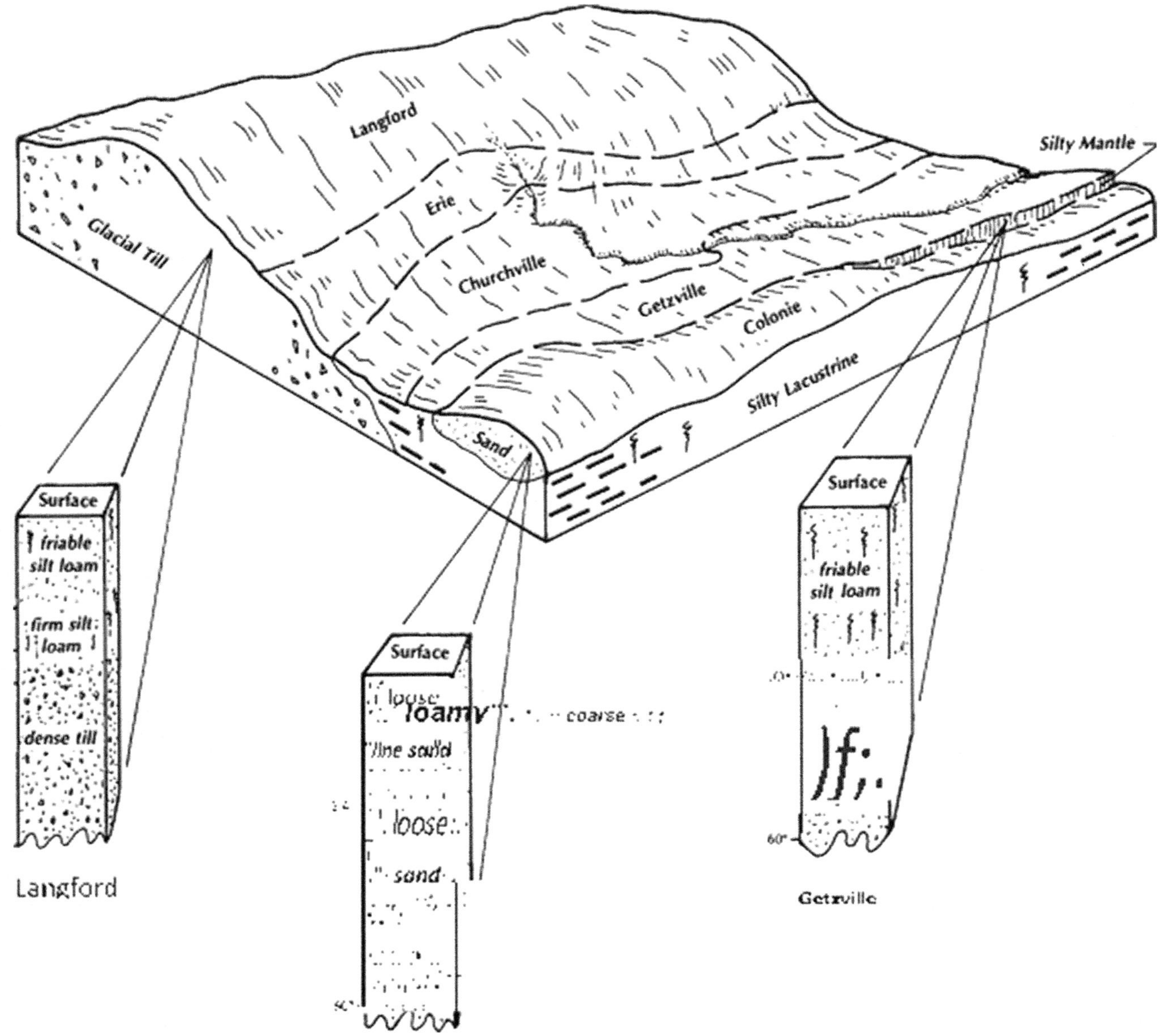

Fig. 9.4 Landforms, parent materials, and soil series of segment 2, eastern Lake Erie. *Source* soil survey of Chautauqua County, NY (Puglia 1994)

Fig. 9.5 The Grayling soil series, a Typic Udipsamments (left) and the Zimmerman soil series, a Lamellic Udispamments (right). The Grayling photo was taken by Dwight Jerome, Resource Soil Scientist, NRCS, Marquette, MI, on Lake Superior. The Zimmerman photo is from the soil survey of Alcona County, MI (Williams 1998)

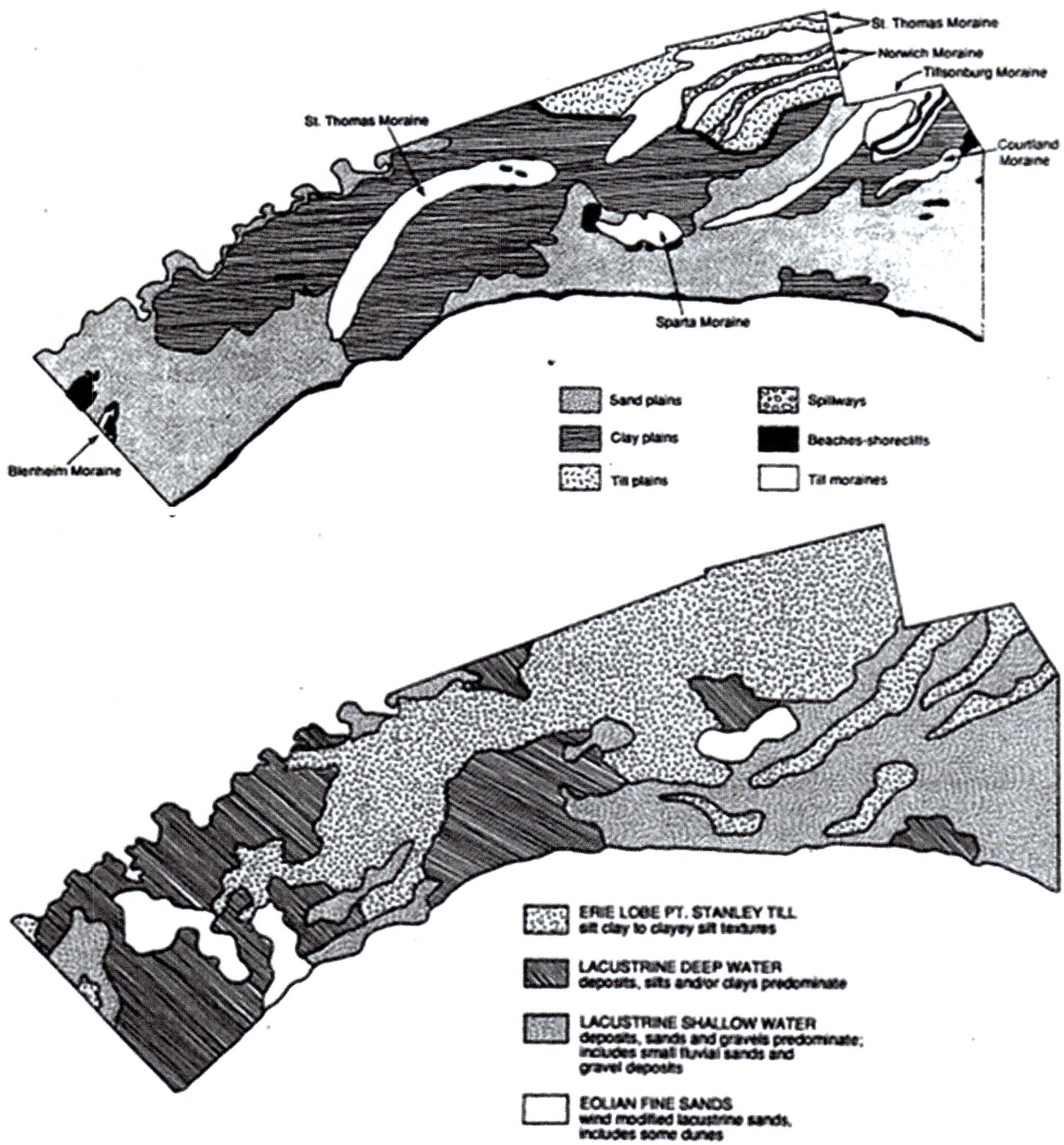

Fig. 9.6 Surficial geology of segment 3, the north shore of Lake Erie. *Source* soil survey of Elgin County, Ontario (Schut 1992)

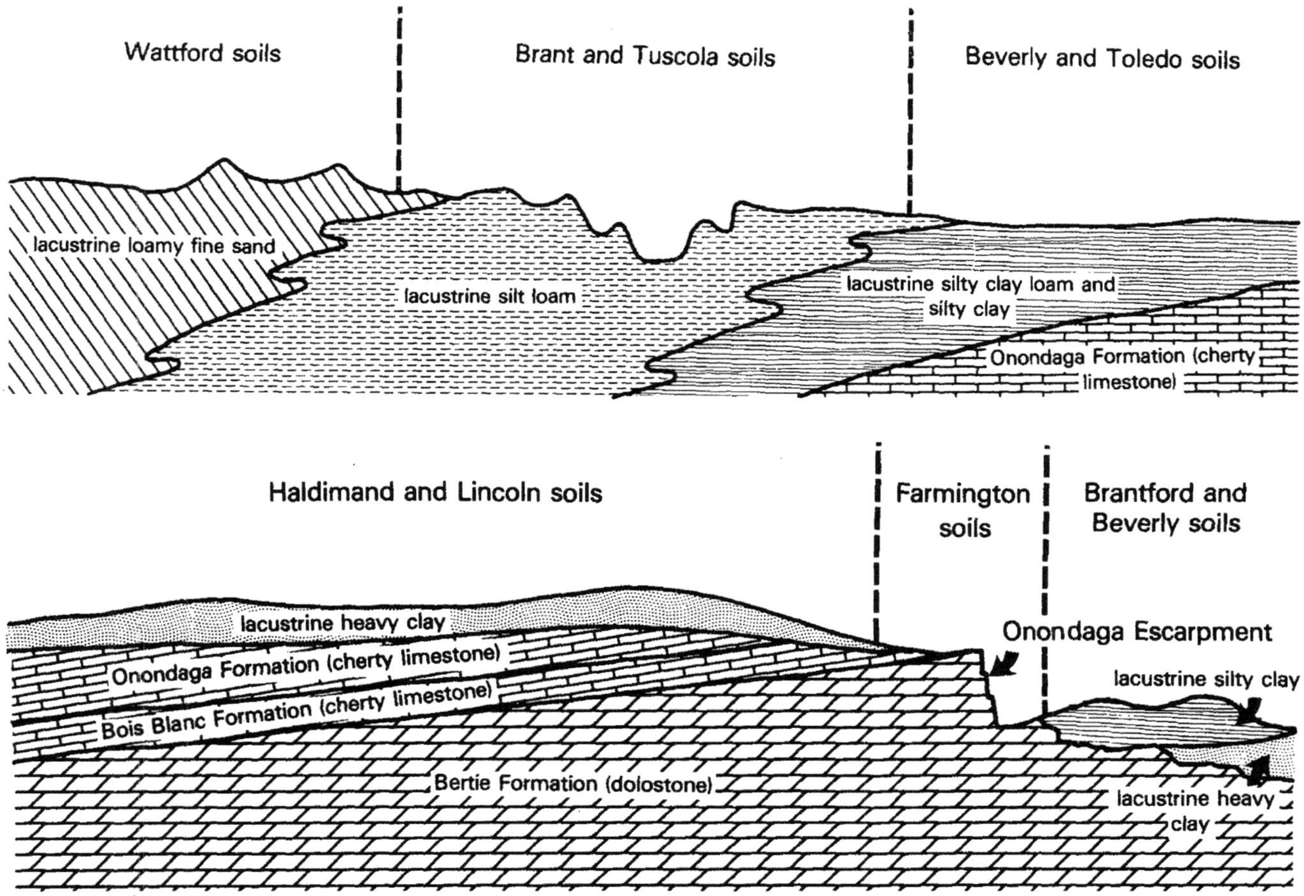

Fig. 9.7 Parent materials of segment 3, the north shore of Lake Erie. *Source* Soils of the Regional Municipality of Haldimand-Norfolk (Presant and Acton 1984)

9.3 Conclusions

The Lake Erie Coastal Zone can be divided into four natural resource segments. Epiaqualfs are the dominant soil great group in the Lake Erie CZ, followed by Epiaquepts, Hapludalfs, and Endoaquepts and Udipsamments. These soils occur on lacustrine deposits with a variety of textural classes and clay loam till. The soils are urbanized in parts, but otherwise are used almost entirely for agriculture.

References

Larson, J.D. 1977. *Soil survey of Wayne county area, Michigan*. US Department of Agriculture, Soil Conservation Service.

Presant, E.W., and Acton, C.J. 1984. *Soils of the regional municipality of Haldimand-Norfolk, Ontario*. Ontario Institute of Pedology, Rep. No. 57.

Puglia, P.S. 1994. *Soil survey of Chautauqua county, New York*. US Department of Agriculture, Soil Conservation Service.

Schut, L.W. 1992. *The soils of Elgin county, Ontario*. Ontario Centre Soil Resource Evalution, Rep. No. 63.

Stone, K.L., E.H. McConoughey, G.D. Bottrell, and D.J. Crowner. 1980. *Soil survey of Lucas county, Ohio*. US Department of Agriculture, Soil Conservation Service.

Williams, T.E. 1998. *Soil survey of Alcona County, Michigan*. Natural Resources Conservation Service.

10 Soils of the Lake Ontario Coastal Zone

10.1 Introduction

Lake Ontario is the smallest (18,960 km^2), easternmost (76.30 W), and has lowest surface elevation (74 m) of the Great Lakes. The northern and eastern shores of Lake Ontario feature rocks derived from dolomitic limestone and sandstone; the southern shore is composed of dolomitic limestone. Whereas the north shore of Lake Ontario features temperate mixed coniferous-broadleaved forest, the north, east, and southern coastline has temperate broadleaved forest. The soils of Lake Ontario are derived from lacustrine sediments and wave-washed till from the Iroquois lake stage approximately 13 kyr BP.

10.2 Soils by Natural Resource Segment

The Lake Ontario coastal zone is composed primarily of Hapludalfs, Endoaqualfs-Epiaqualfs, and Fragiudepts (Table 6.1; Fig. 10.1). Urban areas are extensive around Hamilton and Toronto, Ontario and Rochester, New York. Lake Ontario can be divided into two segments based on MLRAs, parent materials, and dominant soil series. The first segment follows the US Lake Ontario coastal zone is from Youngstown to Cape Vincent, New York and is part of MLRA 101, the Ontario-Erie Plain and Finger Lakes Region. The parent materials are eolian sand, lacustrine silts, calcareous loamy till over shale or dolomitic limestone (Figs. 10.2 and 10.3).

The dominant great groups are Hapludalfs (Collamer, Dunkirk, and Hilton soil series), Endoaqualfs (Niagara and Rhinebeck), and Fragiudepts (Ira, Sodus, and Williamson), with some Udipsamments (Colonie) and Endoaquepts (Canandaigua). Derived from deep, marly organic materials over sandy lacustrine sediments, the Edwards (Fig. 10.4). The Martisco soil series is a Histic Humaquepts that has formed in a thin mantle (10 inches; 25 cm) of highly decomposed, black organic deposits overlying gray marl (Fig. 10.5). The Vergennes soil series, a Glossaquic Hapludalfs, is formed from calcareous estuarine and glaciolacustrine clays in the eastern Lake Ontario basin (Fig. 10.6). The photo shows only the dark grayish brown Ap horizon and the grayish brown, clayey B/E (glossic) horizon. The Churchville soil series, an Aeric Endoaqualfs, occurs along the southern Lake Ontario shores (Fig. 5.3).

The second segment of the Lake Ontario coastal zone extends from Kingston to Niagara-on-the-Lake, Ontario (Iroquois Plain). The parent materials are lacustrine silt, clay, and sand and fine-loamy till (Figs. 10.7, 10.8, and 10.9). The major great groups are Endo-, Epiaqualfs (Chinguacousy, Edenvale, Guerin, Landsdowne, Smithfield, Solmesville, Tecumseth, Tavistock, Vineland soil series), Hapludalfs (Bondhead, Bookton, Colborne, Dundonald, Fox, Grimsby, and Newcastle), Hapludolls (Ameliasburg, Athol, and Farmington), and Eutrudepts (Brighton, Hillier, and Otonabee), with some Endo-, Epiaquepts (Nappanee and Gilford), Endo-, Epiaquolls (Lindsay), Udipsamments (Eastport), Humaquepts (Gerow), and Haplosaprists (i.e., marsh). Sand dunes occur at Sandbanks, from Brighton to Port Hope, from Pickering to Burlington, and near St. Catharines.

Along the US portion of the GLCZ, 56% of the soil series occur on one lake only; 30% occur along two lakes, 13% around three lakes, and 1.3% around four lakes (Fig. 10.10). The dominant great groups containing soil series around three or four lakes are Haplosaprists (13 soil series), Haplorthods (9), Endoaquolls (7), Udipsamments and Hapludalfs (6 soil series each). No soil series occurred around all five lakes.

J. G. Bockheim, *Soils of the Laurentian Great Lakes, USA and Canada*,
https://doi.org/10.1007/978-3-030-52425-8_10

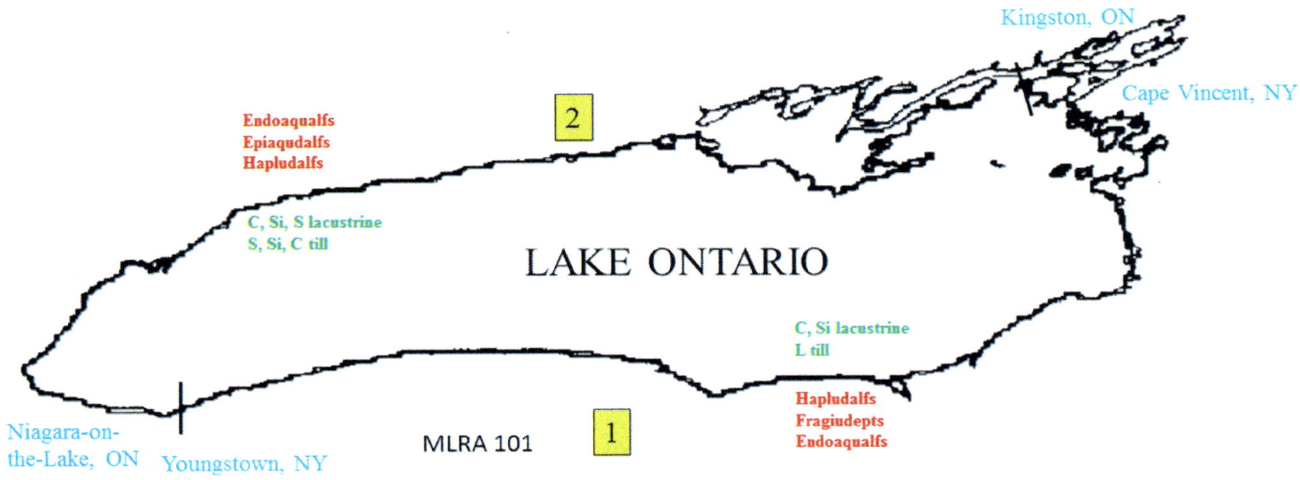

Fig. 10.1 Key segments on the Lake Ontario coastal zone

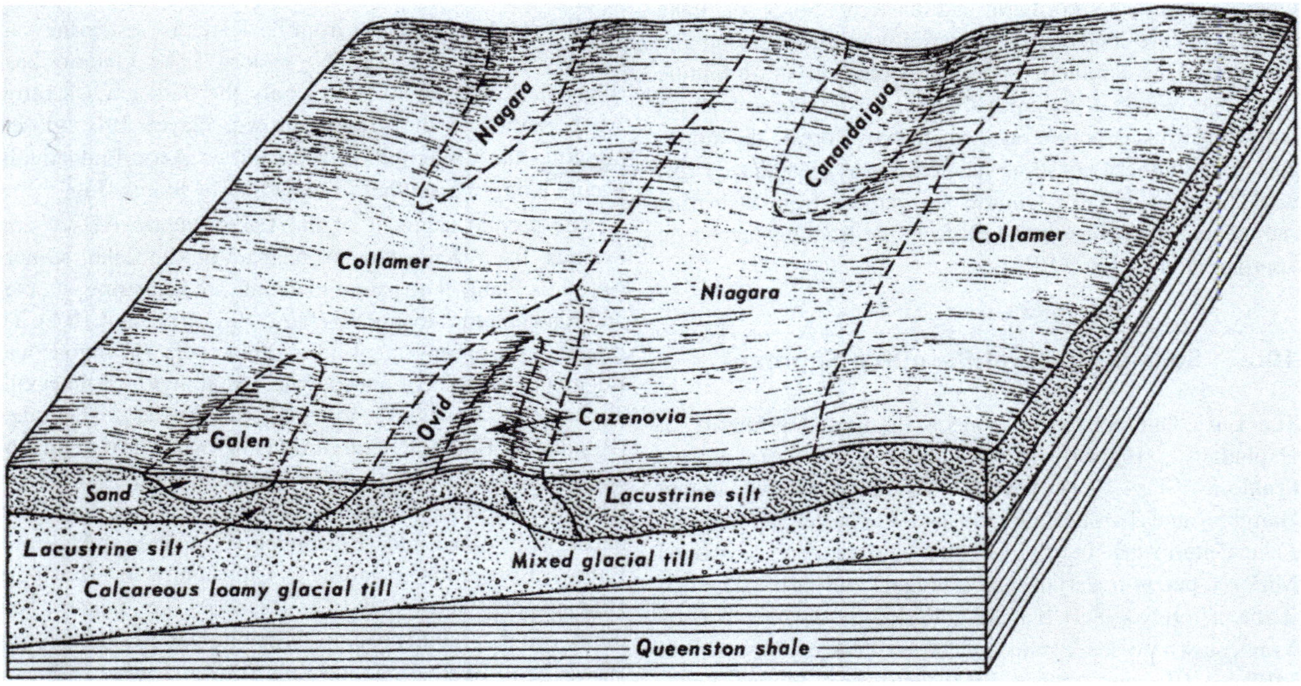

Fig. 10.2 Eolian and lacustrine deposits over till and Queenston shale along segment 1, the south shore of Lake Ontario. *Source* soil survey of Orleans County, NY (Higgins et al. 1977)

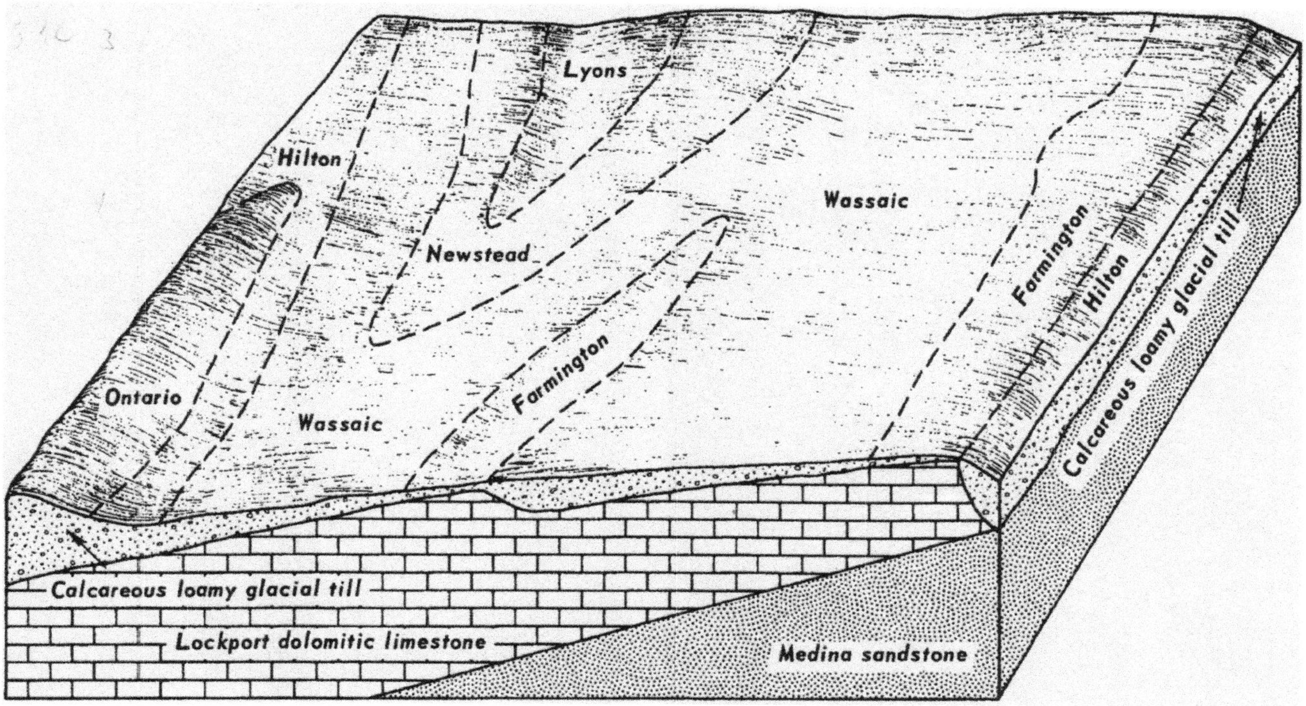

Fig. 10.3 Calcareous till over Lockport dolomitic limestone along segment 1, the south shore of Lake Ontario. *Source* soil survey of Orleans County, NY (Higgins et al. 1977)

Fig. 10.4 The Edwards soil series is a Limnic Haplosaprists derived from marly lacustrine sediments. The scale is in feet. *The photo* is from the soil survey of Pulaski County, IN (Indiana Headwaters MLRA Soils Team 2003)

Fig. 10.5 The Martisco soil series is derived from sapric material overlying marl. The scale is in inches. *The photo* is from the soil survey of St. Joseph County, IN (McBurnett et al. 2004)

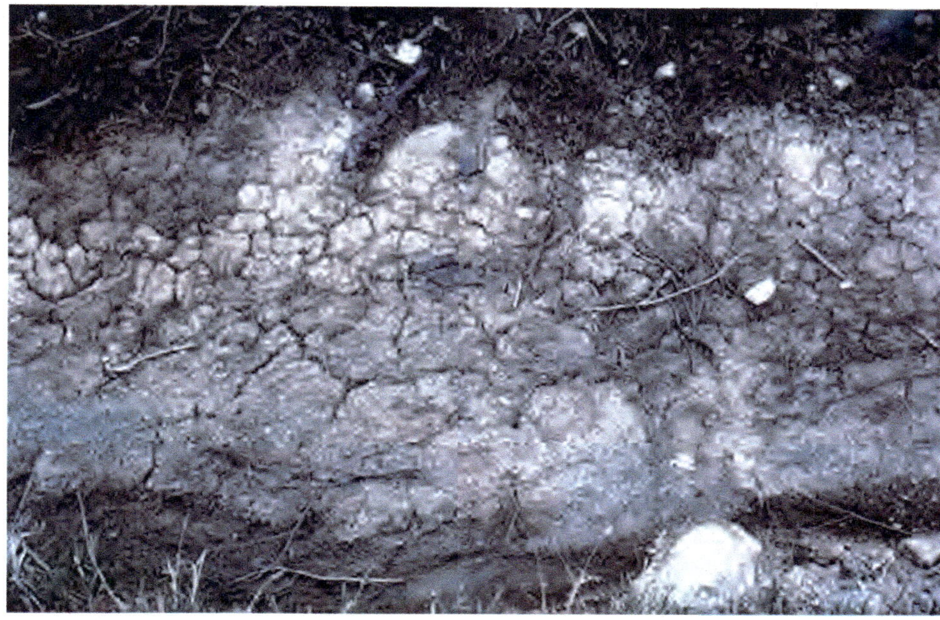

Fig. 10.6 The Vergennes soil series is a Glossaquic Hapludalfs derived from calcareous estuarine and glaciolacustrine clays in the eastern Lake Ontario basin. The exposure is from a road cut of the upper 25 cm.*The photo* is from the soil survey of Essex County, NY (Smith 2010)

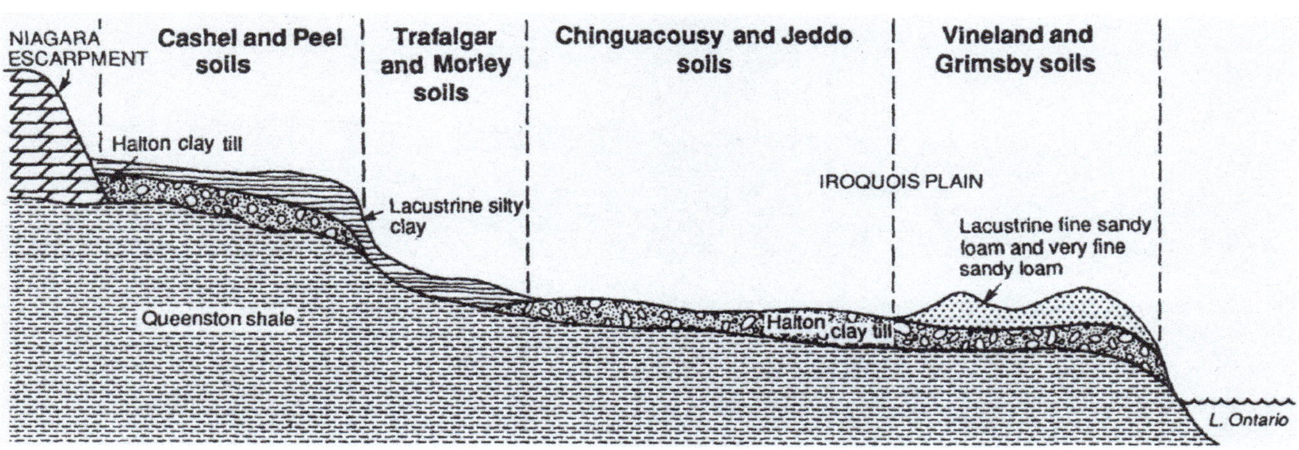

Fig. 10.7 Schematic landscape cross-section showing the relationship of soils on the Iroquois Plain between the Niagara Escarpment and Lake Ontario. *Source* The Soils of the Regional Municipality of Niagara, ON, Vol. 1 (Kingston and Presant 1989)

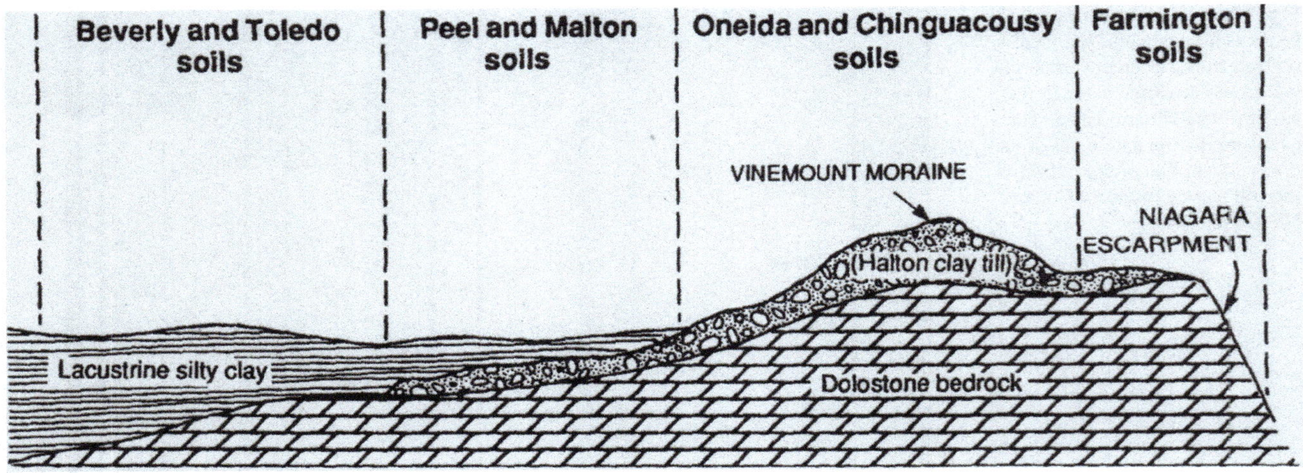

Fig. 10.8 Schematic landscape cross-section showing the relationship of soils in the area of the Vinemount Moraine, between the Niagara Escarpment and the Haldimand Clay Plain. *Source* The Soils of the Regional Municipality of Niagara, ON, Vol. 1 (Kingston and Presant 1989)

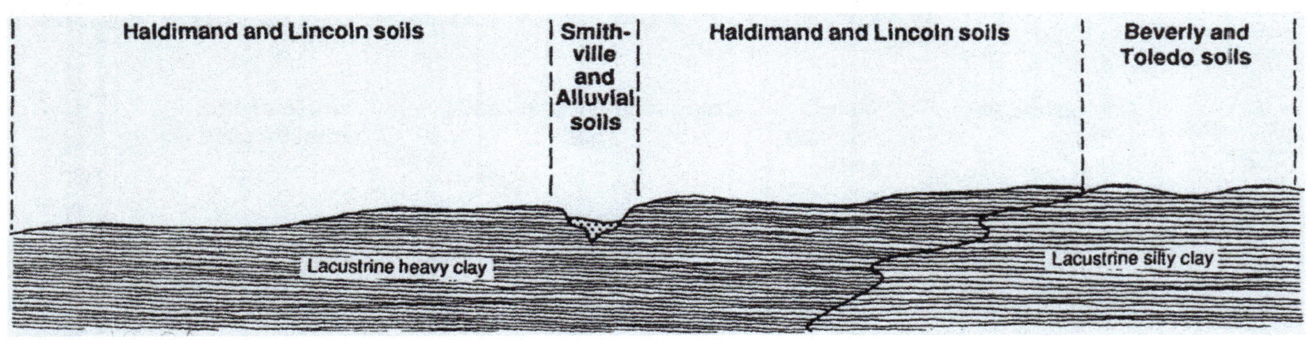

Fig. 10.9 Schematic landscape cross-section showing the relationship of soils on the Haldimand Clay Plain. *Source* The Soils of the Regional Municipality of Niagara, ON, Vol. 1 (Kingston and Presant 1989)

Fig. 10.10 Numbers of soil series occurring on one or more of the Great Lakes. Although 12 soil series occur on four lakes, there is no single soil series that occurs on all five lakes

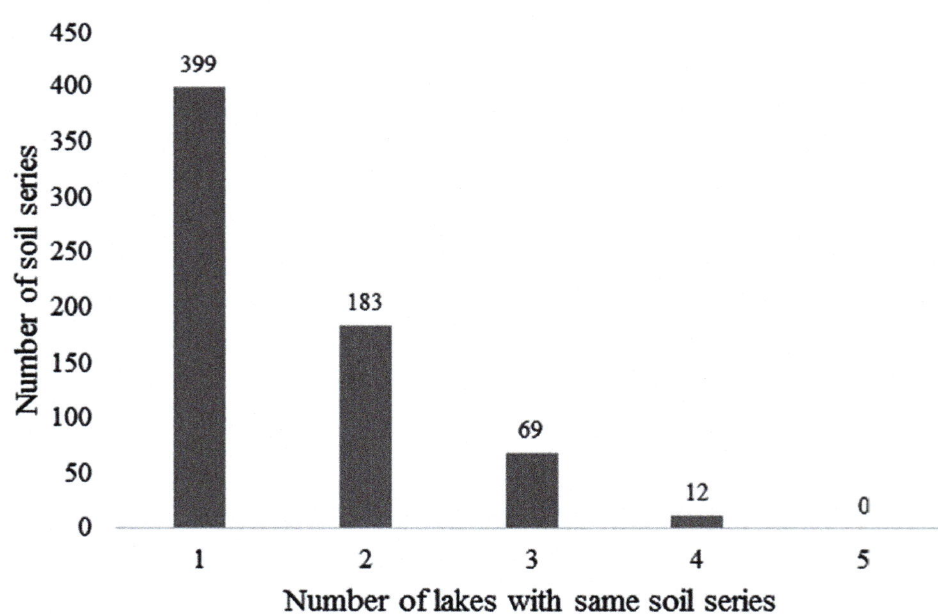

10.3 Conclusions

The Lake Ontario Coastal Zone can be divided into two natural resource segments. The dominant soil great groups are Hapludalfs and Endoaqualfs, along with Fragiudepts along the southeastern and eastern shores. These soils occur on lacustrine silts and clays and till ranging from clayey to loamy in texture. The soils are urbanized in the "Golden Horseshoe" of the eastern shore and along the south-central shore. Otherwise the soils of the Lake Ontario CZ are used for agriculture and recreation.

References

Higgins, et al. 1977. *Soil survey of Orleans county, New York*. US Department of Agriculture, Soil Conservation Service.

Indiana Headwaters MLRA Soils Team. 2003. *Soil survey of Pulaski County, Indian*. Natural Resources Conservation Service.

Kingston, M.S., and E.W. Presant. 1989. *The soils of the regional municipality of Niagara, Ontario*. Ontario Institute of Pedology, Rep. No. 60.

McBurnett, S.L., D.A. Gehring, and R.W. Neilson. 2004. *Soil survey of St. Joseph County, Indiana*. Natural Resources Conservation Service.

Smith, G.W. 2010. *Soil survey of Essex County, New York*. Natural Resources Conservation Service.

11 Soil-Forming Processes in the Great Lakes Coastal Zone

11.1 Introduction

Specific soil processes are determined by the soil-forming factors and are expressed in diagnostic horizons, properties, and materials, which are then used to classify soils: soil-forming factors → soil-forming processes → diagnostic horizons, properties, materials → soil taxonomic system (Bockheim and Gennadiyev 2000). Accordingly, these investigators identified 17 generalized soil-forming processes. Two additional processes are added here, including cambisolization and glossification, and they will be defined forthwith. The dominant soil-forming processes in the GLCZ are gleization, argilluviation, podsolization, cambisolization, melanization, glossification, paludization, and vertization (Fig. 11.1).

11.2 Gleization

Gleization (hydromorphism) refers to the presence of aquic conditions often evidenced by reductimorphic or redoximorphic features such as mottles and gleying. In the Great Lakes region, gleization is favored by proximity to the lakes, abundant precipitation, and parent materials that restrict drainage by virtue of texture, a layer that restricts moisture movement, or the presence of bedrock.

11.3 Argilluviation

Argilluviation (lessivage) refers to the movement of clays in the solum. This process is favored in the Great Lakes region by an abundance of precipitation (udic and aquic soil-moisture regimes), parent materials enriched in carbonate-free clays and coarse fragments, backslopes rather than eroding shoulders, and a time interval of >2,000 yr (Bockheim and Hartemink 2013).

11.4 Podzolization

Podzolization is a complex collection of processes that includes eluviation of base cations, weathering transformation of Fe and Al compounds, mobilization of Fe and Al in surface horizons, and transport of these compounds to the spodic Bs. horizon as Fe and Al complexes with fulvic acids and other complex polyaromatic compounds. This process occurs in soils located primarily along the northern and eastern shores of Lake Superior and the north shore of Lake Huron. Podzolization in the Great Lakes region is favored by abundant precipitation and low temperatures (cryic and frigid soil-temperature regimes), low-base-status, sandy parent materials, the presence of coniferous vegetation, and a time interval of >8,000 yr (Bockheim 1997; 2003).

11.5 Cambisolization

Cambisolization refers to a collection of weak soil-forming processes that leads to the formation of a cambic horizon, which is present largely in Inceptisols and some Mollisols in the Great Lakes region. A cambic horizon may form in <1,000 yr.

11.6 Melanization

Melanization (humification) refers to the accumulation of well-humified organic compounds in the upper mineral soil. This process is favored in the Great Lakes region by

J. G. Bockheim, *Soils of the Laurentian Great Lakes, USA and Canada*,
https://doi.org/10.1007/978-3-030-52425-8_11

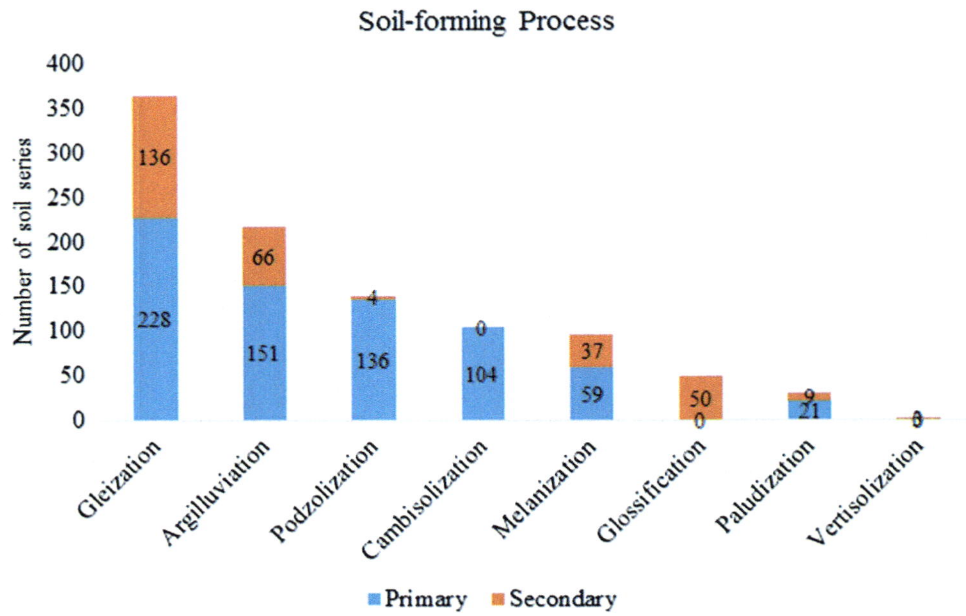

Fig. 11.1 Primary and secondary soil-forming processes in the Great Lakes Coastal Zone

grassland vegetation, base-rich parent materials (in the case of Mollisols), and a time interval of >500 yr.

11.7 Glossification

Glossification, as defined here, refers to tonguing in soils that results from degradation of an argillic horizon from which clay and free iron oxides are removed (Soil Survey Staff 2014). This process leads to the development of a glossic horizon, which occurs in some Alfisols and Spodosols along the Great Lakes shore.

11.8 Paludization

Paludization pertains primarily to the deep (>40 cm) accumulation of organic matter (histic materials) on the landscape usually in marshy areas. This process is dominant in most Histosols and occurs in some Aquepts in the Great Lakes region.

11.9 Vertization

Vertization represents a collection of sub-processes occurring in soils with >60% smectitic clay, which enables soils to undergo shrinking and swelling that leads to cracking on the surface, tilted, wedge-shaped aggregates, and slickensides on aggregate faces. In the Great Lakes region, vertization occurs in a few soils derived from clayey till or lacustrine deposits with a very-fine particle-size class.

11.10 Anthrosolization

In addition to these processes, anthrosolization—the effects of humans on soil development–has affected many of the soils directly through urbanization, industrialization, agricultural practices that have accelerated erosion and supplied excessive N and P, and the introduction of exotic and invasive plant species. Humans affect soils indirectly in the Great Lakes region by air and water pollution and changing the climate.

11.11 Linkage Soil-Forming Processes and Soil Taxa

There is a pronounced linkage among soil-forming processes, soil taxa (great-group level), and landform-parent material (Fig. 4.13). On fine-textured lake plains and till plains, argilluviation leads to the development of Hapludalfs and Glossudalfs, podsolization contributes to the formation of Haplorthods and Fragiorthods, and gleization is dominant in Aqualfs-Aquepts-Aquolls-Aquods. On sandy lake plains and outwash plains, argilluviation forms Hapludalfs, podsolization forms Haplorthods, and gleization forms Aqualfs-Aquepts-Aquolls-Aquods. Beach ridges represent a smaller-scale versions of these interactions. Podzolization is dominant on the ridges of stabilized dunes and gleization and paludization persist in the swales. Active dunes, evidenced by blowouts, feature dune land and Psamments, and stabilized dunes reflect podsolization.

Glacial lake benches comprised of limestone are influenced by melanization and calcification that lead to the development of Haprendolls (Fig. 4.13). Well-drained alluvial fans and floodplains feature cambisolization leading to Udepts; and poorly drained fans and floodplains evidence gleization and Aqu-great groups.

11.12 Conclusions

The dominant soil-forming processes in the Great Lakes Coastal Zone are gleization and argilluviation, followed by podsolization, cambisolization, and melanization. Glossification, paludization, and vertization occur to a limited extent. There is a pronounced linkage among soil-forming processes, soil taxa (great-group level), and landform-parent material.

References

Bockheim, J.G. 1997. Soils in hemlock-hardwood ecosystem-mosaic. *Canadian Journal of Forest Research* 27: 1147–1153.

Bockheim, J.G. 2003. Genesis of bisequal soils on acidic drift in the upper Great Lakes region, USA. *Oil Science Society of America Journal* 67: 612–619.

Bockheim, J.G., and A.N. Gennadiyev. 2000. The role of soil-forming processes in the definition of taxa in soil taxonomy and the world soil reference base. *Geoderma* 95: 53–72.

Bockheim, J.G., and A.E. Hartemink. 2013. Distribution and classification of soils with clay-enriched horizons in the USA. *Geoderma* 209–210: 153–160.

Soil Survey Staff. 2014. *Keys to soil taxonomy* (12th edn.). U.S: Department of Agriculture and National Resources Conservation Service.

12 Pedodiversity of the Great Lakes Coastal Zone

12.1 Soil Series Density

Based on the density of soil series within 5 km of the shorelines, all five of the Great Lakes have a high level of pedodiversity. In the US GLCZ at a scale of 1:24,000, the Lake Ontario coastal zone is most diverse (0.026 soil series per km^2) and Lake Huron is least diverse (0.010 soil series per km^2) (Table 12.1). Most of the mapping in the Canadian Great Lakes coastal zone was conducted at a scale of 1:63,360. The mean density of soil series in the Canadian GLCZ is 0.005.

12.2 Factors Influencing Pedodiversity in the Great Lakes Region

At a scale of 1:24,000, the US GLCZ has a three-fold greater average pedodiversity (0.016) than for Land Resource Regions, Major Land Resource Areas, and state-wide values (Table 12.2). The high pedodiversity of the GLCZ can be attributed to (i) a diversity of landforms and parent materials of different age; (ii) strong climatic gradients; and (iii) a strong vegetation gradient.

12.2.1 Parent Materials

The influence of parent materials on pedodiversity in the GLCZ relates to the composition of the bedrock and the kinds and ages of unconsolidated materials overlying the bedrock. Precambrian granites and gneisses surround the north shores of Lakes Superior and Huron, and Paleozoic sedimentary rocks, largely sandstones, shales, and dolomitic limestones, are dominant in the south (Fig. 12.1).

About half (53%) of the landforms in the GLCZ were identified as lake plains or lake terraces, largely from the Algonquin, Iroquois, and Nipissing lake stages (Hough 1958; Prest 1970; Clark et al. 2012). These lake stages range from 14.1 to 3.2 kyr BP in age, and allow for a considerable range in time for soil development. A third (33%) of the landforms are comprised of till. Several soil series occur on post-glacial shorelines carved into till plains, including the Blount, Ensley, Hoytville, Londo, Mishwabic, Nonesuch, Parkhill, Shebeon, and Tappan series. The Tyner and Elnora series represent relict longshore bars. A number of other soil series are derived from a mantle of lacustrine deposits over till, including the Anton, Augustana, Avoca, Baden, Belleville, Blackhoof, Cayuga, Churchville, Hegberg, Herbster, Normanna, Portwing, and Rimer series. Several soils are in glacial or post-glacial drainageways, including the Glendora, Graycalm, Gull Point, Hettinger, Kerston, Rapson, Richter, Roscommon, and Shingleton series. Beach ridges compose about 5% of the coastal zone and have Udipsamments on ridges and Psammaquents and Haplosaprists in the swales.

Active dunes, particularly the perched type, comprise about 9% of the GLCZ; stabilized dunes comprise another 0.8% of the area. Dominant soils on dunes include Udipsamments (Colonie, Deer Park, Eastport, Morocco, Oakville, Ottokee, Plainfield, Tedrow, Wurtsmith, and Zimmerman soil series) and Haplorthods (Covert, Croswell, Liminga, and Rubicon soil series). The dunes began forming on lake terraces during and after the Nipissing lake stage approximately 3.7–3.0 ka (Arbogast and Loope 1999; Arbogast 2000; Loope et al. 2004, 2010).

Many of the landforms have been affected by glacial isostatic adjustment following recession of the Laurentide ice, particularly those surrounding Lake Superior and the northern portions of Lakes Michigan, Huron, and Ontario. This adjustment varies from near 0 at the hinge line to 300 m along the north shore of Lake Superior.

12.2.2 Climatic Gradient

There are two pronounced climate gradients in the Great Lakes region, including one in temperature that is reflected in the soil-temperature class of the soils and another pertaining to moisture gradients from "lake-effect" snow

J. G. Bockheim, *Soils of the Laurentian Great Lakes, USA and Canada*,
https://doi.org/10.1007/978-3-030-52425-8_12

Table 12.1 Density of soil series in the US and Canadian Great Lakes Coastal Zone

Lake	States	Shoreline length (km)	No. soil series	Density (No./km^2)
US; scale 1:24,000				
Erie	MI, OH, NY, PA	1092	143	0.026
Ontario	NY	547	101	0.037
Superior	MI, WI, MN	3186	233	0.015
Michigan	MI, WI, IN, IL	2673	317	0.024
Huron	MI	1853	263	0.028
Total		9351	691	0.015
Canada; scale 1:63,350				
Ontario	ON	781	27	0.007
Erie	ON	929	26	0.006
Huron	ON	1879	38	0.004
Total		3589	91	0.005

Table 12.2 Soil series density in the United States by region

Region	No. units	Average area (km^2)	Soil series density		
			Avg.	St dev	CV
Land Resource Region	26	3,32,878	0.005	0.007	140
Major Land Resource Area	25	45,107	0.005	0.004	88
State	50	1,82,949	0.005	0.004	80
Great Lakes coastal zone (US)	5	5750	0.015	0.005	33

(Fig. 12.1). Soils in the Lake Superior and northern Lake Michigan and Huron drainage basins are in the frigid soil-temperature class, i.e., the mean annual soil temperature at 50 cm is between 0 and 8 °C. In contrast, soils in the southern Lake Michigan and Huron and Lake Erie and Ontario drainage basins have a mesic soil-temperature class, i.e., the mean annual soil temperature at 50 cm is between 8 and 15 °C.

The Great Lakes are subject to "lake-effect" snow, whereby a cold air mass moves across long expanses of warmer lake water, warms the low layer of air which picks up water vapor from the lake, rises up through the colder air above, freezes and is deposited on the leeward (downwind) shores (see Fig. 2.1). In the Lake Superior drainage basin, average annual snowfall ranges from 200 cm along Minnesota's North Shore to over 500 cm on the Keweenaw Peninsula; in the Lake Michigan basin, snowfall ranges from 100 cm along the south shore to over 375 cm on the northeast shore; in the Lake Huron basin, snowfall ranges from 75 cm along the south shore to over 250 cm along the north shore; on Lake Erie snowfall ranges from 95 cm at Toledo on the southwest shore to 260 cm at Erie, PA, on the northeast shore; and on Lake Ontario snowfall ranges from 120 cm at Hamilton, ON, along the southwest shore to 280 cm at Watertown, NY, on the eastern shore.

12.2.3 Vegetation Gradient

A pronounced bioclimatic tension zone separates the temperate mixed broad-leaved and coniferous forest (boreal and Great Lakes-St. Lawrence Forests) to the north from broad-leaved forest (Carolinian Forest) to the south (Fig. 12.1). The tension zone begins at about 50° N and dips sharply to the south across Minnesota, separates Wisconsin and the Lower Peninsula of Michigan in half, and continues across Ontario along the north shore of Lake Ontario. In Wisconsin the tension zone comprises 13% of the state area but has 40% of the endemic soils, a large number of which are bisequal (Bockheim and Schliemann 2014).

The effect of the Great Lakes on the distribution of Spodosols is of particular interest. In northeastern Minnesota, Spodosols occur only to a limited extent (494 km^2), possibly because of limited snowfall. They occur primarily as forested Orthods downslope of summits where snow accumulates (McHugh, unpublished). The separation between Spodosols and Alfisols in Wisconsin occurs at 45° 00′ N, in Michigan at 43°–44° N, in central Ontario at 44° 30′ N, and in eastern Ontario at 44° 20′ N (Fig. 13.1). The sharp break between Spodosols and Alfisols across the Great Lakes has been attributed to (i) the contact between mixed deciduous and coniferous forest and deciduous forest (Curtis

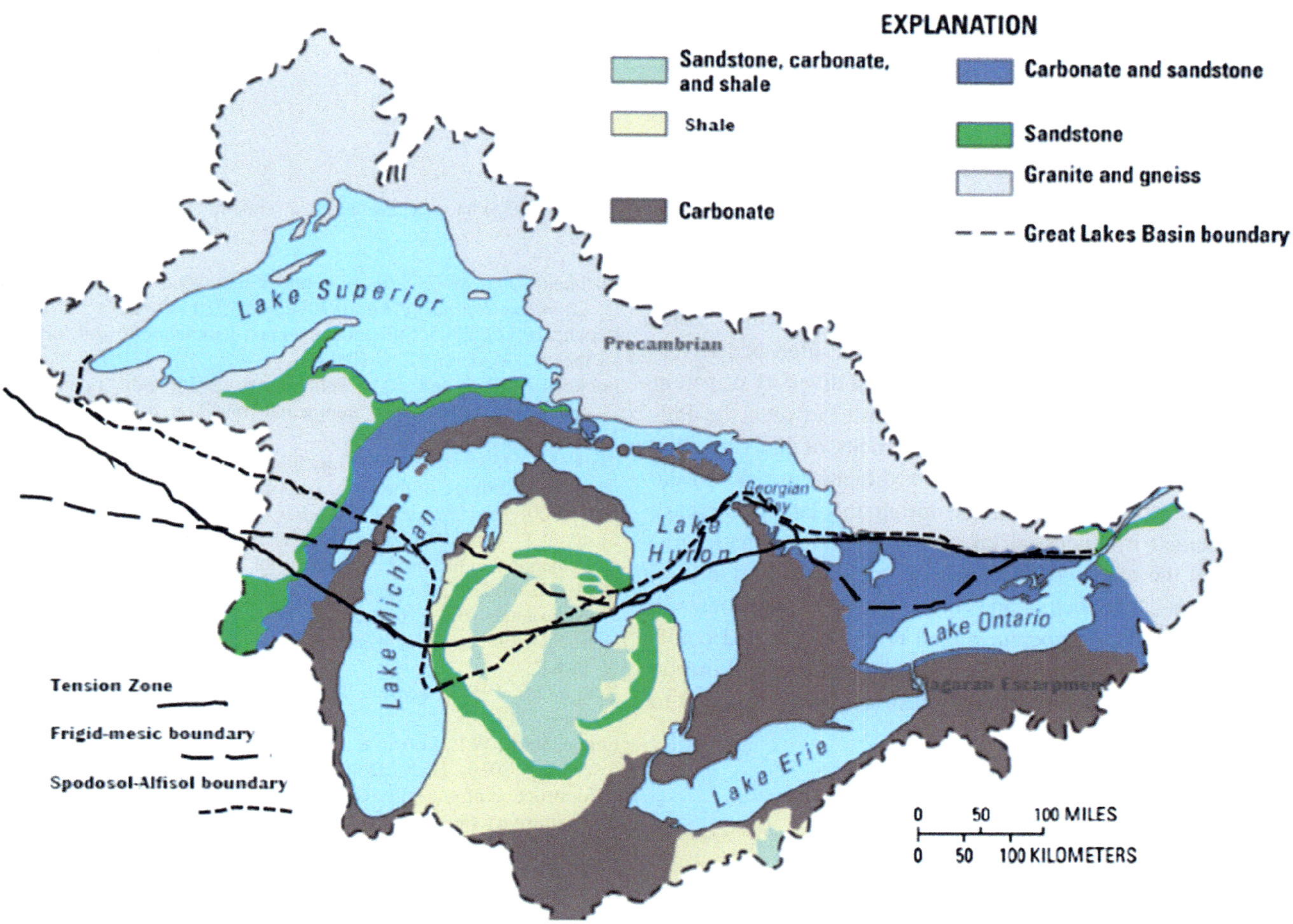

Fig. 12.1 Factors influencing the high pedodiversity in the Great Lakes Coastal Zone, including differences in bedrock, a strong vegetation gradient (tension zone), and pedogenic gradients in soil-temperature class and the Spodosol-Alfisol boundary

1959), (ii) a contact between the cool, temperate (Dbf) and warm temperate (Daf) climatic zones (Hole 1976), (iii) a contact between outwash-dominated and till-dominated parent materials (Bockheim 1997).

12.3 Benchmark, Rare, Endemic, and Endangered Soils of the Great Lakes Coastal Zone

Benchmark soils are those that (i) have a large extent within one or more MLRAs, (ii) hold a key position in *Soil Taxonomy*, (iii) have a large amount of data, (iv) have special importance to one or more significant land uses, or (v) are of significant ecological importance. Whereas only 6.4% of the soil series in the US are benchmark soils, 18% of those in the GLCZ are benchmark soils (Table 12.3). Seven great groups account for two thirds of the benchmark soils in the GLCZ, including Hapludalfs (18%), Haplorthods (12%), Udipsamments (8.9%), Glossudalfs (8.0%), Endoaquolls (8.0%), Epiaqualfs (6.2%), and Endoaqualfs (5.4%). Benchmark soils that are particularly abundant in the GLCZ include the Blount, Carbondale, Collamer, Colonie, Kalkaska, Niagara, Oakville, Plainfield, Roscommon, and Rubicon series.

Plants, animals, and soils that are native or restricted to a certain area are said to be endemic (Bockheim 2005). In the US GLCZ, 35% of the soils are the sole member of their family and are considered here to be endemic, 41% of the soils have an area less than 10,000 ha and can be considered rare (Ditzler 2003); and 18% of the soils are endemic and rare and can be considered endangered (Table 12.3). By way of comparison, in the US, 35% of the soil series established to date are endemic, 62% are rare, and 18% are endangered.

12.4 Conclusions

In the US GLCZ at a scale of 1:24,000, the Lake Ontario coastal zone is most diverse (0.036 soil series per km^2) and Lake Superior is least diverse (0.014 soil series per km^2).

Table 12.3 Benchmark, endemic, rare, and endangered soils of the US Great Lakes coastal zone.[1]

Class	No. of soil series	% of total
Benchmark	121	18
Endemic	245	35
Rare	287	42
Endangered	133	19

[1]*Endemic* only soil series in a family; *rare* <10,000 ha area; *Endangered* endemic and rare

The US GLCZ has a three-fold greater average pedodiversity (0.014) than for Land Resource Regions, Major Land Resource Areas, and individual states. The high pedodiversity of the GLCZ can be attributed to (i) a diversity of parent materials related to a geological contrast between the Pre-cambrian igneous and metamorphic rocks of the Canadian Shield to the north and Paleozoic sedimentary rocks to the south and repeated glaciations during the late Quaternary accompanied by ancestral lakes; (ii) a strong climatic gradient and the effects of the lakes on snow distribution (i.e., "lake-effect" snow belts), and (iii) a bioclimatic tension zone that separates the temperate mixed broad-leaved and coniferous forest (boreal and Great Lakes-St. Lawrence Forests) to the north from broad-leaved forest (Carolinian Forest) to the south.

References

Arbogast, A.F. 2000. Estimating the time since final stabilization of a perched dune field along Lake Superior. *Professor—Department of Geography* 52: 594–606.

Arbogast, A.F., and W.L. Loope. 1999. Maximum-limiting age of Lake Michigan coastal dunes: Their correlation with Holocene lake level history. *Journal of Great Lakes Research* 25: 372–382.

Bockheim, J.G. 1997. Soils in hemlock-hardwood ecosystem-mosaic. *Canadian Journal of Forest Research* 27: 1147–1153.

Bockheim, J.G. 2005. Soil endemism and its relation to soil formation theory. *Geoderma* 129: 109–124.

Bockheim, J.G., and S.A. Schliemann. 2014. Soil richness and endemism across an environmental transition zone in Wisconsin, USA. *Catena* 113: 86–94.

Clark, J.A., K.M. Befus, and G.R. Sharman. 2012. A model of surface water hydrology of the Great Lakes, North America during the past 16,000 years. *Physics & Chemistry Earth* 53–54: 61–71.

Curtis, J.T. 1959. *Vegetation of Wisconsin: An ordination of plant communities*. Madison: University of Wisconsin Press.

Ditzler, C. 2003. Endangered soils. *National Cooperative Soil Survey*. Newsletter No. 25, 1–2.

Hole, F.D. 1976. *Soils of Wisconsin*. Madison: University of Wisconsin Press.

Hough, J.L. 1958. *Geology of the Great Lakes*. University of Illinois Press. 313 pp.

Loope, H.M., W.L. Loope, R.J. Goble, T.G. Fisher, H.M. Jol, and J.C. Seong. 2010. Early Holocene dune activity linked with final destruction of glacial Lake Minong, Eastern Upper Michigan, USA. *Quaternary Research* 74: 73–81.

Loope, W.L., T.G. Fisher, H.M. Jol, R.J. Goble, J.B. Anderton, and W. L. Blewett. 2004. A Holocene history of dune-mediated landscape change along the southeastern shore of Lake Superior. *Geomorphology* 61: 303–322.

Prest, V.K. 1970. Quaternary geology of Canada. In Douglas, R.J.W. (ed.) *Geology and economic minerals of Canada. Geological Survey of Canada Economic Geology* 1: 676–764.

13 Protection of Great Lakes Soils

13.1 Introduction

Key environmental problems in the Great Lakes region are (i) reductions of the water levels from withdrawals and human-caused climate change; (ii) pollution from point-source, nonpoint-source, and atmospheric sources; (iii) the introduction of invasive species; and (iv) degradation of terrestrial and aquatic ecosystems (Smith et al. 2015).

13.2 Key Environmental Problems in the Great Lakes Coastal Zone

The mean levels of Lakes Michigan and Huron have declined by 1 m since the early 1990s (Environmental Protection Agency; Angel and Kunkel 2010), due primarily to water diversion, consumptive use for energy production and drinking water, and from increased evaporation from climate warming. By increasing stream gradients, these activities have the potential to increase shoreline erosion. Indeed, shoreline erosion and bluff retreat are key environmental issues on all of the Great Lakes. Figure 13.1 shows the effects of sediments from the Grand River being deposited in Lake Michigan. Figure 13.2 illustrates bluff retreat on Lake Michigan in Racine County, WI.

The most common pollutants in the Great Lakes are polychlorinated biphenyls (PCBs), mercury, asbestos, copper, DDT, dioxins, heavy metals, excessive nitrogen and phosphorus, and polycyclic aromatic hydrocarbons (PAHs) (Environmental Protection Agency 1995). The most common sources of these contaminants are improper municipal sewage treatment and discharge; contamination from mining, steel production, pulp and paper making; combustion of fossil fuels; and urban and agricultural runoff. While necessary to keep federal harbors open in the Great Lakes, dredging brings up contaminated sediments that are often landfilled nearby.

There are 185 aquatic species and 157 terrestrial species that are exotic to the Great Lakes (Environmental Protection Agency 1995). The sum of four shoreline taxa provides a rapidly measured indicator of anthropogenic stress across the Great Lakes, including *Phragmites australis* (common reed), *Typha* sp. (cattail), *Stuckenia pectinata* (sago pondweed), and the exotic species, *Lythrum salicaria* (purple loosestrife) (Trebitz and Taylor 2007). These exotic and invasive plants were more prevalent in wetlands in Lakes Erie and Ontario than in Lakes Superior and Huron, with Lake Michigan wetlands intermediate.

Coastal development has led to loss in riparian vegetation, dune degradation, and loss of coastal wetlands. There has been extensive harvesting of dune sand along the Great Lakes, especially Lake Michigan, for foundries and fracking (Fig. 13.3). About 2.3 million metric tonnes of sand are removed annually from Michigan alone (Lake Michigan Federation 1999). Beach ridges occur at 138 sites in the Lake Superior, Michigan, and Huron coastal zones (Baedke et al. 2004) and are important bird and fish habitats. However, over two-thirds of the Great Lakes wetlands have been lost, and many of those remaining have been threatened by development or pollution (Environmental Protection Agency 1995; Albert 2003) (Fig. 13.4).

The Great Lakes are protected by the Great Lakes Water Quality Agreement (GLWQA) of 1972, which was updated in 2012. Signed by Canada and the USA, the primary goals of this agreement are to identify areas of concern (AOC), provide guidelines for lake-wide management, identify chemicals and excessive nutrients of concern, prevent discharges from vessels, control aquatic invasive species, preserve habitat and species, protect groundwater, and minimize climate change impacts.

J. G. Bockheim, *Soils of the Laurentian Great Lakes, USA and Canada*,
https://doi.org/10.1007/978-3-030-52425-8_13

Fig. 13.1 Storm runoff into Lake Michigan from erosion upstream on the Grand River (*photo* by http://www.mlive.com/news/grand-rapids/index.ssf/2014/11/drones_could_help_improve_west.html)

Fig. 13.2 Shoreline erosion along Lake Michigan in Racine Co., WI (*photo* from http://journaltimes.com/news/local/erosion-emergency-threatens-mount-pleasant-homes/article_bd600984-48ec-5952-8717-e67f7c03c8f5.html)

In 1987 GLWQA identified 43 areas of concern (AOCs) around the Great Lakes, 26 of which were in the USA, 12 were in Canada, and 5 were binational. Since then four of the US sites have been cleaned up and delisted. Recent budgetary constraints and relaxation of environmental controls threaten the Great Lakes as never before.

13.3 Conclusions

The protection of soils in the GLCZ is an important part of Great Lakes Restoration Initiative by implementing conservation practices that benefit water quality, protect watersheds

Fig. 13.3 Sand mining of dunes along the Lake Michigan shore near Ludington on Lake Michigan (*photo* from http://journaltimes.com/news/local/erosion-emergency-threatens-mount-pleasant-homes/article_bd600984-48ec-5952-8717-e67f7c03c8f5.html)

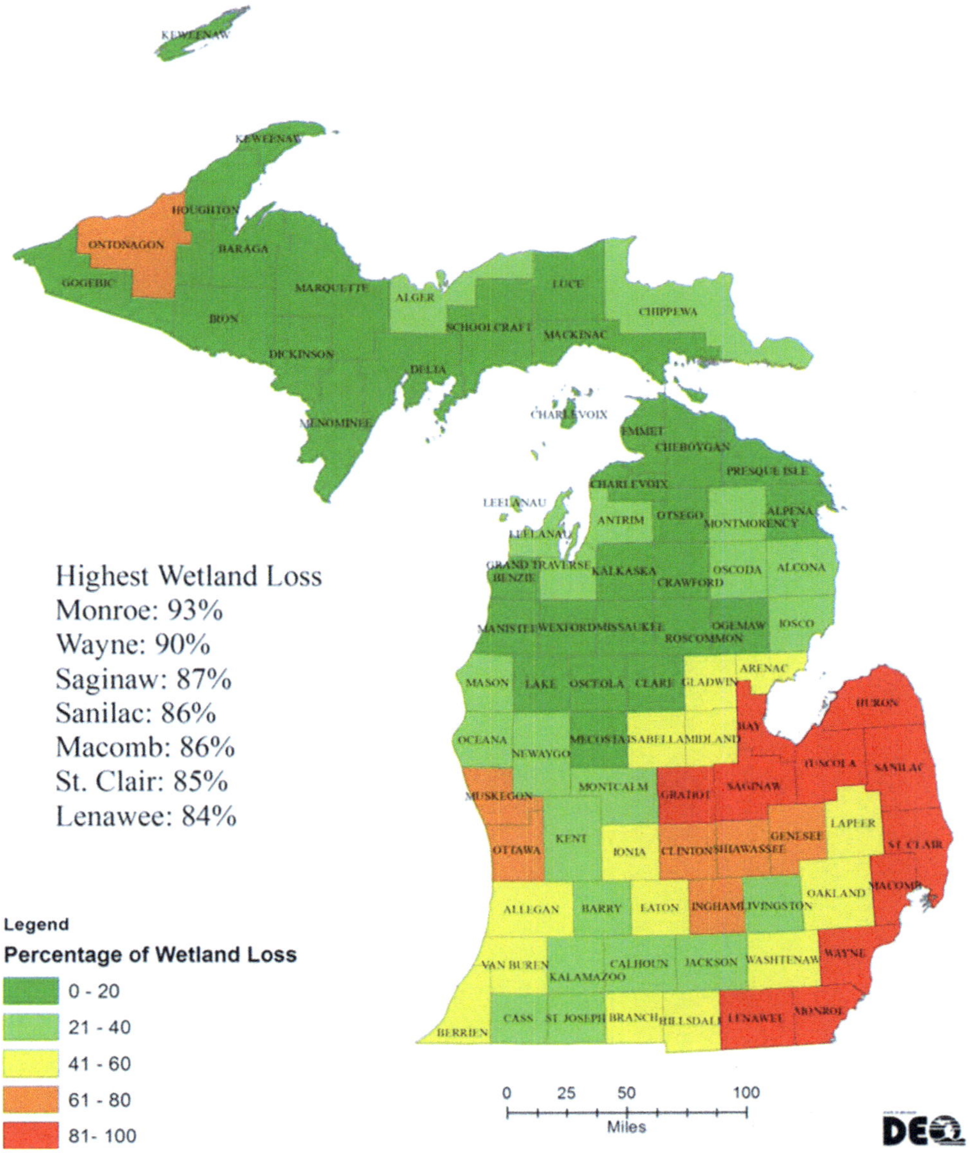

Fig. 13.4 Loss of wetlands in Michigan (Michigan Department of Environmental Quality). The greatest losses are along the Lake Huron and Michigan shores

and the coastal zone from non-point source pollution, and restore wetlands. Some of the coastal zone of the Great Lakes is protected by two national parks, three national lakeshores, and 96 state and provincial parks along the shores of the Great Lakes. However, recent proposed reductions in funding for the Great Lakes Restoration Initiative and the EPA may undermine these activities.

References

Albert, D.A. 2003. Between land the lake: Michigan's Great Lakes coastal wetlands. Michigan Natural Features Inventory, Michigan State University. *Extension bulletin E-2902* 96.

Angel, J.R., and K.E. Kunkel. 2010. The response of Great Lakes water levels to future climate scenarios with an emphasis on Lake Michigan-Huron. *Journal of Great Lakes Research* 36: 51–58.

Baedke, S.J., T.A. Thompson, J.W. Johnston, and D.A. Wilcox. 2004. *Reconstructing paleo lake levels from relict shorelines along the upper Great Lakes*. Envir. Sci. & Biol. Fac. Publ. Paper 36. http://digitalcommons.brockport.edu/env_facpub/36.

Environmental Protection Agency. 1995. *The Great Lakes: an environmental atlas and resource book* (3rd edn.). EPA 9-5-B-95-001.

Smith, S.D.P., P.B. McIntyre, B.S. Halpern, et al. 2015. Rating impacts in a multi-stressor world: a quantitative assessment of 50 stressors affecting the Great Lakes. *Ecological Applications* 25: 717–728.

Trebitz, A.S., and D.L. Taylor. 2007. Exotic and invasive aquatic plants in Great Lakes coastal wetlands: distribution and relation to watershed land use and plant richness and cover. *Journal of Great Lakes Research* 33: 705–721.

14 Some Unusual Relict Features in the Great Lakes Coastal Zone

14.1 Introduction

With the exception of features in the modern surf and swash zones, nearly all of the geomorphic features in the GLCZ are relict, i.e., have survived from an earlier geologic period and are being modified by current geomorphic and pedologic processes. This chapter highlights some of the unusual relict features from the earliest ancestral Great Lakes. These are discussed under three categories, including (i) glacial and periglacial, (ii) lacustrine and glaciolacustrine, and (iii) early people.

14.2 Glacial and Periglacial

Following recession of the Laurentide glaciers from 13.6 to ca. 9.5 ky ago, a periglacial climate existed throughout much of the Great Lakes basin. This is evidenced not only by the fossil pollen record, but also by landscape features attesting to a subarctic climate. Solifluction lobes and sheets are types of slope failure that are common in permafrost environments. Solifluction forms tongue-shaped features due to differential downhill flow rates. Figure 14.1 shows relict solifluction lobes at Ead's Bush 3.6 km southwest of Little Current on northern Manitoulin Island, Ontario. There are at least four southeast-facing terraces with lobes at elevation between 270 and 290 m (885 and 950 ft) on the upper Glacial Lake Algonquin terraces. The individual lobes range from 50 to 670 m (165–2,200 ft) in length, 70 to 90 m (230–295 ft) in width, and 2 to 5 m (6.5–16 ft) in height. The features are developed on the Farmington soil (Orthic Melanic Brunisol; Typic Eutrudepts), which contains 30 cm (12 in.) of calcareous till over dolomitic bedrock. The solifluction lobes likely developed following the Main Algonquin ca. 13.8–ca. 12.4 ka.

Relict patterned ground occurs in lowlands of the Great Lakes, particularly in the Huron Basin (Morgan 1982; Gao 2005; Lusch et al. 2009). Figure 14.2 is a satellite image showing relict flat-centered ice-wedge polygons near Highwood Glen, 25 km (15 mi) northeast of Sarnia, Ontario. The polygons are 70–160 m across and the polygon fissures are 5–10 m across. According to Gao (2005; site 2), the polygons range from 1 to over 3 m (3–10 ft) in depth. Permafrost existed in the area from ca. 13.8 to ca. 12.4 ka.

14.3 Glaciolacustrine and Lacustrine

Raised beaches are one of the more common geomorphic features in the GLCZ. Figure 14.3 shows a "staircase" of raised beaches on Morphets Siding Road south of Little Current, Ontario, on the shore of Georgian Bay (Lake Huron). Five of the eight raised beaches listed in Table 14.1 are shown in the figure, including the Korah (206 m), Sheguiandah (223 m), Payette (240 m), Cedar Point (256 m), and the Penetang (271 m) (Table 14.4). In addition to raised beaches, there are spits, tombolos, beach ridges, sea stacks, cavettos, deltas, and dunes distant from the present Great Lakes shoreline. Many of these features represent Early Lake Algonquin or Iroquois shorelines but some are from Lake Nipissing. Figure 14.4 shows a tombolo surrounded by uplifted beach ridges on the Point Isabella Headland on the eastern shore of the Keweenaw Peninsula (Lake Superior). The current shoreline of Lake Superior is at 183 m (601 ft). The raised beaches range from 186 to 202 m (610–663 ft) and the tombolo is at 206 m (676 ft), all within the limits of the former Nipissing shoreline.

J. G. Bockheim, *Soils of the Laurentian Great Lakes, USA and Canada*,
https://doi.org/10.1007/978-3-030-52425-8_14

Fig. 14.1 Solifluction lobes and terraces on the Highest Algonquin raised beach at Ead's Bush 3.6 km southwest of Little Current, Ontario, on the shores of Georgian Bay (Lake Huron). The individual lobes range from 50 to 670 m in length, 70 to 90 m in width, and 2 to 5 m in height

Dune fields are common on the shores of the Great Lakes. Inland dune fields represent shorelines of the ancestral Great Lakes. Figure 14.5 shows coastal and interior dune fields at the maximum extent of the ancestral Great Lakes.

Schaetzl et al. (2016) identified 49 relict spits along the Glacial Lake Algonquin shoreline in the Lake Michigan and Lake Huron basins (Fig. 14.6). Figure 14.7 is a section of the Ardtrea Island South spit north of the present Lake Simcoe near Barrie, Ontario (Schaetzl et al. 2016; site 47). Ardtrea was an island in the Huron Lowland during the Main Algonquin.

Castle Rock (Fig. 14.8) is an Algonquin sea stack 59 m (194 ft) above the current level of Lake Huron at St. Ignace, Michigan.

Deltas deposited in the ancestral Great Lakes have been uplifted by glacio-isostatic adjustment and now appear above the current lake level. The St. Clair delta is one of the largest in the Great Lakes. Figure 14.9 is a cross-section of a portion of the delta showing the uplifted portion (premodern delta), the modern delta, and the underlying glaciolacustrine deposits. The delta surface is 1.5 m above Lake St. Clair and

Fig. 14.2 This satellite image shows relict flat-centered, ice-wedge polygons at Highwood Glen, 25 km northeast of Sarnia, Ontario, in the Huron Basin. The polygons are 70–160 m across and the fissures are 5–10 m across and attest to the presence of permafrost from ca. 13.8 to ca. 12.4 ka ago

Fig. 14.3 A staircase of raised beaches along Morphets Side Road 3.6 km southwest of Little Current, Ontario, on the north shore of Georgian Bay (Lake Huron). From the bottom to the top, the photographs shows the Korah (210 m), Sheguiandah (233 m), Payette (246 m), Cedar Point (259 m), and Penetang (271 m) raised beaches

Table 14.1 Raised beaches at Little Current, Lake Huron, Ontario

Raised beach	Elevation (m)	Approx. age (ka yr)
Algoma	189	2.9
Nipissing	198	5.3
Korah	210	
Sheguiandah	233	
Payette	246	
Cedar Point	259	
Penetang	271	
Kagawong	277	10.2
Algonquin upper	296	11.2

Source Sly and Lewis (1972)

Fig. 14.4 An uplifted tombolo surrounded by beach ridges from the Nipissing lake stage on the Point Isabella Headland on the eastern shore of the Keweenaw Peninsula (Lake Superior)

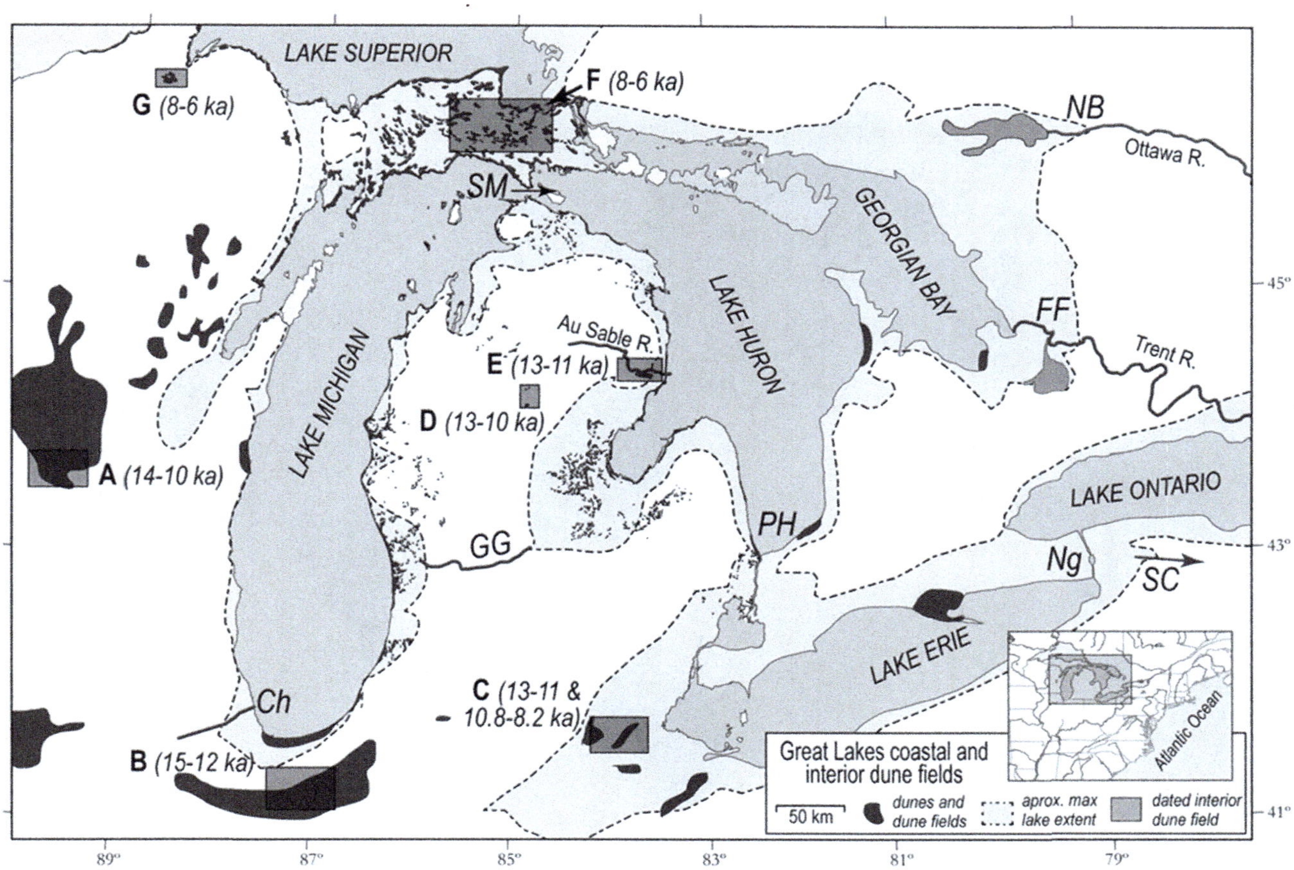

Fig. 14.5 A map of coastal and inland dune fields in the Great Lakes region. The interior dune fields range from Glenwood (15–12 ka) to Nipissing (8.6 ka) in age. *Source* Arbogast et al. (2017)

is from Lake Nipissing 3.5 to 5.0 ka yr BP. The glaciolacustrine deposits are dated at 9.5–7.3 ka yr BP.

14.4 Early People

There have been extensive archaeological investigations in the Great Lakes region (Fig. 14.10). Here, two case studies of early people in the GLCZ briefly are discussed. Sonnenburg et al. (2012) studied Holocene paleoshorelines and the submerged, prehistoric sites near Rice Lake, Ontario. Figure 14.11 shows shoreline extent of Rice Lake in the Lake Erie basin, and known terrestrial occupation sites and areas of archaeological potential.

O'Shea and Meadows (2009) identified a series of stone structures along the Alpena-Amberley ridge (Fig. 14.12 upper), a feature that would have been a dry land corridor crossing Lake Huron during the Lake Stanley low water states 10–7.5 ka BP. The hypothesized that the aligned stones (Fig. 14.12 lower) were used for caribou hunting in

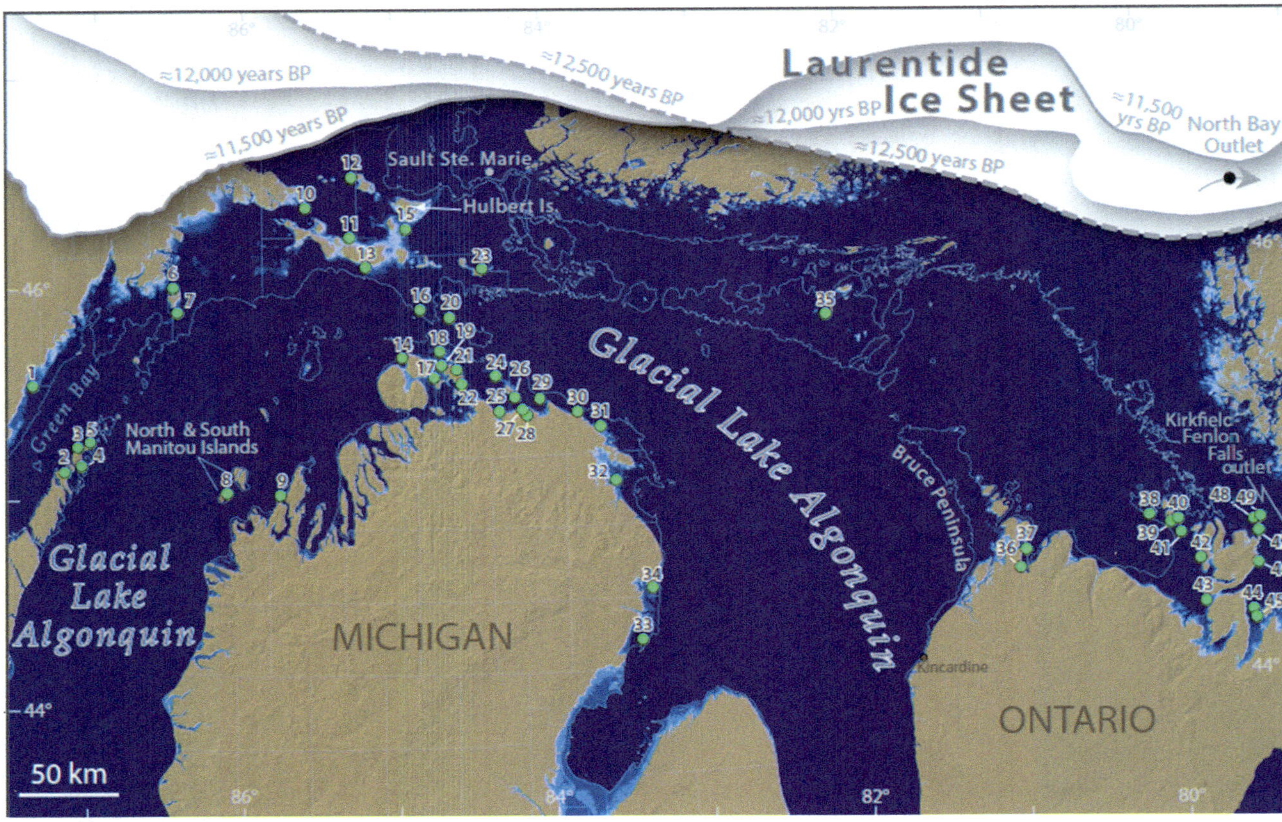

Fig. 14.6 Map of Glacial Lake Algonquin north of 43.5° latitude. Different water planes, associated with lower lake stage, are shown in progressively darker shades of blue. Brown areas were dry land at even the highest lake stage. *Source* Schaetzl et al. (2016)

Fig. 14.7 Spit sediments from Ardtrea Island South along the southeast shore of Georgian Bay near the village of Ardtrea at 44° 39′ N and 79° 26′ W. Photo by Schaetzl et al. (2016)

Fig. 14.8 Castle Rock, a sea stack from Glacial Lake Algonquin, rises 59 m above the current level of Lake Huron at St. Ignace, Michigan. Photo by Michigan.org

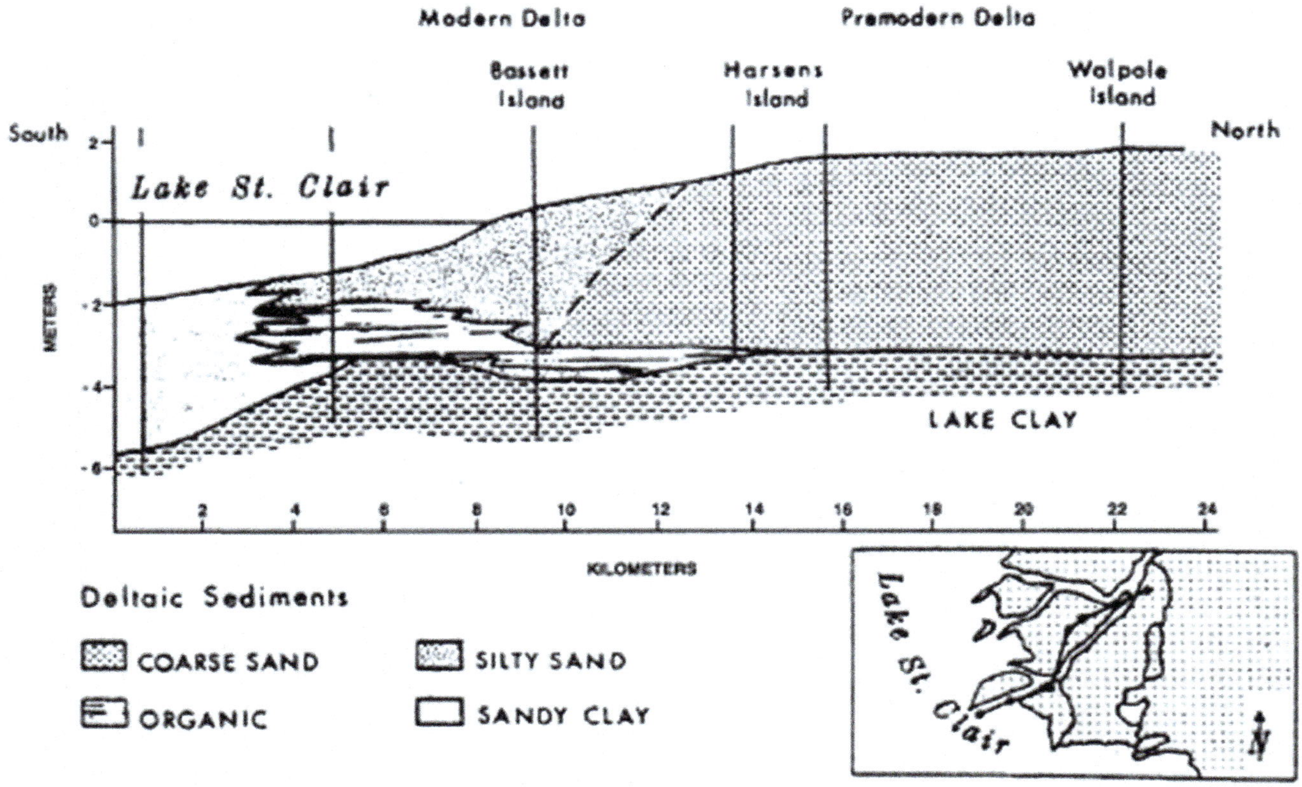

Fig. 14.9 An uplifted delta on Lake St. Clair between Lakes Huron and Erie. The uplifted portion of the delta is 1.5 m above the modern lake and originates from Lake Nipissing 3.5 to 5.0 ka ago. The glaciolacustrine deposits upon which the uplifted and modern deltaic deposits are laid is 9.5–7.3 ka in age. The figure is from Raphael and Jaworski (1982), in Thomas et al. (2006)

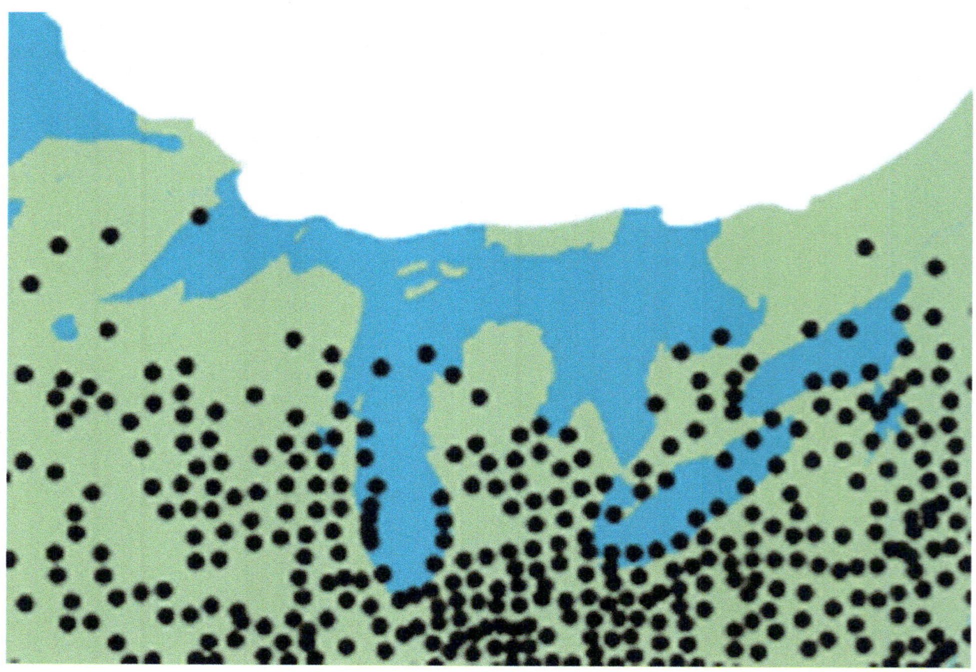

Fig. 14.10 Sites in the Great Lakes region where Clovis points have been found. The white zone at the top is the position of the Laurentide ice 12 ka ago. *Source* Paleoindian Database of the Americas; http://pidba.utk.edu

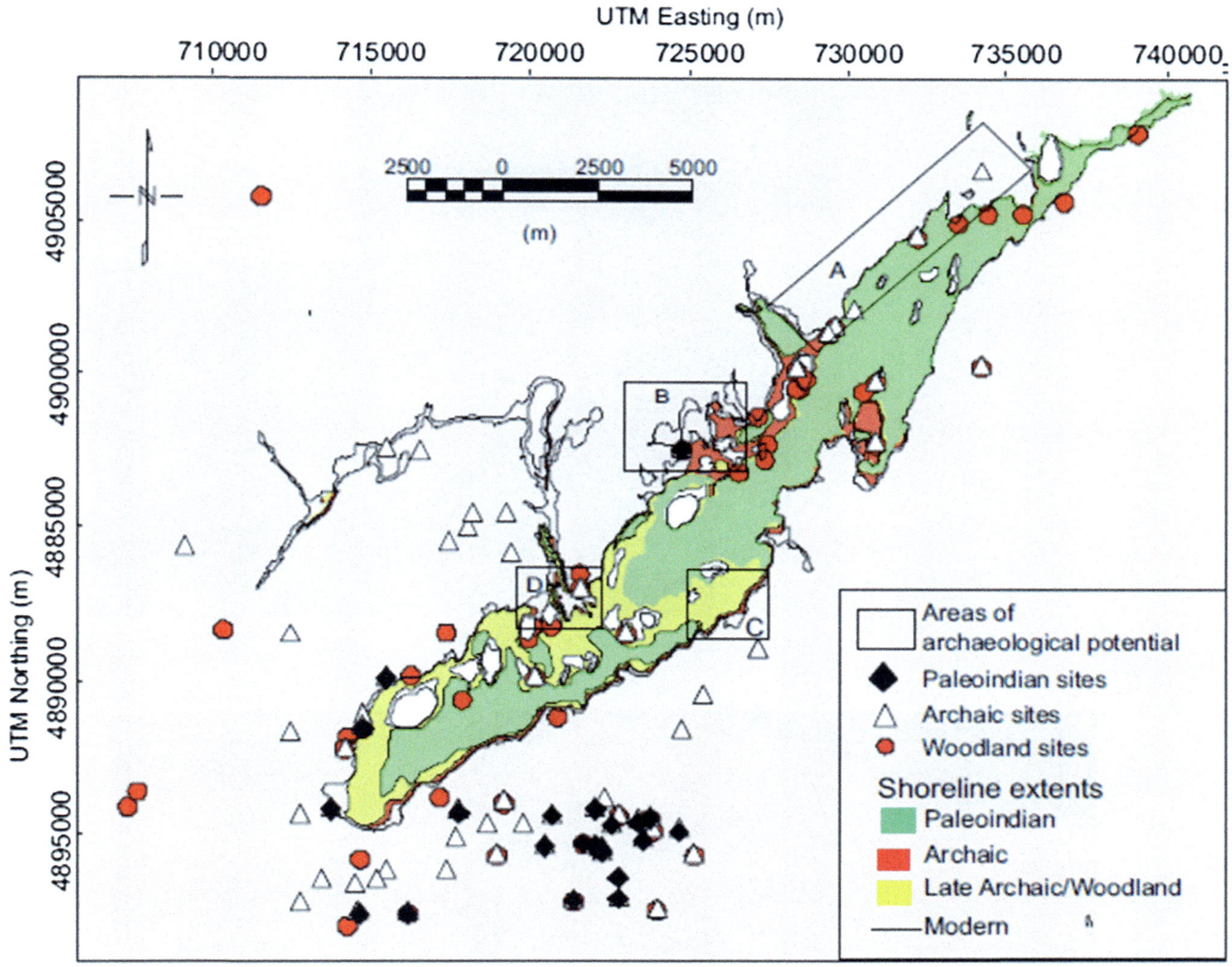

Fig. 14.11 Actual and potential (submerged) occupation sites near Rice Lake in the Lake Ontario basin. *Source* Sonnenberg et al. (2012)

prehistoric and ethnographic times, raising the possibility that intact settlements and ancient landscapes are preserved beneath Lake Huron.

Prior to European settlement, Iroquoian and Algonquian peoples lived around the Great Lakes, along with many other tribes. Europeans settled in GLCZ, beginning in the early 17th century. Figure 14.13 shows a 13-km portion of the former Talbot Trail from Duart to Morpeth, Ontario, along the north shore of Lake Erie. The Talbot Trail extended nearly 260 km (155 mi) between Windsor and Fort Erie, Ontario. It was constructed in the 1820s, and parts of it became Highway 3.

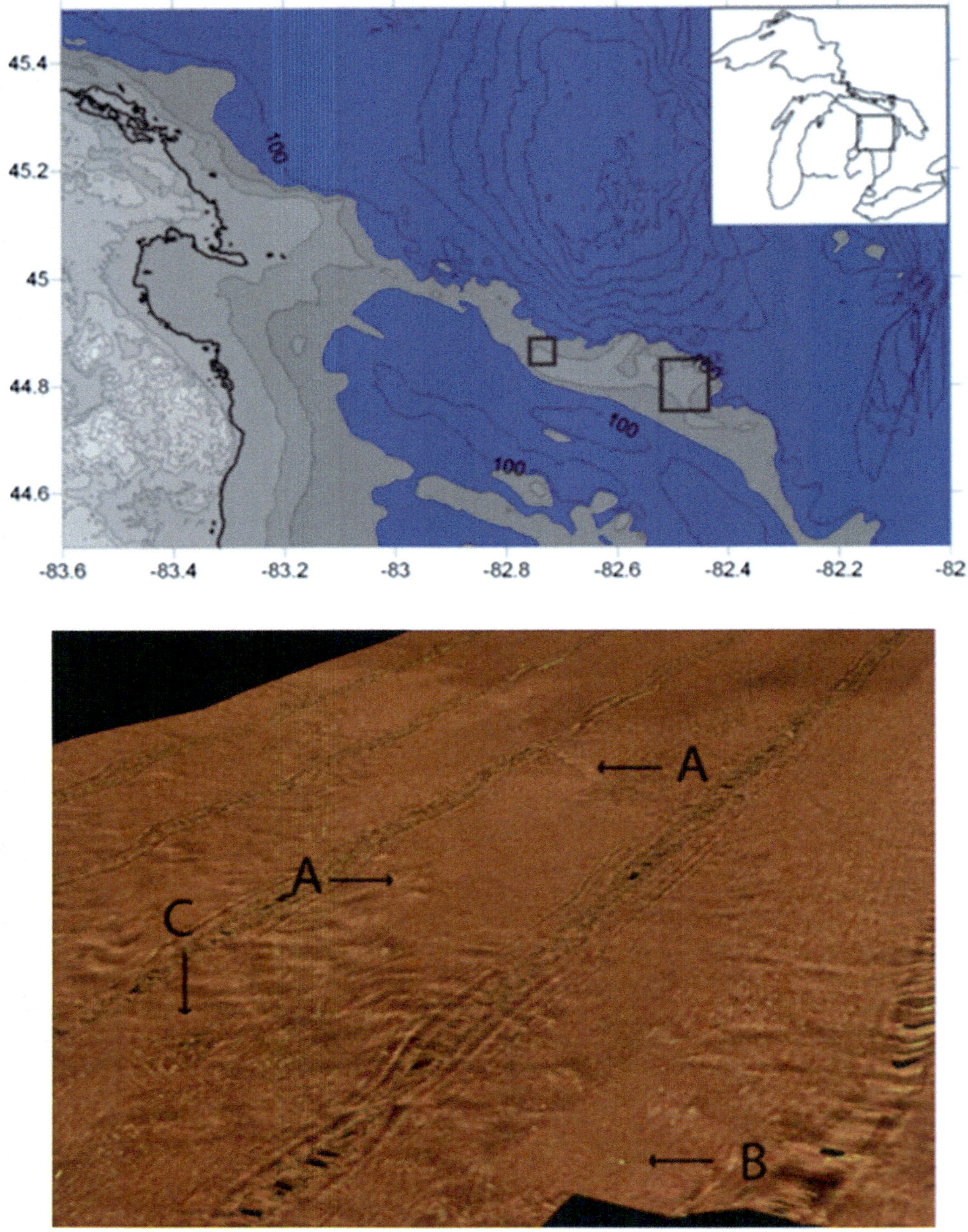

Fig. 14.12 Evidence for early hunters beneath the Great Lake (*Source* O'Shea and Meadows 2009). The top panel shows the Alpena-Amberley ridge as it would have appeared during the Lake Stanley low-water phase (76–121 m below the modern lake level) in Lake Huron ca. 9.9–7.0 ka

Fig. 14.13 Wheel tracks from the Talbot Trail established in the 1820s near Palmyra, Ontario. The Talbot Trail extended from Fort Erie and Niagara Falls in the east to Windsor in the west

14.5 Conclusions

This chapter reviews some of the unusual relict features from the earliest ancestral Great Lakes, including glacial and periglacial, glaciolacustrine and lacustrine, and paleoindian occupation sites. Many of these features represent Early Lake Algonquin or Iroquois shorelines but some are from Lake Nipissing. These features include relict solifluction terraces and patterned ground, staircases of raised beaches, inland dune fields, relict spits, uplifted deltas, potential submerged archaeological sites, and wheel tracks of early European settlers.

References

Arbogast, A.F., M.D. Luehmann, G.W. Monaghan, W.A. Lovis, and H. Wang. 2017. Paleoenvironmental and geomorphic significance of bluff-top dunes along the Au Sable river in northeastern Lower Michigan, USA. *Geomorphology* 297: 112–121.

Gao, C. 2005. Ice-wedge casts in Late Wisconsinan glaciofluvial deposits, southern Ontario, Canada. *Canadian Journal of Earth Sciences* 42: 2117–2126.

Lusch, D.P., K.E. Stanley, R.J. Schaetzl, et al. 2009. Characterization and mapping of patterned ground in the Saginaw Lowlands, Michigan: possible evidence for Late-Wisconsin permafrost. *Annals of the Association of American Geographers* 99: 445–466.

Morgan, A.. 1982. Distribution and probable age of relict permafrost features in south-western Ontario. In *Proceedings of the 4th International Permafrost Conference*, 91–100.

O'Shea, J.M., and G.A. Meadows. 2009. Evidence for early hunters beneath the Great Lakes. In *Proceedings of the National Academy of Sciences* 106: 10120–10123.

Schaetzl, R.J., F.J. Krist, Jr., C.F.M. Lewis, M.D. Luehmann, and M. J. Michalek. 2016. Spits formed in Glacial Lake Algonquin indicate strong easterly winds over the Laurentian Great Lakes during late Pleistocene. *Journal of Paleolimnology* 55: 49–65.

Sly, P.G., and C.F.M. Lewis. 1972. The Great Lakes of Canada–Quaternary geology and limnology. In *XXIV International Geological Congress*, Field Excursion A43 Guidebook, ed. D.J. Glass, 92 pp.

Sonnenburg, E.P., J.I. Boyce, and P. Suttak. 2012. Holocene paleoshorelines, water levels and submerged prehistoric site potential of Rice Lake (Ontario, Canada). *Journal of Archaeological Science* 39: 3553–3567.

Raphael, C.N., and E. Jaworski. 1982. The St. Clair delta. A unique lake delta. *The Geographical Bulletin* 21:7–27.

Thomas, R.L., M.D. Christensen, E. Szalinska, and M. Scarlat. 2006. Formation of the St. Clair river delta in the Laurentian Great Lakes system. *Journal of Great Lakes Research* 32: 738–748.

15 Conclusions

- The Laurentian Great Lakes contain nearly 17,000 km of coastline (includes islands). The Great Lakes Coastal Zone (GLCZ) is defined by the Great Lakes Environmental Assessment and Mapping Project (GLEAM) as all land and water within 5 km of the shorelines.
- Soils are recognized here as "a natural body comprised of solids (minerals and organic matter), liquid, and gases that occurs on the land surface, occupies space, and is characterized by horizons, or layers, that are distinguishable from the initial material as a result of additions, losses, transfers, and transformation of energy and matter.
- Soils have been described and mapped in the Great Lakes Coastal Zone since 1901 and include 82 counties in the USA and 26 counties and districts in Canada.
- More than 800 soil series have been identified in the GLCZ in Canada and the USA; soils have been mapped 79% of the GLCZ. Areas remaining to be mapped include the north and east shores of Lakes Superior and Huron.
- Soils in the GLCZ form from a variety of climates due to latitude, longitude, elevation, and proximity to the lakes themselves; four broad vegetation types including boreal forest, temperate mixed coniferous and broad-leaved forest, broad-leaved forest, and prairie; rocks derived from Precambrian granites and gneisses and surficial deposits from glaciers, former lakes, eolian deposits; and a variety of times during which the sediments were deposited from more than 14 kyr to a few hundreds of years.
- Two natural soil classification systems are employed in this study, including The *Canadian System of Soil Classification* (1998) and *Soil Taxonomy* (1999; 2014).
- In the US GLCZ, there are soils representing six orders, 16 suborders, 39 great groups, 140 subgroups, 507 families, and 691 soil series. In the Canadian GLCZ, there are soils representing five orders, 8 great groups, 17 soil subgroups, and 144 soil series.
- The ranking of soil orders by numbers of soil series and soil area in the GLCZ include Alfisols (Luvisols) > Spodosols (Podzols) > Inceptisols (some Brunisols) > Entisols (Regosols) > Mollisols (some Brunisols) > Histosols (Organic soils).
- Each of the Great Lakes was divided into segments based on Major Land Resource Area, dominant shoreline parent materials, and dominant great groups and soil series, including six for Lake Superior, six for Lake Michigan, eight for Lake Huron, four for Lake Erie, and two for Lake Ontario.
- Dominant soil-forming processes in the GLCZ include gleization (redoximorphic features from restricted drainage), argilluviation (clay translocation), podsolization (translocation of Fe and Al in combination with organic C), cambisolization (weak development of color and structure), melanization (accumulation of mineralized organic matter at the surface), glossification (degradation of an argillic horizon), paludization (accumulation of organic materials), vertization (seasonal cracking), and anthrosolization (human influences). These process are linked to great groups and landforms.
- The dominant landforms in the GLCZ include lake plains-lake terraces (48%), till plains-moraines (26%), outwash plains (12%), dunes, wind-modified beach ridges, and longshore bars (5.0%), glacial lake benches (4.4%), and alluvial fans and floodplains (3.9%). The thickness of lacustrine and glaciolacustrine sediments ranges from a decimeter to hundreds of meters. Till plains and moraines have been modified in the Saginaw Bay area and in eastern Lake Huron by wave-washing. Some former coastal features occur as much as 150 km inland.
- Eighteen percent of the soils in the US GLCZ are benchmark soils; 35% are endemic and occur nowhere else in the USA, 41% are rare (occupy less than 10,000 ha), and 18% are endangered (endemic and rare).

J. G. Bockheim, *Soils of the Laurentian Great Lakes, USA and Canada*,
https://doi.org/10.1007/978-3-030-52425-8_15

- Based on the density of soil series within 5 km of the shorelines, all five of the Great Lakes have a high level of pedodiversity. The pedodiversity is related to a variety of parent materials, strong climate gradients, and pronounced vegetation gradients.
- From an examination of seven published soil chronosequences from the GLCZ, soils derived from sandy dunes, beach ridges, and lake terraces begin as dune lands and upon dune stabilization (i.e., no blowouts) evolve to Typic Udipsamments and after 4,000 yr develop a weak spodic-like horizon, becoming Spodic Udipsamments. Spodosols require approximately 7,000 years to form.
- Key environmental problems in the Great Lakes region are (i) reductions of the water levels from withdrawals and human-caused climate change; (ii) pollution from point-source, nonpoint-source, and atmospheric sources; (iii) the introduction of invasive species; and (iv) degradation of terrestrial and aquatic ecosystems.
- The protection of soils in the GLCZ is an important part of the Great Lakes Restoration Initiative by implementing conservation practices that benefit water quality, protect watersheds and the coastal zone from non-point source pollution, and restore wetlands.

Appendix A
Classification of soil series in the Great Lakes Coastal Zone.

Series name	Order	Suborder	Great group	Subgroup	Particle-size or substitute class	Mineralogy class	CEC actiivity class	Reaction class	Soil temperature class	Other family class
ABBAYE	Spodosols	Orthods	Haplorthods	Alfic Oxyaquic Haplorthods	Coarse-loamy	Mixed	Active		Frigid	
ABSCOTA	Entisols	Psamments	Udipsamments	Oxyaquic Udipsamments	Sandy	Mixed			Mesic	
ADAMS	Spodosols	Orthods	Haplorthods	Typic Haplorthods	Sandy	Isotic			Frigid	
ADRIAN	Histosols	Saprists	Haplosaprists	Terric Haplosaprists	Sandy or sandy-skeletal	Mixed		Euic	Mesic	
AHMEEK	Inceptisols	Udepts	Eutrudepts	Dystric Eutrudepts	Coarse-loamy	Isotic			Frigid	
ALCONA	Spodosols	Orthods	Haplorthods	Alfic Haplorthods	Coarse-loamy	Mixed	Active		Frigid	
ALDENLAKE	Inceptisols	Udepts	Eutrudepts	Dystric Eutrudepts	Coarse-loamy	Isotic			Frigid	
ALGANSEE	Entisols	Psamments	Udipsamments	Aquic Udipsamments	Sandy	Mixed			Mesic	
ALGONQUIN	Alfisols	Udalfs	Hapludalfs	Aquic Hapludalfs	Fine	Mixed	Semiactive		Frigid	
ALLENDALE	Spodosols	Aquods	Epiaquods	Alfic Epiaquods	Sandy over clayey	Mixed	Semiactive		Frigid	
ALLIS	Inceptisols	Aquepts	Endoaquepts	Typic Endoaquepts	Fine	Illitic		Acid	Mesic	
ALLOUEZ	Inceptisols	Udepts	Dystrudepts	Humic Dystrudepts	Sandy-skeletal	Mixed			Frigid	
ALPENA	Mollisols	Udolls	Hapludolls	Entic Hapludolls	Sandy-skeletal	Mixed			Frigid	
ALSTAD	Alfisols	Udalfs	Glossudalfs	Aquic Glossudalfs	Fine-loamy	Mixed	Superactive		Frigid	
ALTMAR	Entisols	Psamments	Udipsamments	Aquic Udipsamments	Sandy	Mixed			Mesic	
ALTON	Inceptisols	Udepts	Eutrudepts	Dystric Eutrudepts	Loamy-skeletal	Mixed	Active		Mesic	
AMADON	Spodosols	Orthods	Haplorthods	Lithic Haplorthods	Loamy	Mixed	Active		Frigid	
AMASA	Spodosols	Orthods	Haplorthods	Typic Haplorthods	Coarse-loamy over sandy or sandy-skeletal	Mixed	Superactive		Frigid	
AMBOY	Inceptisols	Udepts	Fragiudepts	Typic Fragiudepts	Coarse-silty	Mixed	Active		Mesic	
AMENIA	Inceptisols	Udepts	Eutrudepts	Aquic Eutrudepts	Coarse-loamy	Mixed	Active		Mesic	
AMNICON	Alfisols	Udalfs	Glossudalfs	Oxyaquic Vertic Glossudalfs	Very-fine	Mixed	Active		Frigid	
ANGELICA	Inceptisols	Aquepts	Endoaquepts	Aeric Endoaquepts	Fine-loamy	Mixed	Active	Nonacid	Frigid	
ANGOLA	Alfisols	Aqualfs	Endoaqualfs	Aeric Endoaqualfs	Fine-loamy	Mixed	Active		Mesic	

(continued)

J. G. Bockheim, *Soils of the Laurentian Great Lakes, USA and Canada*,
https://doi.org/10.1007/978-3-030-52425-8

Series name	Order	Suborder	Great group	Subgroup	Particle-size or substitute class	Mineralogy class	CEC actiivity class	Reaction class	Soil temperature class	Other family class
ANNALAKE	Spodosols	Orthods	Haplorthods	Alfic Oxyaquic Haplorthods	Coarse-loamy	Mixed	Superactive		Frigid	
ANNANIAS	Spodosols	Aquods	Endoaquods	Typic Endoaquods	Coarse-silty	Isotic			Frigid	
APPLETON	Alfisols	Aqualfs	Endoaqualfs	Aeric Endoaqualfs	Fine-loamy	Mixed	Active		Mesic	
APTAKISIC	Alfisols	Aqualfs	Endoaqualfs	Aeric Endoaqualfs	Fine-silty	Mixed	Superactive		Mesic	
ARCADIAN	Spodosols	Orthods	Haplorthods	Lithic Haplorthods	Loamy-skeletal	Mixed	Active		Frigid	
ARKONA	Spodosols	Aquods	Epiaquods	Alfic Epiaquods	Sandy over clayey	Mixed	Semiactive		Mesic	
ARKPORT	Alfisols	Udalfs	Hapludalfs	Lamellic Hapludalfs	Coarse-loamy	Mixed	Active		Mesic	
ARNHEIM	Entisols	Aquents	Fluvaquents	Typic Fluvaquents	Coarse-loamy	Mixed	Superactive	Nonacid	Frigid	
ASHKUM	Mollisols	Aquolls	Endoaquolls	Typic Endoaquolls	Fine	Mixed	Superactive		Mesic	
ASHWABAY	Spodosols	Orthods	Haplorthods	Alfic Oxyaquic Haplorthods	Sandy	Isotic			Frigid	
ASSININS	Spodosols	Aquods	Endoaquods	Argic Endoaquods	Coarse-loamy	Mixed	Active		Frigid	
AU GRES	Spodosols	Aquods	Endoaquods	Typic Endoaquods	Sandy	Mixed			Frigid	
AU TRAIN	Spodosols	Orthods	Haplorthods	Oxyaquic Haplorthods	Sandy	Isotic			Frigid	Shallcw
AUBARQUE	Inceptisols	Aquepts	Epiaquepts	Aeric Epiaquepts	Coarse-loamy	Mixed	Semiactive	Calcareous	Mesic	
AUGER	Spodosols	Orthods	Haplorthods	Oxyaquic Haplorthods	Coarse-silty	Isotic			Frigid	
AUGUSTANA	Alfisols	Udalfs	Hapludalfs	Mollic Oxyaquic Hapludalfs	Fine-loamy	Mixed	Superactive		Frigid	
AVOCA	Entisols	Orthents	Udorthents	Aquic Udorthents	Sandy over loamy	Mixed	Semiactive	Nonacid	Mesic	
AZTALAN	Mollisols	Udolls	Argiudolls	Aquic Argiudolls	Fine-loamy	Mixed	Superactive		Mesic	
BACH	Inceptisols	Aquepts	Endoaquepts	Mollic Endoaquepts	Coarse-silty	Mixed	Semiactive	Calcareous	Mesic	
BADAXE	Alfisols	Udalfs	Hapludalfs	Aquic Hapludalfs	Coarse-loamy	Mixed	Semiactive		Mesic	
BADRIVER	Alfisols	Aqualfs	Glossaqualfs	Aeric Glossaqualfs	Fine	Mixed	Active		Frigid	
BAMFIELD	Alfisols	Udalfs	Glossudalfs	Haplic Glossudalfs	Fine-loamy	Mixed	Active		Frigid	
BANAT	Alfisols	Aqualfs	Endoaqualfs	Aeric Endoaqualfs	Loamy-skeletal	Mixed	Active		Frigid	
BARCELONA	Alfisols	Aqualfs	Endoaqualfs	Aeric Endoaqualfs	Fine-silty	Mixed	Active		Mesic	
BARRE	Alfisols	Aqualfs	Endoaqualfs	Udollic Endoaqualfs	Fine	Illitic			Mesic	
BARRINGTON	Mollisols	Udolls	Argiudolls	Oxyaquic Argiudolls	Fine-silty	Mixed	Superactive		Mesic	
BARTO	Inceptisols	Udepts	Eutrudepts	Lithic Eutrudepts	Loamy	Isotic			Frigid	
BATTLEFIELD	Spodosols	Aquods	Endoaquods	Typic Endoaquods	Sandy	Mixed			Frigid	
BATTYDOE	Spodosols	Orthods	Haplorthods	Typic Haplorthods	Coarse-loamy	Mixed	Active		Frigid	
BEAVERTAIL	Inceptisols	Aquepts	Humaquepts	Histic Humaquepts	Loamy-skeletal	Carbonatic			Frigid	
BEECHER	Alfisols	Aqualfs	Epiaqualfs	Udollic Epiaqualfs	Fine	Illitic			Mesic	
BELDING	Spodosols	Aquods	Epiaquods	Alfic Epiaquods	Coarse-loamy	Mixed	Superactive		Frigid	
BELLEVILLE	Mollisols	Aquolls	Endoaquolls	Typic Endoaquolls	Sandy over loamy	Mixed	Active		Mesic	
BENNINGTON	Alfisols	Aqualfs	Epiaqualfs	Aeric Epiaqualfs	Fine	Illitic			Mesic	
BENONA	Spodosols	Orthods	Haplorthods	Lamellic Haplorthods	Sandy	Mixed			Mesic	
BENSON	Inceptisols	Udepts	Eutrudepts	Lithic Eutrudepts	Loamy-skeletal	Mixed	Active		Mesic	

(continued)

Series name	Order	Suborder	Great group	Subgroup	Particle-size or substitute class	Mineralogy class	CEC actiivity class	Reaction class	Soil temperature class	Other family class
BENZONIA	Spodosols	Orthods	Haplorthods	Lamellic Haplorthods	Sandy	Isotic			Mesic	
BERGLAND	Alfisols	Aqualfs	Epiaqualfs	Aeric Vertic Epiaqualfs	Very-fine	Mixed	Semiactive		Frigid	
BESEMAN	Histosols	Saprists	Haplosaprists	Terric Haplosaprists	Loamy	Mixed		Dysic	Frigid	
BETE GRISE	Spodosols	Aquods	Endoaquods	Typic Endoaquods	Sandy-skeletal	Mixed			Frigid	
BETSY BAY	Entisols	Aquents	Psammaquents	Typic Psammaquents	Sandy	Mixed			Frigid	
BIG IRON	Alfisols	Udalfs	Glossudalfs	Aquic Glossudalfs	Fine-loamy	Mixed	Superactive		Frigid	
BISCUIT	Spodosols	Aquods	Epiaquods	Alfic Epiaquods	Coarse-silty over clayey	Mixed	Superactive		Frigid	
BIXLER	Alfisols	Udalfs	Hapludalfs	Aquic Arenic Hapludalfs	Loamy	Mixed	Active		Mesic	
BLAKESLEE	Alfisols	Udalfs	Hapludalfs	Oxyaquic Hapludalfs	Fine-loamy	Mixed	Active		Mesic	
BLASDELL	Inceptisols	Udepts	Dystrudepts	Typic Dystrudepts	Loamy-skeletal	Mixed	Active		Mesic	
BLOUNT	Alfisols	Aqualfs	Epiaqualfs	Aeric Epiaqualfs	Fine	Illitic			Mesic	
BLUE LAKE	Spodosols	Orthods	Haplorthods	Lamellic Haplorthods	Sandy	Mixed			Frigid	
BOGART	Alfisols	Udalfs	Hapludalfs	Aquic Hapludalfs	Fine-loamy	Mixed	Active		Mesic	
BOHEMIAN	Spodosols	Orthods	Haplorthods	Alfic Haplorthods	Fine-loamy	Mixed	Active		Frigid	
BOMBAY	Alfisols	Udalfs	Hapludalfs	Oxyaquic Hapludalfs	Coarse-loamy	Mixed	Active		Mesic	
BONDUEL	Alfisols	Udalfs	Hapludalfs	Aquollic Hapludalfs	Fine-loamy	Mixed	Active		Frigid	
BONO	Mollisols	Aquolls	Endoaquolls	Typic Endoaquolls	Fine	Illitic			Mesic	
BORGS TROM	Spodosols	Orthods	Durorthods	Typic Durorthods	Sandy	Mixed			Frigid	Shallow, ortstein
BOURBON	Alfisols	Udalfs	Hapludalfs	Aquultic Hapludalfs	Coarse-loamy	Mixed	Active		Mesic	
BOWERS	Alfisols	Udalfs	Glossudalfs	Aquic Glossudalfs	Fine	Mixed	Semiactive		Frigid	
BOYER	Alfisols	Udalfs	Hapludalfs	Typic Hapludalfs	Coarse-loamy	Mixed	Semiactive		Mesic	
BRADY	Alfisols	Udalfs	Hapludalfs	Aquollic Hapludalfs	Coarse-loamy	Mixed	Active		Mesic	
BRECKENRIDGE	Inceptisols	Aquepts	Endoaquepts	Mollic Endoaquepts	Coarse-loamy	Mixed		Nonacid	Frigid	
BRECKSVILLE	Inceptisols	Udepts	Dystrudepts	Typic Dystrudepts	Fine-loamy	Mixed	Active		Mesic	
BREMS	Entisols	Psamments	Udipsamments	Aquic Udipsamments	Sandy	Mixed			Mesic	
BRETHREN	Spodosols	Orthods	Haplorthods	Oxyaquic Haplorthods	Sandy	Isotic			Mesic	
BREVORT	Entisols	Aquents	Endoaquents	Mollic Endoaquents	Sandy over loamy	Mixed	Active	Nonacid	Frigid	
BRIGGSVILLE	Alfisols	Udalfs	Hapludalfs	Typic Hapludalfs	Fine	Mixed	Superactive		Mesic	
BRIMLEY	Spodosols	Aquods	Endoaquods	Argic Endoaquods	Fine-loamy	Mixed	Superactive		Frigid	
BROCKPORT	Alfisols	Aqualfs	Endoaqualfs	Aeric Endoaqualfs	Fine	Illitic			Mesic	
BROOKSTON	Mollisols	Aquolls	Argiaquolls	Typic Argiaquolls	Fine-loamy	Mixed	Superactive		Mesic	
BROWNSTONE	Spodosols	Orthods	Haplorthods	Typic Haplorthods	Sandy-skeletal	Mixed			Frigid	
BRUCE	Inceptisols	Aquepts	Endoaquepts	Mollic Endoaquepts	Fine-loamy	Mixed	Superactive	Nonacid	Frigid	
BRYCE	Mollisols	Aquolls	Endoaquolls	Vertic Endoaquolls	Fine	Mixed	Superactive		Mesic	
BUCKROE	Entisols	Orthents	Udorthents	Lithic Udorthents	Sandy-skeletal	Mixed			Frigid	
BURLEIGH	Entisols	Aquents	Endoaquents	Mollic Endoaquents	Sandy over loamy	Mixed	Active	Nonacid	Frigid	

(continued)

Series name	Order	Suborder	Great group	Subgroup	Particle-size or substitute class	Mineralogy class	CEC actiivity class	Reaction class	Soil temperature class	Other family class
BURSAW	Entisols	Orthents	Udorthents	Oxyaquic Udorthents	Sandy-skeletal	Mixed			Frigid	
BURT	Entisols	Aquents	Psammaquents	Lithic Psammaquents	Sandy	Siliceous			Frigid	
BUSTI	Inceptisols	Aquepts	Endoaquepts	Aeric Endoaquepts	Coarse-loamy	Mixed	Active	Nonacid	Mesic	
CAFFEY	Entisols	Aquents	Endoaquents	Aeric Endoaquents	Sandy over loamy	Mixed	Semiactive	Nonacid	Frigid	
CANADICE	Alfisols	Aqualfs	Endoaqualfs	Typic Endoaqualfs	Fine	Illitic			Mesic	
CANANDAIGUA	Inceptisols	Aquepts	Endoaquepts	Mollic Endoaquepts	Fine-silty	Mixed	Active	Nonacid	Mesic	
CANOSIA	Inceptisols	Aquepts	Epiaquepts	Typic Epiaquepts	Coarse-loamy	Isotic		Nonacid	Frigid	
CAPAC	Alfisols	Udalfs	Glossudalfs	Aquic Glossudalfs	Fine-loamy	Mixed	Active		Mesic	
CARBONDALE	Histosols	Saprists	Haplosaprists	Hemic Haplosaprists				Euic	Frigid	
CARLISLE	Histosols	Saprists	Haplosaprists	Typic Haplosaprists				Euic	Mesic	
CARLSHEND	Spodosols	Orthods	Haplorthods	Oxyaquic Haplorthods	Loamy	Mixed	Superactive	Shallow	Frigid	
CASCO	Alfisols	Udalfs	Hapludalfs	Inceptic Hapludalfs	Fine-loamy over sandy or sandy-skeletal	Mixed	Superactive		Mesic	
CASTALIA	Mollisols	Rendolls	Haprendolls	Inceptic Haprendolls	Loamy-skeletal	Carbonatic			Mesic	
CATHRO	Histosols	Saprists	Haplosaprists	Terric Haplosaprists	Loamy	Mixed		Euic	Frigid	
CAYUGA	Alfisols	Udalfs	Glossudalfs	Oxyaquic Glossudalfs	Fine	Illitic			Mesic	
CAZENOVIA	Alfisols	Udalfs	Hapludalfs	Oxyaquic Hapludalfs	Fine-loamy	Mixed	Active		Mesic	
CERESCO	Mollisols	Udolls	Hapludolls	Fluvaquentic Hapludolls	Coarse-loamy	Mixed	Superactive		Mesic	
CHABENEAU	Spodosols	Orthods	Haplorthods	Oxyaquic Haplorthods	Coarse-loamy over sandy or sandy-skeletal	Mixed	Superactive		Frigid	
CHADAKOIN	Inceptisols	Udepts	Dystrudepts	Typic Dystrudepts	Coarse-loamy	Mixed	Superactive		Mesic	
CHANNING	Spodosols	Aquods	Endoaquods	Typic Endoaquods	Coarse-loamy over sandy or sandy-skeletal	Mixed	Superactive		Frigid	
CHARITY	Mollisols	Aquolls	Epiaquolls	Typic Epiaquolls	Fine-silty	Mixed	Superactive	Calcareous	Frigid	
CHARLEVOIX	Spodosols	Aquods	Endoaquods	Argic Endoaquods	Coarse-loamy	Mixed	Semiactive		Frigid	
CHAUMONT	Alfisols	Aqualfs	Endoaqualfs	Aeric Endoaqualfs	Very-fine	Mixed	Active		Mesic	
CHEBOYGAN	Spodosols	Orthods	Haplorthods	Alfic Haplorthods	Coarse-loamy	Mixed	Active		Frigid	
CHEEKTOWAGA	Mollisols	Aquolls	Epiaquolls	Typic Epiaquolls	Sandy over clayey	Mixed	Active		Mesic	
CHELSEA	Entisols	Psamments	Udipsamments	Lamellic Udipsamments	Sandy	Mixed			Mesic	
CHENANGO	Inceptisols	Udepts	Dystrudepts	Typic Dystrudepts	Loamy-skeletal	Mixed	Superactive		Mesic	
CHILI	Alfisols	Udalfs	Hapludalfs	Typic Hapludalfs	Fine-loamy	Mixed	Active		Mesic	
CHINWHISKER	Spodosols	Orthods	Haplorthods	Lamellic Oxyaquic Haplorthods	Sandy	Mixed			Frigid	
CHIPPENY	Histosols	Saprists	Haplosaprists	Lithic Haplosaprists				Euic	Frigid	
CHIPPEWA HARBOR	Spodosols	Orthods	Haplorthods	Fragic Haplorthods	Coarse-loamy	Isotic			Frigid	
CHOCOLAY	Spodosols	Orthods	Haplorthods	Oxyaquic Haplorthods	Loamy-skeletal	Mixed	Superactive		Frigid	
CHURCHVILLE	Alfisols	Aqualfs	Endoaqualfs	Aeric Endoaqualfs	Fine	Illitic			Mesic	

(continued)

Series name	Order	Suborder	Great group	Subgroup	Particle-size or substitute class	Mineralogy class	CEC actiivity class	Reaction class	Soil temperature class	Other family class
CLAVERACK	Entisols	Orthents	Udorthents	Aquic Udorthents	Sandy over clayey	Mixed	Superactive	Nonacid	Mesic	
COHOCTAH	Mollisols	Aquolls	Endoaquolls	Fluvaquentic Endoaquolls	Coarse-loamy	Mixed	Active		Mesic	
COLLAMER	Alfisols	Udalfs	Hapludalfs	Glossaquic Hapludalfs	Fine-silty	Mixed	Semiactive		Mesic	
COLOMA	Entisols	Psamments	Udipsamments	Lamellic Udipsamments	Sandy	Mixed			Mesic	
COLONIE	Entisols	Psamments	Udipsamments	Lamellic Udipsamments	Sandy	Mixed			Mesic	
COLWOOD	Mollisols	Aquolls	Endoaquolls	Typic Endoaquolls	Fine-loamy	Mixed	Active		Mesic	
CONDIT	Alfisols	Aqualfs	Epiaqualfs	Typic Epiaqualfs	Fine	Illitic			Mesic	
CONNEAUT	Inceptisols	Aquepts	Epiaquepts	Aeric Epiaquepts	Fine-silty	Mixed	Active	Nonacid	Mesic	
CONOTTON	Alfisols	Udalfs	Hapludalfs	Typic Hapludalfs	Loamy-skeletal	Mixed	Active		Mesic	
COOKSON	Spodosols	Orthods	Haplorthods	Alfic Haplorthods	Coarse-loamy	Mixed	Active		Frigid	
COPEMISH	Spodosols	Orthods	Haplorthods	Entic Haplorthods	Sandy	Mixed			Mesic	Ortstein
COPPER HARBOR	Spodosols	Orthods	Haplorthods	Oxyaquic Haplorthods	Sandy-skeletal	Isotic			Frigid	
CORUNNA	Mollisols	Aquolls	Endoaquolls	Typic Endoaquolls	Coarse-loamy	Mixed	Semiactive		Mesic	
COSAD	Entisols	Orthents	Udorthents	Aquic Udorthents	Sandy over clayey	Mixed	Superactive	Nonacid	Mesic	
COVERT	Spodosols	Orthods	Haplorthods	Oxyaquic Haplorthods	Sandy	Mixed			Mesic	
COVINGTON	Alfisols	Aqualfs	Endoaqualfs	Mollic Endoaqualfs	Very-fine	Mixed	Active		Mesic	
COZY	Spodosols	Orthods	Haplorthods	Oxyaquic Haplorthods	Loamy	Mixed	Active		Frigid	Shallow
CROSWELL	Spodosols	Orthods	Haplorthods	Oxyaquic Haplorthods	Sandy	Isotic			Frigid	
CROWELL	Spodosols	Orthods	Haplorthods	Entic Haplorthods	Sandy	Mixed			Frigid	Ortstein
CUBLAKE	Spodosols	Orthods	Haplorthods	Oxyaquic Haplorthods	Sandy	Mixed			Frigid	
CUNARD	Alfisols	Udalfs	Hapludalfs	Typic Hapludalfs	Coarse-loamy	Mixed	Active		Frigid	
CURTISVILLE	Alfisols	Udalfs	Glossudalfs	Haplic Glossudalfs	Fine	Mixed	Semiactive		Frigid	
CUSINO	Spodosols	Orthods	Haplorthods	Typic Haplorthods	Sandy	Isotic			Frigid	
CUTTRE	Alfisols	Aquepts	Glossaqualfs	Aeric Glossaqualfs	Very-fine	Mixed	Active		Frigid	
DAIR	Entisols	Aquents	Psammaquents	Typic Psammaquents	Sandy	Mixed			Mesic	
DANLEY	Alfisols	Udalfs	Hapludalfs	Glossaquic Hapludalfs	Fine-loamy	Mixed	Active		Mesic	
DARIEN	Alfisols	Aqualfs	Endoaqualfs	Aeric Endoaqualfs	Fine-loamy	Mixed	Active		Mesic	
DAVIES	Inceptisols	Aquepts	Endoaquepts	Typic Endoaquepts	Sandy-skeletal	Mixed			Frigid	
DAWSON	Histosols	Saprists	Haplosaprists	Terric Haplosaprists	Sandy or sandy-skeletal	Mixed		Dysic	Frigid	
DEER PARK	Entisols	Psamments	Udipsamments	Spodic Udipsamments	Sandy	Mixed			Frigid	
DEERFIELD	Entisols	Psamments	Udipsamments	Aquic Udipsamments	Sandy	Mixed			Mesic	
DEERTON	Spodosols	Orthods	Haplorthods	Typic Haplorthods	Sandy	Mixed			Frigid	
DEFORD	Entisols	Aquents	Psammaquents	Typic Psammaquents	Sandy	Mixed			Frigid	

(continued)

Series name	Order	Suborder	Great group	Subgroup	Particle-size or substitute class	Mineralogy class	CEC actiivity class	Reaction class	Soil temperature class	Other family class
DEL REY	Alfisols	Aqualfs	Epiaqualfs	Aeric Epiaqualfs	Fine	Illitic			Mesic	
DENOMIE	Alfisols	Udalfs	Glossudalfs	Haplic Glossudalfs	Fine-silty	Mixed	Active		Frigid	
DERB	Inceptisols	Aquepts	Epiaquepts	Aeric Epiaquepts	Fine-silty	Mixed	Active	Acid	Mesic	
DETOUR	Inceptisols	Udepts	Eutrudepts	Aquic Eutrudepts	Fine-loamy	Mixed	Active		Frigid	
DILLINGHAM	Spodosols	Orthods	Fragiorthods	Typic Fragiorthods	Sandy	Isotic			Frigid	
DISHNO	Spodosols	Orthods	Haplorthods	Oxyaquic Haplorthods	Coarse-loamy over sandy or sandy-skeletal	Mixed	Superactive		Frigid	
DIXBORO	Alfisols	Udalfs	Hapludalfs	Aquollic Hapludalfs	Coarse-loamy	Mixed	Active		Mesic	
DORA	Histosols	Saprists	Haplosaprists	Terric Haplosaprists	Clayey	Smectitic		Euic	Frigid	
DORVAL	Histosols	Saprists	Haplosaprists	Terric Haplosaprists	Clayey	Mixed		Euic	Frigid	
DUANE	Spodosols	Orthods	Haplorthods	Typic Haplorthods	Sandy-skeletal	Mixed			Frigid	Ortste n
DUEL	Spodosols	Orthods	Haplorthods	Entic Haplorthods	Sandy	Mixed			Frigid	
DUNBRIDGE	Alfisols	Udalfs	Hapludalfs	Mollic Hapludalfs	Fine-loamy	Mixed	Active		Mesic	
DUNKIRK	Alfisols	Udalfs	Hapludalfs	Glossic Hapludalfs	Fine-silty	Mixed	Active		Mesic	
EAST LAKE	Spodosols	Orthods	Haplorthods	Entic Haplorthods	Sandy	Mixed			Frigid	
EASTPORT	Entisols	Psamments	Udipsamments	Spodic Udipsamments	Sandy	Mixed			Frigid	
EDMORE	Entisols	Aquents	Endoaquents	Mollic Endoaquents	Sandy	Mixed			Mesic	
EDWARDS	Histosols	Saprists	Haplosaprists	Limnic Haplosaprists		Marly		Euic	Mesic	
ELCAJON	Alfisols	Udalfs	Glossudalfs	Aquic Glossudalfs	Fine-loamy	Mixed	Active		Frigid	
ELDES	Alfisols	Aqualfs	Epiaqualfs	Typic Epiaqualfs	Fine-loamy	Mixed	Superactive		Frigid	
ELMRIDGE	Inceptisols	Udepts	Eutrudepts	Aquic Dystric Eutrudepts	Coarse-loamy over clayey	Mixed	Semiactive		Mesic	
ELNORA	Entisols	Psamments	Udipsamments	Aquic Udipsamments		Mixed			Mesic	
EMMET	Alfisols	Udalfs	Hapludalfs	Inceptic Hapludalfs	Coarse-loamy	Mixed	Active		Frigid	
ENGADINE	Spodosols	Aquods	Epiaquods	Alfic Epiaquods	Coarse-loamy over clayey	Mixed	Active		Frigid	
ENSIGN	Inceptisols	Udepts	Eutrudepts	Lithic Eutrudepts	Loamy	Mixed	Superactive		Frigid	
ENSLEY	Inceptisols	Aquepts	Endoaquepts	Aeric Endoaquepts	Coarse-loamy	Mixed	Active	Nonacid	Frigid	
EPOUFETTE	Alfisols	Aqualfs	Endoaqualfs	Mollic Endoaqualfs	Coarse-loamy	Mixed	Superactive		Frigid	
EPWORTH	Spodosols	Orthods	Haplorthods	Entic Haplorthods	Sandy	Siliceous			Mesic	
ERMATINGER	Entisols	Aquents	Fluvaquents	Aeric Fluvaquents	Coarse-silty	Mixed	Superactive	Nonacid	Frigid	
ESAU	Entisols	Orthents	Udorthents	Aquic Udorthents	Sandy-skeletal	Mixed			Frigid	
ESCANABA	Spodosols	Orthods	Haplorthods	Alfic Haplorthods	Sandy over loamy	Mixed	Superactive		Frigid	
ESSEXVILLE	Mollisols	Aquolls	Endoaquolls	Typic Endoaquolls	Sandy over loamy	Mixed	Active	Calcareous	Mesic	
EVART	Mollisols	Aquolls	Endoaquolls	Fluvaquentic Endoaquolls	Sandy	Mixed			Frigid	
FABIUS	Mollisols	Udolls	Argiudolls	Aquic Argiudolls	Fine-loamy over sandy or sandy-skeletal	Mixed	Semiactive		Mesic	
FARMINGTON	Inceptisols	Udepts	Eutrudepts	Lithic Eutrudepts	Loamy	Mixed	Active		Mesic	
FARNHAM	Inceptisols	Udepts	Dystrudepts	Aquic Dystrudepts	Loamy-skeletal	Mixed	Active		Mesic	

(continued)

Series name	Order	Suborder	Great group	Subgroup	Particle-size or substitute class	Mineralogy class	CEC actiivity class	Reaction class	Soil temperature class	Other family class
FARQUAR	Spodosols	Orthods	Haplorthods	Oxyaquic Haplorthods	Sandy-skeletal	Mixed			Frigid	
FELDTMANN	Entisols	Psamments	Udipsamments	Typic Udipsamments	Sandy	Mixed			Frigid	
FENCE	Spodosols	Orthods	Haplorthods	Alfic Oxyaquic Haplorthods	Coarse-silty	Mixed	Superactive		Frigid	
FERN	Alfisols	Udalfs	Glossudalfs	Arenic Oxyaquic Glossudalfs	Loamy	Mixed	Active		Mesic	
FIBRE	Spodosols	Aquods	Epiaquods	Alfic Epiaquods	Sandy over clayey	Mixed	Semiactive		Frigid	
FILER	Alfisols	Udalfs	Glossudalfs	Haplic Glossudalfs	Fine-loamy	Mixed	Semiactive		Mesic	
FILION	Inceptisols	Aquepts	Epiaquepts	Typic Epiaquepts	Fine-loamy	Mixed	Semiactive	Calcareous	Mesic	
FINCH	Spodosols	Aquods	Duraquods	Typic Duraquods	Sandy	Mixed			Frigid	Shallow, ortstein
FITCHVILLE	Alfisols	Aqualfs	Endoaqualfs	Aeric Endoaqualfs	Fine-silty	Mixed	Superactive		Mesic	
FLINK	Spodosols	Aquods	Epiaquods	Typic Epiaquods	Sandy	Mixed			Frigid	
FLINTSTEEL	Alfisols	Udalfs	Glossudalfs	Oxyaquic Glossudalfs	Fine-loamy	Mixed	Superactive		Frigid	
FOGG	Spodosols	Orthods	Haplorthods	Alfic Haplorthods	Sandy	Mixed			Mesic	
FONDA	Inceptisols	Aquepts	Endoaquepts	Mollic Endoaquepts	Fine	Illitic		Nonacid	Mesic	
FORBAY	Alfisols	Udalfs	Hapludalfs	Mollic Hapludalfs	Fine-loamy	Mixed	Superactive		Frigid	
FORTRESS (also LIVONIA)	Entisols	Psamments	Udipsamments	Aquic Udipsamments	Dredgic	Mixed			Mesic	
FOX	Alfisols	Udalfs	Hapludalfs	Typic Hapludalfs	Fine-loamy over sandy or sandy-skeletal	Mixed	Superactive		Mesic	
FOXPAW	Spodosols	Aquods	Endoaquods	Typic Endoaquods	Coarse-loamy	Isotic			Frigid	
FREDA	Spodosols	Orthods	Haplorthods	Lithic Haplorthods	Loamy	Mixed	Active		Frigid	
FREDON	Inceptisols	Aquepts	Endoaquepts	Aeric Endoaquepts	Coarse-loamy over sandy or sandy-skeletal	Mixed	Active	Nonacid	Mesic	
FREESOIL	Inceptisols	Udepts	Eutrudepts	Aquic Eutrudepts	Coarse-loamy	Mixed	Semiactive		Mesic	
FRIES	Mollisols	Aquolls	Endoaquolls	Typic Endoaquolls	Fine	Illitic			Mesic	
FROHLING	Spodosols	Orthods	Fragiorthods	Alfic Haplorthods	Coarse-loamy	Mixed	Active		Frigid	
FULTON	Alfisols	Aqualfs	Epiaqualfs	Aeric Epiaqualfs	Fine	Illitic			Mesic	
FURLONG	Spodosols	Orthods	Haplorthods	Typic Haplorthods	Sandy	Mixed			Frigid	
GAASTRA	Spodosols	Aquods	Endoaquods	Argic Endoaquods	Coarse-loamy	Mixed	Active		Frigid	
GAGETOWN	Mollisols	Udolls	Hapludolls	Oxyaquic Hapludolls	Coarse-silty	Carbonatic			Mesic	
GAGEVILLE	Alfisols	Udalfs	Hapludalfs	Oxyaquic Hapludalfs	Fine-loamy	Mixed	Semiactive		Mesic	
GALEN	Alfisols	Udalfs	Hapludalfs	Oxyaquic Hapludalfs	Coarse-loamy	Mixed	Active		Mesic	
GALOO	Entisols	Orthents	Udorthents	Lithic Udorthents	Loamy	Mixed	Active	Nonacid	Mesic	
GALWAY	Inceptisols	Udepts	Eutrudepts	Typic Eutrudepts	Coarse-loamy	Mixed	Superactive		Mesic	
GARLIC	Spodosols	Orthods	Haplorthods	Typic Haplorthods	Sandy	Mixed			Frigid	Ortstein
GAY	Inceptisols	Aquepts	Endoaquepts	Aeric Endoaquepts	Coarse-loamy	Mixed	Active	Nonacid	Frigid	
GETZVILLE	Inceptisols	Aquepts	Endoaquepts	Aeric Endoaquepts	Fine-silty over sandy or sandy-skeletal	Mixed	Active	Nonacid	Mesic	
GICHIGAMI	Alfisols	Udalfs	Glossudalfs	Oxyaquic Glossudalfs	Fine-silty	Mixed	Superactive		Frigid	

(continued)

Series name	Order	Suborder	Great group	Subgroup	Particle-size or substitute class	Mineralogy class	CEC actiivity class	Reaction class	Soil temperature class	Other family class
GIESE	Inceptisols	Aquepts	Humaquepts	Typic Humaquepts	Coarse-loamy	Isotic		Nonacid	Frigid	
GILCHRIST	Spodosols	Orthods	Haplorthods	Oxyaquic Haplorthods	Sandy	Mixed			Frigid	
GILFORD	Mollisols	Aquolls	Endoaquolls	Typic Endoaquolls	Coarse-loamy	Mixed	Superactive		Mesic	
GLADWIN	Spodosols	Aquods	Endoaquods	Argic Endoaquods	Sandy	Mixed			Frigid	
GLAWE	Mollisols	Aquolls	Endoaquolls	Typic Endoaquolls	Coarse-silty	Mixed	Semiactive	Calcareous	Frigid	
GLENDORA	Entisols	Aquents	Psammaquents	Mollic Psammaquents		Mixed			Mesic	
GLENFORD	Alfisols	Udalfs	Hapludalfs	Aquic Hapludalfs	Fine-silty	Mixed	Superactive		Mesic	
GLYNWOOD	Alfisols	Udalfs	Hapludalfs	Aquic Hapludalfs	Fine	Illitic			Mesic	
GOGEBIC	Spodosols	Orthods	Fragiorthods	Alfic Oxyaquic Fragiorthods	Coarse-loamy	Isotic			Frigid	
GOGOMAIN	Inceptisols	Aquepts	Epiaquepts	Aeric Epiaquepts	Coarse-loamy over clayey	Mixed	Active	Nonacid	Frigid	
GONGEAU	Entisols	Aquents	Psammaquents	Typic Psammaquents	Sandy	Siliceous			Frigid	Shallow
GOODHARBOR	Entisols	Psamments	Quartzipsamments	Typic Quartzipsamments	Sandy				Mesic	Uncoated
GORVAN	Mollisols	Aquolls	Endoaquolls	Fluvaquentic Endoaquolls	Fine-loamy over sandy or sandy-skeletal	Mixed	Semiactive		Mesic	
GOWDY	Alfisols	Udalfs	Glossudalfs	Arenic Glossudalfs	Clayey	Mixed	Active		Mesic	
GRACE	Alfisols	Udalfs	Glossudalfs	Oxyaquic Glossudalfs	Coarse-silty	Mixed	Semiactive		Frigid	
GRANBY	Mollisols	Aquolls	Endoaquolls	Typic Endoaquolls	Sandy	Mixed			Mesic	
GRAND SABLE	Entisols	Orthents	Udorthents	Typic Udorthents	Sandy	Isotic			Frigid	
GRATIOT	Spodosols	Aquods	Fragiaquods	Typic Fragiaquods	Loamy-skeletal	Mixed	Superactive		Frigid	
GRATTAN	Spodosols	Orthods	Haplorthods	Entic Haplorthods	Sandy	Mixed			Mesic	
GRAVERAET	Spodosols	Orthods	Fragiorthods	Alfic Oxyaquic Fragiorthods	Coarse-loamy	Mixed	Active		Frigid	
GRAYCALM	Entisols	Psamments	Udipsamments	Lamellic Udipsamments	Sandy	Isotic			Frigid	
GRAYLING	Entisols	Psamments	Udipsamments	Typic Udipsamments	Sandy	Isotic			Frigid	
GRAYS	Alfisols	Udalfs	Hapludalfs	Mollic Oxyaquic Hapludalfs	Fine-silty	Mixed	Superactive		Mesic	
GREENSTONE	Alfisols	Udalfs	Hapludalfs	Fragiaquic Hapludalfs	Loamy	Isotic			Frigid	Shallow
GREENWOOD	Histosols	Hemists	Haplohemists	Typic Haplohemists				Dysic	Frigid	
GREYLOCK	Spodosols	Orthods	Haplorthods	Alfic Haplorthods	Coarse-loamy	Mixed	Active		Frigid	
GREYSOLON	Inceptisols	Udepts	Eutrudepts	Oxyaquic Eutrudepts	Coarse-loamy	Isotic			Frigid	
GRINDSTONE	Alfisols	Udalfs	Hapludalfs	Glossaquic Hapludalfs	Fine-loamy	Mixed	Active		Mesic	
GROTON	Inceptisols	Udepts	Eutrudepts	Typic Eutrudepts	Sandy-skeletal	Mixed			Mesic	
GUARDLAKE	Spodosols	Orthods	Haplorthods	Typic Haplorthods	Sandy-skeletal	Mixed			Frigid	
GUELPH	Alfisols	Udalfs	Glossudalfs	Haplic Glossudalfs	Fine-loamy	Mixed	Active		Mesic	
GUFFIN	Inceptisols	Aquepts	Endoaquepts	Mollic Endoaquepts	Very-fine	Mixed	Active	Nonacid	Mesic	
GULL POINT	Mollisols	Aquolls	Argiaquolls	Typic Argiaquolls	Fine-loamy	Mixed	Superactive		Frigid	

(continued)

Series name	Order	Suborder	Great group	Subgroup	Particle-size or substitute class	Mineralogy class	CEC actiivity class	Reaction class	Soil temperature class	Other family class
GUTPORT	Inceptisols	Aquepts	Humaquepts	Histic Humaquepts	Fine	Mixed	Active	Nonacid	Frigid	
HAGENSVILLE	Mollisols	Udolls	Hapludolls	Aquic Hapludolls	Coarse-loamy	Mixed	Semiactive		Frigid	
HALFADAY	Spodosols	Orthods	Haplorthods	Oxyaquic Haplorthods	Sandy	Mixed			Frigid	
HALSEY	Inceptisols	Aquepts	Humaquepts	Typic Humaquepts	Coarse-loamy over sandy or sandy-skeletal	Mixed	Active	Nonacid	Mesic	
HARBOR	Alfisols	Udalfs	Hapludalfs	Aquic Hapludalfs	Coarse-loamy	Mixed	Active		Mesic	
HASKINS	Alfisols	Aqualfs	Epiaqualfs	Aeric Epiaqualfs	Fine-loamy	Mixed	Active		Mesic	
HEALYLAKE	Spodosols	Orthods	Haplorthods	Oxyaquic Haplorthods	Sandy	mixed			mesic	ortstein
HEBRON	Alfisols	Udalfs	Hapludalfs	Oxyaquic Hapludalfs	Fine-loamy	Mixed	Superactive		Mesic	
HEGBERG	Alfisols	Udalfs	Hapludalfs	Aquollic Hapludalfs	Fine-loamy	Mixed	Superactive		Frigid	
HEINZ	Spodosols	Orthods	Haplorthods	Oxyaquic Haplorthods	Sandy	Mixed			Frigid	
HENDRIE	Inceptisols	Aquepts	Endoaquepts	Aeric Endoaquepts	Coarse-silty	Mixed	Superactive	Nonacid	Frigid	
HERBSTER	Alfisols	Aqualfs	Glossaqualfs	Aeric Glossaqualfs	Fine	Mixed	Active		Frigid	
HERMANTOWN	Inceptisols	Udepts	Eutrudepts	Aquic Dystric Eutrudepts	Coarse-loamy	Isotic			Frigid	
HESSEL	Mollisols	Aquolls	Epiaquolls	Typic Epiaquolls	Coarse-loamy	Mixed	Semiactive	Calcareous	Frigid	
HETTINGER	Inceptisols	Aquepts	Epiaquepts	Mollic Epiaquepts	Fine-loamy	Mixed	Active	Nonacid	Frigid	
HILTON	Alfisols	Udalfs	Hapludalfs	Oxyaquic Hapludalfs	Fine-loamy	Mixed	Active		Mesic	
HINCKLEY	Entisols	Orthents	Udorthents	Typic Udorthents	Sandy-skeletal	Mixed			Mesic	
HOCHHEIM	Mollisols	Aquolls	Argiaquolls	Typic Argiaquolls	Fine-loamy	Mixed	Active		Mesic	
HOIST	Alfisols	Udalfs	Glossudalfs	Oxyaquic Glossudalfs	Coarse-loamy	Mixed	Semiactive		Frigid	
HORNELL	Inceptisols	Aquepts	Endoaquepts	Aeric Endoaquepts	Fine	Illitic		Acid	Mesic	
HORTONVILLE	Alfisols	Udalfs	Glossudalfs	Haplic Glossudalfs	Fine-loamy	Mixed	Active		Mesic	
HOUGHTON	Histosols	Saprists	Haplosaprists	Typic Haplosaprists				Euic	Mesic	
HOWARD	Alfisols	Udalfs	Hapludalfs	Glossic Hapludalfs	Loamy-skeletal	Mixed	Active		Mesic	
HOYTVILLE	Alfisols	Aqualfs	Epiaqualfs	Mollic Epiaqualfs	Fine	Illitic			Mesic	
HUDSON	Alfisols	Udalfs	Hapludalfs	Glossaquic Hapludalfs	Fine	Illitic			Mesic	
HULLIGAN	Inceptisols	Aquepts	Humaquepts	Typic Humaquepts	Coarse-loamy	Isotic		Nonacid	Frigid	
IARGO	Alfisols	Udalfs	Hapludalfs	Oxyaquic Hapludalfs	Fine	Mixed	Semiactive		Frigid	
INGALLS	Spodosols	Aquods	Endoaquods	Typic Endoaquods	Sandy over loamy	Mixed	Active		Frigid	
IOSCO	Spodosols	Aquods	Endoaquods	Argic Endoaquods	Sandy over loamy	Mixed	Active		Frigid	
IRA	Inceptisols	Udepts	Fragiudepts	Typic Fragiudepts	Coarse-loamy	Mixed	Active		Mesic	
ISABELLA	Spodosols	Orthods	Haplorthods	Alfic Haplorthods	Fine-loamy	Mixed	Active		Frigid	
ISHPEMING	Spodosols	Orthods	Haplorthods	Entic Haplorthods	Sandy	Mixed			Frigid	
ITHACA	Alfisols	Udalfs	Glossudalfs	Aquic Glossudalfs	Fine	Mixed	Semiactive		Mesic	
JACOBSVILLE	Inceptisols	Aquepts	Endoaquepts	Aeric Endoaquepts	Coarse-loamy	Mixed	Active	Nonacid	Frigid	
JEBAVY	Spodosols	Aquods	Duraquods	Typic Duraquods	Sandy	Mixed			Mesic	Shallow, ortstein

(continued)

Series name	Order	Suborder	Great group	Subgroup	Particle-size or substitute class	Mineralogy class	CEC actiivity class	Reaction class	Soil temperature class	Other family class
JESKE	Entisols	Aquents	Psammaquents	Typic Psammaquents	Sandy	Siliceous		Acid	Frigid	Shallow
JIMTOWN	Alfisols	Aqualfs	Endoaqualfs	Aeric Endoaqualfs	Fine-loamy	Mixed	Superactive		Mesic	
JOHNSWOOD	Mollisols	Udolls	Argiudolls	Oxyaquic Argiudolls	Loamy-skeletal	Mixed	Semiactive		Frigid	
JUNIUS	Entisols	Aquents	Psammaquents	Typic Psammaquents	Sandy	Mixed			Mesic	
KALEVA	Spodosols	Orthods	Haplorthods	Typic Haplorthods	Sandy	Isotic			Mesic	
KALKASKA	Spodosols	Orthods	Haplorthods	Typic Haplorthods	Sandy	Isotic			Frigid	
KANOTIN	Spodosols	Aquods	Epiaquods	Histic Epiaquods	Sandy	Mixed			Frigid	
KARLIN	Spodosols	Orthods	Haplorthods	Entic Haplorthods	Sandy	Mixed			Frigid	
KAWKAWLIN	Alfisols	Udalfs	Glossudalfs	Aquic Glossudalfs	Fine	Mixed	Semiactive		Mesic	
KELLOGG	Spodosols	Orthods	Haplorthods	Alfic Oxyaquic Haplorthods	Sandy over clayey	Mixed	Active		Frigid	
KENT	Alfisols	Udalfs	Glossudalfs	Oxyaquic Glossudalfs	Fine	Mixed	Semiactive		Frigid	
KEOWNS	Inceptisols	Aquepts	Endoaquepts	Mollic Endoaquepts	Coarse-loamy	Mixed	Superactive	Nonacid	Mesic	
KERSTON	Histosols	Saprists	Haplosaprists	Fluvaquentic Haplosaprists				Euic	Mesic	
KEWAUNEE	Alfisols	Udalfs	Hapludalfs	Typic Hapludalfs	Fine	Mixed	Active		Mesic	
KEWEENAW	Spodosols	Orthods	Haplorthods	Alfic Haplorthods	Sandy	Mixed			Frigid	
KIBBIE	Alfisols	Udalfs	Hapludalfs	Aquollic Hapludalfs	Fine-loamy	Mixed	Active		Mesic	
KILMANAGH	Inceptisols	Aquepts	Epiaquepts	Aeric Epiaquepts	Fine-loamy	Mixed	Active	Nonacid	Mesic	
KINGSBURY	Alfisols	Aqualfs	Endoaqualfs	Aeric Endoaqualfs	Very-fine	Mixed	Active		Mesic	
KINGSVILLE	Entisols	Aquents	Psammaquents	Mollic Psammaquents	Sandy	Mixed			Mesic	
KINROSS	Spodosols	Aquods	Endoaquods	Typic Endoaquods	Sandy	Mixed			Frigid	
KIVA	Spodosols	Orthods	Haplorthods	Entic Haplorthods	Sandy	Mixed			Frigid	
KLACKING	Alfisols	Udalfs	Glossudalfs	Arenic Glossudalfs	Loamy	Mixed	Semiactive		Frigid	
KOLBERG	Alfisols	Udalfs	Glossudalfs	Haplic Glossudalfs	Fine	Mixed	Active		Frigid	
KRAKOW	Alfisols	Udalfs	Hapludalfs	Inceptic Hapludalfs	Loamy-skeletal	Mixed	Semiactive		Frigid	
LACHINE	Mollisols	Udolls	Hapludolls	Lithic Hapludolls	Loamy	Mixed	Superactive		Frigid	
LACOTA	Mollisols	Aquolls	Endoaquolls	Typic Endoaquolls	Fine-loamy over sandy or sandy-skeletal	Mixed	Semiactive		Frigid	
LAGROSS	Inceptisols	Udepts	Dystrudepts	Typic Dystrudepts	Loamy-skeletal	Mixed	Active		Frigid	
LAKEMONT	Alfisols	Aqualfs	Endoaqualfs	Udollic Endoaqualfs	Fine	Illitic			Mesic	
LAMSON	Inceptisols	Aquepts	Endoaquepts	Aeric Endoaquepts	Coarse-loamy	Mixed	Active	Nonacid	Mesic	
LAPOIN	Spodosols	Orthods	Haplorthods	Alfic Oxyaquic Haplorthods	Fine	Mixed	Active		Frigid	
LATTY	Inceptisols	Aquepts	Endoaquepts	Typic Endoaquepts	Fine	Illitic		Nonacid	Mesic	
LEAFRIVER	Inceptisols	Aquepts	Humaquepts	Histic Humaquepts	Sandy	Mixed			Frigid	
LEELANAU	Spodosols	Orthods	Haplorthods	Alfic Haplorthods	Sandy	Mixed			Frigid	
LENAWEE	Inceptisols	Aquepts	Epiaquepts	Mollic Epiaquepts	Fine	Mixed	Semiactive	Nonacid	Mesic	
LEVASSEUR	Entisols	Aquents	Endoaquents	Aeric Endoaquents	Sandy-skeletal	Mixed			Frigid	Shallow
LIMINGA	Spodosols	Orthods	Haplorthods	Typic Haplorthods	Sandy	Isotic			Frigid	

(continued)

Series name	Order	Suborder	Great group	Subgroup	Particle-size or substitute class	Mineralogy class	CEC actiivity class	Reaction class	Soil temperature class	Other family class
LINWOOD	Histosols	Saprists	Haplosaprists	Terric Haplosaprists	Loamy	Mixed		Euic	Mesic	
LIVINGSTON	Inceptisols	Aquepts	Endoaquepts	Mollic Endoaquepts	Very-fine	Mixed	Active	Nonacid	Mesic	
LOGGERHEAD	Spodosols	Orthods	Haplorthods	Alfic Oxyaquic Haplorthods	Coarse-loamy	Mixed	Superactive		Frigid	
LONDO	Alfisols	Aqualfs	Glossaqualfs	Aeric Glossaqualfs	Fine-loamy	Mixed	Semiactive		Mesic	
LONGRIE	Spodosols	Orthods	Haplorthods	Typic Haplorthods	Coarse-loamy	Mixed	Superactive		Frigid	
LORAIN	Alfisols	Aqualfs	Epiaqualfs	Mollic Epiaqualfs	Fine	Illitic			Mesic	
LOXLEY	Histosols	Saprists	Haplosaprists	Typic Haplosaprists				Dysic	Frigid	
LUPTON	Histosols	Saprists	Haplosaprists	Typic Haplosaprists				Euic	Frigid	
LURAY	Mollisols	Aquolls	Argiaquolls	Typic Argiaquolls	Fine-silty	Mixed	Superactive		Mesic	
LYONS	Inceptisols	Aquepts	Endoaquepts	Mollic Endoaquepts	Fine-loamy	Mixed	Active	Nonacid	Mesic	
MACKINAC	Spodosols	Aquods	Endoaquods	Argic Endoaquods	Fine-loamy	Mixed	Superactive		Frigid	
MADALIN	Alfisols	Aqualfs	Endoaqualfs	Mollic Endoaqualfs	Fine	Illitic			Mesic	
MADAUS	Inceptisols	Aquepts	Humaquepts	Histic Humaquepts	Coarse-silty over sandy or sandy-skeletal	Carbonatic over mixed			Mesic	
MADRID	Alfisols	Udalfs	Glossudalfs	Haplic Glossudalfs	Coarse-loamy	Mixed	Active		Mesic	
MAHONING	Alfisols	Aqualfs	Epiaqualfs	Aeric Epiaqualfs	Fine	Illitic			Mesic	
MANARY	Mollisols	Udolls	Argiudolls	Aquic Argiudolls	Fine	Mixed	Active		Frigid	
MANAWA	Alfisols	Udalfs	Hapludalfs	Aquollic Hapludalfs	Fine	Mixed	Active		Mesic	
MANCELONA	Spodosols	Orthods	Haplorthods	Alfic Haplorthods	Sandy	Mixed			Frigid	
MANIDO	Spodosols	Orthods	Haplorthods	Lamellic Oxyaquic Haplorthods	Sandy	Isotic			Frigid	
MANISTEE	Spodosols	Orthods	Haplorthods	Alfic Haplorthods	Sandy over clayey	Mixed	Active		Frigid	
MANISTIQUE	Entisols	Aquents	Psammaquents	Spodic Psammaquents	Sandy	Mixed			Frigid	
MANLIUS	Inceptisols	Udepts	Dystrudepts	Typic Dystrudepts	Loamy-skeletal	Mixed	Active		Mesic	
MARBLEHEAD	Mollisols	Udolls	Hapludolls	Lithic Hapludolls	Loamy	Mixed	Superactive		Mesic	
MARKEY	Histosols	Saprists	Haplosaprists	Terric Haplosaprists	Sandy or sandy-skeletal	Mixed		Euic	Frigid	
MARLETTE	Alfisols	Udalfs	Glossudalfs	Oxyaquic Glossudalfs	Fine-loamy	Mixed	Semiactive		Mesic	
MARTINSVILLE	Alfisols	Udalfs	Hapludalfs	Typic Hapludalfs	Fine-loamy	Mixed	Active		Mesic	
MARTINTON	Mollisols	Udolls	Argiudolls	Aquic Argiudolls	Fine	Illitic			Mesic	
MARTISCO	Inceptisols	Aquepts	Humaquepts	Histic Humaquepts	Fine-silty	Carbonatic			Mesic	
MASSENA	Inceptisols	Aquepts	Endoaquepts	Aeric Endoaquepts	Coarse-loamy	Mixed	Active	Nonacid	Mesic	
MATTIX	Spodosols	Orthods	Haplorthods	Oxyaquic Haplorthods	Sandy	Mixed			Frigid	
MAUMEE	Mollisols	Aquolls	Endoaquolls	Typic Endoaquolls	Sandy	Mixed			Mesic	
MCGINN	Alfisols	Udalfs	Glossudalfs	Haplic Glossudalfs	Coarse-loamy	Mixed	Semiactive		Frigid	
MCIVOR	Spodosols	Aquods	Duraquods	Typic Duraquods	Sandy	Mixed			Frigid	Shallow, ortstein
MCMASTER	Spodosols	Orthods	Haplorthods	Oxyaquic Haplorthods	Sandy-skeletal	Mixed			Frigid	

(continued)

Series name	Order	Suborder	Great group	Subgroup	Particle-size or substitute class	Mineralogy class	CEC actiivity class	Reaction class	Soil temperature class	Other family class
MCMILLAN	Spodosols	Orthods	Haplorthods	Lamellic Haplorthods	Sandy	Mixed			Frigid	
MEEHAN	Entisols	Psamments	Udipsamments	Aquic Udipsamments		Mixed			Frigid	
MELITA	Spodosols	Orthods	Haplorthods	Alfic Haplorthods	Sandy	Mixed			Frigid	
MENOMINEE	Spodosols	Orthods	Haplorthods	Alfic Haplorthods	Sandy over loamy	Mixed	Active		Frigid	
MENTOR	Alfisols	Udalfs	Hapludalfs	Typic Hapludalfs	Fine-silty	Mixed	Active		Mesic	
MEQUON	Alfisols	Aqualfs	Endoaqualfs	Udollic Endoaqualfs	Fine	Mixed	Superactive		Mesic	
MERMILL	Alfisols	Aqualfs	Epiaqualfs	Mollic Epiaqualfs	Fine-loamy	Mixed	Active		Mesic	
MERWIN	Histosols	Hemists	Haplohemists	Terric Haplohemists	Loamy	Mixed		Dysic	Frigid	
MESABA	Inceptisols	Udepts	Eutrudepts	Dystric Eutrudepts	Coarse-loamy	Isotic			Frigid	
METAMORA	Alfisols	Aqualfs	Epiaqualfs	Udollic Epiaqualfs	Fine-loamy	Mixed	Semiactive		Mesic	
METEA	Alfisols	Udalfs	Hapludalfs	Arenic Hapludalfs	Loamy	Mixed	Active		Mesic	
MICHIGAMME	Spodosols	Orthods	Haplorthods	Fragic Haplorthods	Coarse-loamy	Mixed	Superactive		Frigid	
MIDDLEBURY	Inceptisols	Udepts	Eutrudepts	Fluvaquentic Eutrudepts	Coarse-loamy	Mixed	Superactive		Mesic	
MILFORD	Mollisols	Aquolls	Endoaquolls	Typic Endoaquolls	Fine	Mixed	Superactive		Mesic	
MILL	Inceptisols	Aquepts	Epiaquepts	Aeric Epiaquepts	Fine-loamy	Mixed	Superactive	Nonacid	Mesic	
MILLGROVE	Mollisols	Aquolls	Argiaquolls	Typic Argiaquolls	Fine-loamy	Mixed	Superactive		Mesic	
MILLSDALE	Mollisols	Aquolls	Argiaquolls	Typic Argiaquolls	Fine	Mixed	Active		Mesic	
MILNICHOL	Spodosols	Aquods	Epiaquods	Typic Epiaquods	Sandy	Mixed			Mesic	
MILTON	Alfisols	Udalfs	Hapludalfs	Typic Hapludalfs	Fine	Mixed	Active		Mesic	
MINER	Alfisols	Aqualfs	Epiaqualfs	Mollic Epiaqualfs	Fine	Illitic			Mesic	
MINOA	Inceptisols	Udepts	Eutrudepts	Aquic Dystric Eutrudepts	Coarse-loamy	Mixed	Active		Mesic	
MINOCQUA	Inceptisols	Aquepts	Endoaquepts	Typic Endoaquepts	Coarse-loamy over sandy or sandy-skeletal	Mixed	Superactive	Nonacid	Frigid	
MINONG	Histosols	Folists	Udifolists	Lithic Udifolists				Euic	Frigid	
MISERY	Spodosols	Aquods	Fragiaquods	Argic Fragiaquods	Coarse-loamy	Mixed	Active		Frigid	
MISHWABIC	Inceptisols	Aquepts	Epiaquepts	Typic Epiaquepts	Coarse-loamy	Isotic		Nonacid	Frigid	
MISKOAKI	Alfisols	Udalfs	Glossudalfs	Vertic Glossudalfs	Very-fine	Mixed	Active		Frigid	
MITIWANGA	Alfisols	Aqualfs	Endoaqualfs	Aeric Endoaqualfs	Fine-loamy	Mixed	Active		Mesic	
MOLLINEAUX	Spodosols	Orthods	Haplorthods	Lamellic Haplorthods	Sandy over loamy	Mixed	Active		Mesic	
MOLTKE	Alfisols	Udalfs	Glossudalfs	Aquic Glossudalfs	Coarse-silty	Mixed	Semiactive		Frigid	
MONGO	Alfisols	Udalfs	Glossudalfs	Haplic Glossudalfs	Fine	Mixed	Semiactive		Frigid	
MONITOR	Alfisols	Aqualfs	Endoaqualfs	Udollic Endoaqualfs	Fine-loamy	Mixed	Semiactive		Mesic	
MONTCALM	Spodosols	Orthods	Haplorthods	Alfic Haplorthods	Coarse-loamy	Mixed	Semiactive		Frigid	
MONTGOMERY	Mollisols	Aquolls	Endoaquolls	Vertic Endoaquolls	Fine	Mixed	Active		Mesic	
MONTREAL	Spodosols	Orthods	Fragiorthods	Oxyaquic Fragiorthods	Coarse-loamy	Isotic			Frigid	
MOQUAH	Entisols	Fluvents	Udifluvents	Typic Udifluvents	Coarse-loamy	Mixed	Superactive	Nonacid	Frigid	
MORGANLAKE	Spodosols	Orthods	Haplorthods	Alfic Oxyaquic Haplorthods	Sandy over loamy	Mixed	Active		Frigid	
MORLEY	Alfisols	Udalfs	Hapludalfs	Oxyaquic Hapludalfs	Fine	Illitic			Mesic	
MOROCCO	Entisols	Psamments	Udipsamments	Aquic Udipsamments	Sandy	Mixed			Mesic	

(continued)

Series name	Order	Suborder	Great group	Subgroup	Particle-size or substitute class	Mineralogy class	CEC actiivity class	Reaction class	Soil temperature class	Other family class
MOSEL	Alfisols	Udalfs	Hapludalfs	Aquollic Hapludalfs	Fine-loamy	Mixed	Superactive		Mesic	
MUNDELEIN	Mollisols	Udolls	Argiudolls	Aquic Argiudolls	Fine-silty	Mixed	Superactive		Mesic	
MUNISING	Spodosols	Orthods	Fragiorthods	Alfic Oxyaquic Fragiorthods	Coarse-loamy	Mixed	Active		Frigid	
MUNUSCONG	Inceptisols	Aquepts	Epiaquepts	Mollic Epiaquepts	Coarse-loamy over clayey	Mixed	Active	Nonacid	Frigid	
MUSSEY	Mollisols	Aquolls	Argiaquolls	Typic Argiaquolls	Fine-loamy over sandy or sandy-skeletal	Mixed	Semiactive		Mesic	
NADEAU	Alfisols	Udalfs	Hapludalfs	Typic Hapludalfs	Coarse-loamy	Mixed	Active		Frigid	
NAHMA	Inceptisols	Aquepts	Humaquepts	Histic Humaquepts	Coarse-loamy	Mixed	Active	Nonacid	Frigid	
NAMUR	Mollisols	Udolls	Hapludolls	Lithic Hapludolls	Loamy	Mixed	Active		Frigid	
NAPPANEE	Alfisols	Aqualfs	Epiaqualfs	Aeric Epiaqualfs	Fine	Illitic			Mesic	
NAUMBURG	Spodosols	Aquods	Endoaquods	Typic Endoaquods	Sandy	Isotic			Frigid	
NAVAN	Mollisols	Aquolls	Argiaquolls	Typic Argiaquolls	Fine-loamy	Mixed	Superactive		Mesic	
NECONISH	Spodosols	Orthods	Haplorthods	Oxyaquic Haplorthods	Sandy	Isotic			Frigid	
NEGWEGON	Alfisols	Udalfs	Glossudalfs	Oxyaquic Glossudalfs	Fine	Mixed	Semiactive		Frigid	
NENNO	Mollisols	Udolls	Argiudolls	Aquic Argiudolls	Fine-loamy	Mixed	Active		Mesic	
NESSEN	Spodosols	Orthods	Haplorthods	Typic Haplorthods	Sandy	Mixed			Mesic	
NESTER	Alfisols	Udalfs	Glossudalfs	Oxyaquic Glossudalfs	Fine	Mixed	Semiactive		Frigid	
NET	Spodosols	Aquods	Fragiaquods	Typic Fragiaquods	Coarse-loamy	Mixed	Superactive		Frigid	
NEVENS	Inceptisols	Aquepts	Epiaquepts	Typic Epiaquepts	Coarse-loamy	Isotic		Nonacid	Frigid	
NEWSTEAD	Inceptisols	Aquepts	Endoaquepts	Aeric Endoaquepts	Coarse-loamy	Mixed	Active	Nonacid	Mesic	
NEWTON	Inceptisols	Aquepts	Humaquepts	Typic Humaquepts	Sandy	Mixed			Mesic	
NIAGARA	Alfisols	Aqualfs	Endoaqualfs	Aeric Endoaqualfs	Fine-silty	Mixed	Active		Mesic	
NICHOLS	Inceptisols	Udepts	Eutrudepts	Oxyaquic Eutrudepts	Coarse-silty	Mixed	Superactive		Mesic	
NIPISSING	Spodosols	Orthods	Haplorthods	Typic Haplorthods	Loamy-skeletal	Mixed	Active		Frigid	
NONESUCH	Spodosols	Orthods	Fragiorthods	Alfic Oxyaquic Fragiorthods	Loamy-skeletal	Isotic			Frigid	
NORDHOUSE	Entisols	Psamments	Quartzipsamments	Spodic Quartzipsamments	Sandy				Mesic	Uncoated
NORMANNA	Inceptisols	Udepts	Eutrudepts	Oxyaquic Eutrudepts	Coarse-loamy	Isotic			Frigid	
NOSEUM	Spodosols	Orthods	Haplorthods	Oxyaquic Haplorthods	Sandy	Isotic			Frigid	
NUNICA	Alfisols	Udalfs	Glossudalfs	Haplic Glossudalfs	Fine-silty	Mixed	Superactive		Frigid	
NYKANEN	Inceptisols	Udepts	Eutrudepts	Lithic Eutrudepts	Loamy	Mixed	Superactive		Frigid	
OAKVILLE	Entisols	Psamments	Udipsamments	Typic Udipsamments	Sandy	Mixed			Mesic	
OCQUEOC	Spodosols	Orthods	Haplorthods	Entic Haplorthods	Sandy over loamy	Mixed	Active		Frigid	
ODANAH	Alfisols	Udalfs	Glossudalfs	Haplic Glossudalfs	Fine	Mixed	Active		Frigid	
ODESSA	Alfisols	Aqualfs	Endoaqualfs	Aeric Endoaqualfs	Fine	Illitic			Mesic	
OGEMAW	Spodosols	Orthods	Durorthods	Typic Durorthods	Sandy over loamy	Mixed	Semiactive		Frigid	Ortstein

(continued)

Series name	Order	Suborder	Great group	Subgroup	Particle-size or substitute class	Mineralogy class	CEC actiivity class	Reaction class	Soil temperature class	Other family class
OGONTZ	Alfisols	Udalfs	Hapludalfs	Aquic Hapludalfs	Fine-silty	Mixed	Active		Mesic	
OLDMAN	Spodosols	Orthods	Fragiorthods	Alfic Oxyaquic Fragiorthods	Loamy-skeletal	Mixed	Active		Frigid	
OLMSTED	Alfisols	Aqualfs	Endoaqualfs	Mollic Endoaqualfs	Fine-loamy	Mixed	Active		Mesic	
OMENA	Alfisols	Udalfs	Glossudalfs	Haplic Glossudalfs	Coarse-loamy	Mixed	Active		Frigid	
OMRO	Alfisols	Udalfs	Hapludalfs	Typic Hapludalfs	Clayey over loamy	Mixed	Active		Mesic	
ONAWAY	Alfisols	Udalfs	Hapludalfs	Inceptic Hapludalfs	Fine-loamy	Mixed	Active		Frigid	
ONEKAMA	Alfisols	Udalfs	Glossudalfs	Haplic Glossudalfs	Fine	Mixed	Active		Mesic	
ONOTA	Spodosols	Orthods	Haplorthods	Typic Haplorthods	Coarse-loamy	Mixed	Superactive		Frigid	
ONTARIO	Alfisols	Udalfs	Hapludalfs	Glossic Hapludalfs	Fine-loamy	Mixed	Active		Mesic	
ONTONAGON	Alfisols	Udalfs	Glossudalfs	Haplic Glossudalfs	Very-fine	Mixed	Semiactive		Frigid	
ORONTO	Alfisols	Aqualfs	Glossaqualfs	Aeric Glossaqualfs	Fine-silty	Mixed	Active		Frigid	
ORPARK	Inceptisols	Aquepts	Endoaquepts	Aeric Endoaquepts	Fine-loamy	Mixed	Semiactive	Acid	Mesic	
ORRVILLE	Inceptisols	Aquepts	Endoaquepts	Fluventic Endoaquepts	Fine-loamy	Mixed	Active	Nonacid	Mesic	
OSHKOSH	Alfisols	Udalfs	Hapludalfs	Typic Hapludalfs	Very-fine	Mixed	Active		Mesic	
OSHTEMO	Alfisols	Udalfs	Hapludalfs	Typic Hapludalfs	Coarse-loamy	Mixed	Active		Mesic	
OSSINEKE	Alfisols	Udalfs	Glossudalfs	Oxyaquic Glossudalfs	Fine-loamy	Mixed	Semiactive		Frigid	
OTEGO	Inceptisols	Udepts	Dystrudepts	Fluvaquentic Dystrudepts	Coarse-silty	Mixed	Superactive		Mesic	
OTISCO	Spodosols	Aquods	Endoaquods	Argic Endoaquods	Sandy	Mixed			Frigid	
OTISVILLE	Entisols	Orthents	Udorthents	Typic Udorthents	Sandy-skeletal	Mixed			Mesic	
OTTOKEE	Entisols	Psamments	Udipsamments	Aquic Udipsamments	Sandy	Mixed			Mesic	
OVID	Alfisols	Aqualfs	Endoaqualfs	Aeric Endoaqualfs	Fine-loamy	Mixed	Active		Mesic	
OZAUKEE	Alfisols	Udalfs	Hapludalfs	Oxyaquic Hapludalfs	Fine	Illitic			Mesic	
PAAVOLA	Spodosols	Orthods	Fragiorthods	Alfic Oxyaquic Fragiorthods	Sandy-skeletal	Mixed			Frigid	
PAINESVILLE	Inceptisols	Aquepts	Epiaquepts	Aeric Epiaquepts	Coarse-loamy	Mixed	Active	Nonacid	Mesic	
PALMERS	Inceptisols	Aquepts	Epiaquepts	Vertic Epiaquepts	Very-fine	Mixed	Active	Nonacid	Frigid	
PALMS	Histosols	Saprists	Haplosaprists	Terric Haplosaprists	Loamy	Mixed		Euic	Mesic	
PALMYRA	Alfisols	Udalfs	Hapludalfs	Glossic Hapludalfs	Fine-loamy over sandy or sandy-skeletal	Mixed	Active		Mesic	
PAQUIN	Spodosols	Orthods	Durorthods	Typic Durorthods	Sandy	Isotic			Frigid	Shallow, ortstein
PARKHILL	Inceptisols	Aquepts	Epiaquepts	Mollic Epiaquepts	Fine-loamy	Mixed	Semiactive	Nonacid	Mesic	
PELISSIER	Spodosols	Orthods	Haplorthods	Entic Haplorthods	Sandy-skeletal	Mixed			Frigid	
PELLA	Mollisols	Aquolls	Endoaquolls	Typic Endoaquolls	Fine-silty	Mixed	Superactive		Mesic	
PENCE	Spodosols	Orthods	Haplorthods	Typic Haplorthods	Sandy	Isotic			Frigid	
PEOTONE	Mollisols	Aquolls	Endoaquolls	Cumulic Vertic Endoaquolls	Fine	Smectitic			Mesic	
PEQUAYWAN	Inceptisols	Udepts	Eutrudepts	Aquic Dystric Eutrudepts	Coarse-loamy	Isotic			Frigid	
PERRINTON	Alfisols	Udalfs	Glossudalfs	Oxyaquic Glossudalfs	Fine	Mixed	Active		Mesic	

(continued)

Series name	Order	Suborder	Great group	Subgroup	Particle-size or substitute class	Mineralogy class	CEC actiivity class	Reaction class	Soil temperature class	Other family class
PESHEKEE	Spodosols	Orthods	Haplorthods	Lithic Haplorthods	Loamy	Mixed	Semiactive		Frigid	
PEWAMO	Mollisols	Aquolls	Argiaquolls	Typic Argiaquolls	Fine	Mixed	Active		Mesic	
PHELPS	Alfisols	Udalfs	Hapludalfs	Glossaquic Hapludalfs	Fine-loamy over sandy or sandy-skeletal	Mixed	Active		Mesic	
PICKFORD	Inceptisols	Aquepts	Epiaquepts	Aeric Epiaquepts	Fine	Mixed	Active	Nonacid	Frigid	
PIERPONT	Alfisols	Udalfs	Fragiudalfs	Aquic Fragiudalfs	Fine-silty	Mixed	Active		Mesic	
PINCONNING	Entisols	Aquents	Epiaquents	Mollic Epiaquents	Sandy over clayey	Mixed	Semiactive	Nonacid	Frigid	
PIPESTONE	Spodosols	Aquods	Endoaquods	Typic Endoaquods	Sandy	Mixed			Mesic	
PLAINFIELD	Entisols	Psamments	Udipsamments	Typic Udipsamments	Sandy	Mixed			Mesic	
PLATEA	Alfisols	Aqualfs	Fragiaqualfs	Aeric Fragiaqualfs	Fine-silty	Mixed	Active		Mesic	
PLATTERIVER	Entisols	Psamments	Udipsamments	Oxyaquic Udipsamments	Sandy	Mixed			Mesic	
POMPTON	Inceptisols	Udepts	Dystrudepts	Aquic Dystrudepts	Coarse-loamy	Mixed	Active		Mesic	
PORKIES	Spodosols	Orthods	Haplorthods	Fragic Haplorthods	Loamy-skeletal	Isotic			Frigid	
PORTWING	Alfisols	Udalfs	Glossudalfs	Oxyaquic Glossudalfs	Fine	Mixed	Active		Frigid	
POSEN	Inceptisols	Udepts	Eutrudepts	Typic Eutrudepts	Loamy-skeletal	Mixed	Superactive		Frigid	
POSEYVILLE	Alfisols	Udalfs	Hapludalfs	Aquollic Hapludalfs	Coarse-loamy	Mixed	Active		Mesic	
POTAGANNISSING	Mollisols	Udolls	Hapludolls	Lithic Hapludolls	Loamy	Mixed	Superactive		Frigid	
POY	Mollisols	Aquolls	Endoaquolls	Typic Endoaquolls	Clayey over sandy or sandy-skeletal	Mixed	Active		Mesic	
POYGAN	Mollisols	Aquolls	Epiaquolls	Typic Epiaquolls	Fine	Mixed	Active		Mesic	
PROPER	Spodosols	Orthods	Haplorthods	Oxyaquic Haplorthods	Sandy	Mixed			Frigid	Ortstein
PULLUP	Spodosols	Orthods	Durorthods	Typic Durorthods	Sandy	Mixed			Frigid	Ortstein
QUETICO	Entisols	Orthents	Udorthents	Lithic Udorthents	Loamy	Isotic		Acid	Frigid	
RAPSON	Entisols	Orthents	Udorthents	Aquic Udorthents	Sandy over loamy	Mixed	Semiactive	Nonacid	Mesic	
RAWSON	Alfisols	Udalfs	Hapludalfs	Oxyaquic Hapludalfs	Fine-loamy	Mixed	Active		Mesic	
RAYNHAM	Inceptisols	Aquepts	Epiaquepts	Aeric Epiaquepts	Coarse-silty	Mixed	Active	Nonacid	Mesic	
READE	Spodosols	Orthods	Haplorthods	Aqualfic Haplorthods	Coarse-loamy	Mixed	Superactive		Frigid	
RED HOOK	Inceptisols	Aquepts	Endoaquepts	Aeric Endoaquepts	Coarse-loamy	Mixed	Superactive	Nonacid	Mesic	
REDRIM	Spodosols	Orthods	Haplorthods	Lithic Haplorthods	Sandy-skeletal	Mixed			Frigid	
REMSEN	Alfisols	Aqualfs	Endoaqualfs	Aeric Endoaqualfs	Fine	Illitic			Mesic	
REMUS	Alfisols	Udalfs	Glossudalfs	Haplic Glossudalfs	Fine-loamy	Mixed	Semiactive		Mesic	
RHINEBECK	Alfisols	Aqualfs	Endoaqualfs	Aeric Endoaqualfs	Fine	Illitic			Mesic	
RHODY	Mollisols	Aquolls	Endoaquolls	Typic Endoaquolls	Coarse-silty over sandy or sandy-skeletal	Mixed	Active		Frigid	
RICHTER	Spodosols	Aquods	Endoaquods	Argic Endoaquods	Coarse-loamy	Mixed	Semiactive		Frigid	
RIFLE	Histosols	Hemists	Haplohemists	Typic Haplohemists				Euic	Frigid	
RIGGSVILLE	Spodosols	Aquods	Epiaquods	Alfic Epiaquods	Coarse-loamy	Mixed	Active		Frigid	
RIMER	Alfisols	Udalfs	Hapludalfs	Aquic Arenic Hapludalfs	Loamy	Mixed	Active		Mesic	

(continued)

Series name	Order	Suborder	Great group	Subgroup	Particle-size or substitute class	Mineralogy class	CEC actiivity class	Reaction class	Soil temperature class	Other family class
ROBAGO	Spodosols	Aquods	Endoaquods	Argic Endoaquods	Coarse-loamy	Mixed	Superactive		Frigid	
ROCKBOTTOM	Alfisols	Udalfs	Hapludalfs	Oxyaquic Hapludalfs	Loamy-skeletal	Mixed	Superactive		Frigid	
ROCKCUT	Entisols	Aquents	Fluvaquents	Aeric Fluvaquents	Sandy-skeletal over loamy	Mixed	Active	Nonacid	Frigid	
ROCKLAND	Inceptisols	Udepts	Eutrudepts	Typic Eutrudepts	Fine-loamy	Mixed	Active		Frigid	
ROLLINS	Inceptisols	Udepts	Dystrudepts	Typic Dystrudepts	Sandy-skeletal	Isotic			Frigid	
ROSCOMMON	Entisols	Aquents	Psammaquents	Mollic Psammaquents		Mixed			Frigid	
ROUSSEAU	Spodosols	Orthods	Haplorthods	Entic Haplorthods	Sandy	Mixed			Frigid	
RUBICON	Spodosols	Orthods	Haplorthods	Entic Haplorthods	Sandy	Mixed			Frigid	
RUDYARD	Alfisols	Udalfs	Glossudalfs	Aquic Glossudalfs	Very-fine	Mixed	Active		Frigid	
RUSE	Mollisols	Aquolls	Endoaquolls	Lithic Endoaquolls	Loamy	Mixed	Active		Frigid	
SABATTIS	Inceptisols	Aquepts	Humaquepts	Histic Humaquepts	Coarse-loamy	Mixed	Active	Nonacid	Frigid	
SAGANING	Inceptisols	Aquepts	Endoaquepts	Aeric Endoaquepts	Sandy	Mixed			Frigid	
SANBORG	Alfisols	Udalfs	Glossudalfs	Oxyaquic Glossudalfs	Fine	Mixed	Active		Frigid	
SANDUSKY	Mollisols	Aquolls	Endoaquolls	Fluvaquentic Endoaquolls	Fine-loamy	Carbonatic			Mesic	
SANILAC	Inceptisols	Aquepts	Endoaquepts	Aeric Endoaquepts	Coarse-silty	Mixed	Semiactive	Calcareous	Mesic	
SARANAC	Mollisols	Aquolls	Endoaquolls	Fluvaquentic Endoaquolls	Fine	Mixed	Active		Mesic	
SATAGO	Mollisols	Udolls	Hapludolls	Typic Hapludolls	Fine-silty	Carbonatic			Frigid	
SAUGATUCK	Spodosols	Aquods	Duraquods	Typic Duraquods	Sandy	Mixed			Mesic	Shallow, ortstein
SAUXHEAD	Entisols	Orthents	Udorthents	Lithic Udorthents	Sandy-skeletal	Mixed			Frigid	
SAYLESVILLE	Alfisols	Udalfs	Hapludalfs	Typic Hapludalfs	Fine	Illitic			Mesic	
SAYNER	Spodosols	Orthods	Haplorthods	Entic Haplorthods	Sandy	Mixed			Frigid	
SCHOHARIE	Alfisols	Udalfs	Hapludalfs	Oxyaquic Hapludalfs	Fine	Illitic			Mesic	
SCHWEITZER	Spodosols	Orthods	Fragiorthods	Alfic Fragiorthods	Coarse-loamy	Mixed	Superactive		Frigid	
SCIO	Inceptisols	Udepts	Dystrudepts	Aquic Dystrudepts	Coarse-silty	Mixed	Active		Mesic	
SCRIBA	Inceptisols	Aquepts	Fragiaquepts	Aeric Fragiaquepts	Coarse-loamy	Mixed	Active		Mesic	
SEARCH	Mollisols	Rendolls	Haprendolls	Typic Haprendolls	Coarse-silty	Carbonatic			Frigid	
SEBEWA	Mollisols	Aquolls	Argiaquolls	Typic Argiaquolls	Fine-loamy over sandy or sandy-skeletal	Mixed	Superactive		Mesic	
SEDGWICK	Spodosols	Aquods	Epiaquods	Alfic Epiaquods	Coarse-loamy over clayey	Mixed	Active		Frigid	
SELFRIDGE	Alfisols	Udalfs	Hapludalfs	Aquic Arenic Hapludalfs	Loamy	Mixed	Active		Mesic	
SELKIRK	Alfisols	Udalfs	Glossudalfs	Aquic Glossudalfs	Fine	Mixed	Semiactive		Frigid	
SEWARD	Alfisols	Udalfs	Hapludalfs	Arenic Oxyaquic Hapludalfs	Loamy	Mixed	Active		Mesic	
SHAG	Mollisols	Aquolls	Endoaquolls	Typic Endoaquolls	Coarse-silty	Mixed	Active		Frigid	
SHAKER	Inceptisols	Aquepts	Epiaquepts	Aeric Epiaquepts	Coarse-loamy over clayey	Mixed	Semiactive	Nonacid	Mesic	
SHAVENAUGH	Alfisols	Udalfs	Hapludalfs	Psammentic Hapludalfs	Sandy	Mixed			Mesic	
SHAWANO	Entisols	Psamments	Udipsamments	Typic Udipsamments	Sandy	Mixed			Frigid	
SHEBEON	Alfisols	Aqualfs	Epiaqualfs	Aeric Epiaqualfs	Fine-loamy	Mixed	Active		Mesic	

(continued)

Series name	Order	Suborder	Great group	Subgroup	Particle-size or substitute class	Mineralogy class	CEC actiivity class	Reaction class	Soil temperature class	Other family class
SHELLDRAKE	Entisols	Psamments	Quartzipsamments	Typic Quartzipsamments	Sandy				Frigid	Uncoated
SHELTER	Mollisols	Rendolls	Haprendolls	Inceptic Haprendolls	Loamy-skeletal	Carbonatic			Frigid	
SHINGLETON	Spodosols	Orthods	Haplorthods	Lithic Haplorthods	Sandy	Isotic			Frigid	
SHINROCK	Alfisols	Udalfs	Hapludalfs	Aquic Hapludalfs	Fine	Illitic			Mesic	
SHIOCTON	Mollisols	Udolls	Hapludolls	Aquic Hapludolls	Coarse-silty	Mixed	Superactive		Frigid	
SHUBERTS	Entisols	Aquents	Psammaquents	Typic Psammaquents	Sandy	Mixed			Frigid	
SICKLES	Entisols	Aquents	Epiaquents	Mollic Epiaquents	Sandy over clayey	Mixed	Semiactive	Nonacid	Mesic	
SIMS	Inceptisols	Aquepts	Epiaquepts	Mollic Epiaquepts	Fine	Mixed	Semiactive	Nonacid	Frigid	
SISKIWIT	Spodosols	Orthods	Haplorthods	Alfic Oxyaquic Haplorthods	Sandy	Mixed			Frigid	
SISSON	Alfisols	Udalfs	Hapludalfs	Typic Hapludalfs	Fine-loamy	Mixed	Semiactive		Mesic	
SKANDIA	Histosols	Saprists	Haplosaprists	Lithic Haplosaprists				Dysic	Frigid	
SKANEE	Spodosols	Aquods	Fragiaquods	Argic Fragiaquods	Coarse-loamy	Mixed	Active		Frigid	
SKEEL	Spodosols	Orthods	Haplorthods	Alfic Oxyaquic Haplorthods	Sandy over loamy	Mixed	Semiactive		Frigid	Ortstein
SLADE	Alfisols	Udalfs	Glossudalfs	Aquic Glossudalfs	Fine-loamy	Mixed	Active		Frigid	
SLOAN	Mollisols	Aquolls	Endoaquolls	Fluvaquentic Endoaquolls	Fine-loamy	Mixed	Superactive		Mesic	
SODUS	Inceptisols	Udepts	Fragiudepts	Typic Fragiudepts	Coarse-loamy	Mixed	Active		Mesic	
SOLONA	Mollisols	Udolls	Argiudolls	Aquic Argiudolls	Coarse-loamy	Mixed	Superactive		Frigid	
SOO	Entisols	Aquents	Epiaquents	Aeric Epiaquents	Fine-silty	Mixed	Superactive	Nonacid	Frigid	
SOUTHWELLS	Spodosols	Orthods	Haplorthods	Alfic Haplorthods	Sandy	Mixed			Frigid	
SPARTA	Mollisols	Udolls	Hapludolls	Entic Hapludolls	Sandy	Mixed			Mesic	
SPEAR	Alfisols	Udalfs	Glossudalfs	Aquic Glossudalfs	Coarse-silty	Mixed	Superactive		Frigid	
SPINKS	Alfisols	Udalfs	Hapludalfs	Lamellic Hapludalfs	Sandy	Mixed			Mesic	
SPORLEY	Spodosols	Orthods	Haplorthods	Alfic Haplorthods	Coarse-silty	Mixed	Active		Frigid	
SPOT	Spodosols	Aquods	Duraquods	Typic Duraquods	Sandy	Mixed			Frigid	Shallow, ortstein
SPRINGLAKE	Spodosols	Orthods	Haplorthods	Typic Haplorthods	Sandy	Mixed			Frigid	
SPRINGPORT	Mollisols	Aquolls	Epiaquolls	Typic Epiaquolls	Fine	Mixed	Semiactive		Frigid	
ST. CLAIR	Alfisols	Udalfs	Hapludalfs	Oxyaquic Hapludalfs	Fine	Illitic			Mesic	
ST. IGNACE	Mollisols	Rendolls	Haprendolls	Lithic Haprendolls	Loamy	Carbonatic			Frigid	
STAFFORD	Entisols	Aquents	Psammaquents	Typic Psammaquents	Sandy	Mixed			Mesic	
STANHOPE	Inceptisols	Aquepts	Endoaquepts	Fluvaquentic Endoaquepts	Fine-silty	Mixed	Active	Nonacid	Mesic	
STEUBEN	Spodosols	Orthods	Fragiorthods	Alfic Fragiorthods	Coarse-loamy	Mixed	Active		Frigid	
STURGEON	Entisols	Fluvents	Udifluvents	Aquic Udifluvents	Coarse-silty over sandy or sandy-skeletal	Mixed	Superactive	Nonacid	Frigid	
STUTTS	Spodosols	Orthods	Haplorthods	Typic Haplorthods	Sandy	Isotic			Frigid	
SUGAR	Spodosols	Orthods	Haplorthods	Alfic Oxyaquic Haplorthods	Coarse-silty over clayey	Mixed	Superactive		Frigid	
SULTZ	Spodosols	Orthods	Haplorthods	Entic Haplorthods	Sandy	Mixed			Frigid	
SUMMERVILLE	Inceptisols	Udepts	Eutrudepts	Lithic Eutrudepts	Loamy	Mixed	Active		Frigid	
SUN	Inceptisols	Aquepts	Epiaquepts	Aeric Epiaquepts	Coarse-loamy	Mixed	Active	Nonacid	Mesic	
SUNDELL	Mollisols	Udolls	Hapludolls	Aquic Hapludolls	Coarse-loamy	Mixed	Superactive		Frigid	

(continued)

Series name	Order	Suborder	Great group	Subgroup	Particle-size or substitute class	Mineralogy class	CEC actiivity class	Reaction class	Soil temperature class	Other family class
SUPERIOR	Spodosols	Orthods	Haplorthods	Alfic Oxyaquic Haplorthods	Coarse-loamy over clayey	Mixed	Active		Frigid	
SWANTON	Inceptisols	Aquepts	Epiaquepts	Aeric Epiaquepts	Coarse-loamy over clayey	Mixed over illitic	Superactive	Nonacid	Frigid	
SWORMVILLE	Alfisols	Aqualfs	Endoaqualfs	Aeric Endoaqualfs	Fine-silty over sandy or sandy-skeletal	Mixed	Active		Mesic	
SYMCO	Alfisols	Udalfs	Hapludalfs	Aquollic Hapludalfs	Fine-loamy	Mixed	Active		Mesic	
SYMERTON	Mollisols	Udolls	Argiudolls	Oxyaquic Argiudolls	Fine-loamy	Mixed	Superactive		Mesic	
TACODA	Spodosols	Aquods	Epiaquods	Typic Epiaquods	Sandy	Mixed			Frigid	
TACOOSH	Histosols	Hemists	Haplohemists	Terric Haplohemists	Loamy	Mixed		Euic	Frigid	
TAPPAN	Mollisols	Aquolls	Epiaquolls	Typic Epiaquolls	Fine-loamy	Mixed	Active	Calcareous	Mesic	
TATCHES	Inceptisols	Udepts	Dystrudepts	Lamellic Dystrudepts	Sandy	Isotic			Mesic	
TAWAS	Histosols	Saprists	Haplosaprists	Terric Haplosaprists	Sandy or sandy-skeletal	Mixed		Euic	Frigid	
TEDROW	Entisols	Psamments	Udipsamments	Aquic Udipsamments	Sandy	Mixed			Mesic	
TEEL	Inceptisols	Udepts	Eutrudepts	Fluvaquentic Eutrudepts	Coarse-silty	Mixed	Active		Mesic	
TEKENINK	Alfisols	Udalfs	Glossudalfs	Typic Glossudalfs	Coarse-loamy	Mixed	Semiactive		Mesic	
THETFORD	Alfisols	Udalfs	Hapludalfs	Aquic Arenic Hapludalfs	Sandy	Mixed			Mesic	
THOMAS	Inceptisols	Aquepts	Humaquepts	Histic Humaquepts	Fine-loamy	Mixed	Semiactive	Calcareous	Mesic	
THOMPSONVILLE	Spodosols	Orthods	Haplorthods	Alfic Oxyaquic Haplorthods	Sandy	Mixed			Mesic	
TOBICO	Entisols	Aquents	Psammaquents	Mollic Psammaquents	Sandy	Mixed			Mesic	
TOIVOLA	Spodosols	Orthods	Haplorthods	Lamellic Haplorthods	Sandy	Isotic			Frigid	Ortstein
TOKIAHOK	Spodosols	Orthods	Fragiorthods	Alfic Fragiorthods	Sandy	Mixed			Frigid	
TOLEDO	Inceptisols	Aquepts	Endoaquepts	Mollic Endoaquepts	Fine	Illitic		Nonacid	Mesic	
TONKEY	Inceptisols	Aquepts	Endoaquepts	Mollic Endoaquepts	Coarse-loamy	Mixed	Semiactive	Nonacid	Frigid	
TOWERVILLE	Inceptisols	Udepts	Dystrudepts	Aquic Dystrudepts	Fine-loamy	Mixed	Semiactive		Mesic	
TOWES	Mollisols	Aquolls	Endoaquolls	Typic Endoaquolls	Coarse-silty over sandy or sandy-skeletal	Mixed	Active		Frigid	
TRIMOUNTAIN	Spodosols	Orthods	Fragiorthods	Ultic Fragiorthods	Coarse-loamy	Mixed	Superactive		Frigid	
TROUT BAY	Histosols	Saprists	Haplosaprists	Lithic Haplosaprists				Euic	Frigid	
TULA	Spodosols	Aquods	Fragiaquods	Argic Fragiaquods	Coarse-loamy	Mixed	Superactive		Frigid	
TUSCOLA	Alfisols	Udalfs	Hapludalfs	Aquic Hapludalfs	Fine-loamy	Mixed	Active		Mesic	
TUSTIN	Alfisols	Udalfs	Hapludalfs	Arenic Hapludalfs	Clayey	Mixed	Active		Mesic	
TWIG	Inceptisols	Aquepts	Humaquepts	Histic Humaquepts	Coarse-loamy	Isotic		Nonacid	Frigid	
TWINING	Spodosols	Orthods	Haplorthods	Aqualfic Haplorthods	Fine-loamy	Mixed	Semiactive		Frigid	
TYNER	Entisols	Psamments	Udipsamments	Typic Udipsamments	Sandy	Mixed			Mesic	
TYRE	Entisols	Aquents	Psammaquents	Typic Psammaquents	Sandy	Mixed			Mesic	
UBLY	Spodosols	Orthods	Haplorthods	Alfic Haplorthods	Coarse-loamy	Mixed	Semiactive		Frigid	

(continued)

Series name	Order	Suborder	Great group	Subgroup	Particle-size or substitute class	Mineralogy class	CEC actiivity class	Reaction class	Soil temperature class	Other family class
UNADILLA	Inceptisols	Udepts	Dystrudepts	Typic Dystrudepts	Coarse-silty	Mixed	Active		Mesic	
VALOIS	Inceptisols	Udepts	Dystrudepts	Typic Dystrudepts	Coarse-loamy	Mixed	Superactive		Mesic	
VARYSBURG	Alfisols	Udalfs	Hapludalfs	Glossaquic Hapludalfs	Loamy-skeletal over clayey	Mixed	Semiactive		Mesic	
VELVET	Spodosols	Orthods	Fragiorthods	Oxyaquic Fragiorthods	Sandy-skeletal	Mixed			Frigid	
VERGENNES	Alfisols	Udalfs	Hapludalfs	Glossaquic Hapludalfs	Very-fine	Mixed	Active		Mesic	
VILAS	Spodosols	Orthods	Haplorthods	Entic Haplorthods	Sandy	Isotic			Frigid	
VOELKER	Spodosols	Orthods	Durorthods	Typic Durorthods	Sandy	Mixed			Frigid	Shallow, ortstein
VOLUSIA	Inceptisols	Aquepts	Fragiaquepts	Aeric Fragiaquepts	Fine-loamy	Mixed	Active		Mesic	
WABUN	Entisols	Aquents	Psammaquents	Mollic Psammaquents		Mixed			Frigid	
WAHBEGON	Inceptisols	Aquepts	Humaquepts	Typic Humaquepts	Fine-loamy	Mixed	Superactive	Nonacid	Frigid	
WAINOLA	Spodosols	Aquods	Endoaquods	Typic Endoaquods	Sandy	Mixed			Frigid	
WAISKA	Spodosols	Orthods	Haplorthods	Typic Haplorthods	Sandy-skeletal	Mixed			Frigid	
WAKELEY	Entisols	Aquents	Epiaquents	Aeric Epiaquents	Sandy over clayey	Mixed	Semiactive	Nonacid	Frigid	
WAKEVILLE	Inceptisols	Aquepts	Endoaquepts	Fluvaquentic Endoaquepts	Coarse-silty	Mixed	Active	Nonacid	Mesic	
WALLACE	Spodosols	Orthods	Durorthods	Typic Durorthods	Sandy	Mixed			Frigid	Shallow, ortstein
WALLINGTON	Inceptisols	Aquepts	Fragiaquepts	Aeric Fragiaquepts	Coarse-silty	Mixed	Active		Mesic	
WALLKILL	Inceptisols	Aquepts	Humaquepts	Fluvaquentic Humaquepts	Fine-loamy	Mixed	Superactive	Nonacid	Mesic	
WARNERS	Mollisols	Aquolls	Endoaquolls	Fluvaquentic Endoaquolls	Fine-silty	Carbonatic			Mesic	
WASEPI	Alfisols	Udalfs	Hapludalfs	Aquollic Hapludalfs	Coarse-loamy	Mixed	Semiactive		Mesic	
WATSEKA	Mollisols	Udolls	Hapludolls	Aquic Hapludolls	Sandy	Mixed			Mesic	
WATTON	Alfisols	Udalfs	Glossudalfs	Haplic Glossudalfs	Fine-loamy	Mixed	Semiactive		Frigid	
WAUCONDA	Alfisols	Aqualfs	Endoaqualfs	Udollic Endoaqualfs	Fine-silty	Mixed	Superactive		Mesic	
WAUSEON	Mollisols	Aquolls	Epiaquolls	Typic Epiaquolls	Coarse-loamy over clayey	Mixed OVER illitic	Superactive		Mesic	
WAYLAND	Inceptisols	Aquepts	Endoaquepts	Fluvaquentic Endoaquepts	Fine-silty	Mixed	Active	Nonacid	Mesic	
WEGA	Entisols	Fluvents	Udifluvents	Aquic Udifluvents	Coarse-silty	Mixed	Superactive	Nonacid	Frigid	
WESTBURY	Spodosols	Aquods	Fragiaquods	Typic Fragiaquods	Coarse-loamy	Isotic			Frigid	
WHEATLEY	Entisols	Aquents	Psammaquents	Mollic Psammaquents	Sandy	Mixed			Frigid	
WHITTEMORE	Spodosols	Aquods	Duraquods	Typic Duraquods	Sandy	Mixed			Frigid	Shallow, ortstein
WICK	Inceptisols	Aquepts	Endoaquepts	Fluvaquentic Endoaquepts	Fine-silty	Mixed	Superactive	Nonacid	Mesic	
WILLETTE	Histosols	Saprists	Haplosaprists	Terric Haplosaprists	Clayey	Illitic		Euic	Mesic	
WILLIAMSON	Inceptisols	Udepts	Fragiudepts	Typic Fragiudepts	Coarse-silty	Mixed	Active		Mesic	
WILPOINT	Alfisols	Udalfs	Hapludalfs	Aquic Hapludalfs	Very-fine	Mixed			Mesic	
WINDSOR	Entisols	Psamments	Udipsamments	Typic Udipsamments	Sandy	Mixed			Mesic	
WINTERFIELD	Entisols	Psamments	Udipsamments	Aquic Udipsamments	Sandy	Mixed			Frigid	

(continued)

Series name	Order	Suborder	Great group	Subgroup	Particle-size or substitute class	Mineralogy class	CEC actiivity class	Reaction class	Soil temperature class	Other family class
WISNER	Mollisols	Aquolls	Endoaquolls	Typic Endoaquolls	Fine-loamy	Mixed	Semiactive	Calcareous	Frigid	
WITBECK	Inceptisols	Aquepts	Humaquepts	Histic Humaquepts	Coarse-loamy	Mixed	Semiactive	Nonacid	Frigid	
WIXOM	Spodosols	Aquods	Epiaquods	Alfic Epiaquods	Sandy over loamy	Mixed	Semiactive		Mesic	
WORMET	Spodosols	Aquods	Endoaquods	Typic Endoaquods	Sandy	Mixed			Frigid	
WURTSMITH	Entisols	Psamments	Udipsamments	Oxyaquic Udipsamments	Sandy				Frigid	
YALMER	Spodosols	Orthods	Fragiorthods	Alfic Oxyaquic Fragiorthods	Sandy	Mixed			Frigid	
YELLOWDOG	Entisols	Orthents	Udorthents	Typic Udorthents	Sandy-skeletal	Mixed			Frigid	
ZANDI	Spodosols	Orthods	Haplorthods	Lamellic Haplorthods	Coarse-loamy	Isotic			Frigid	
ZEBA	Spodosols	Aquods	Endoaquods	Argic Endoaquods	Coarse-loamy	Mixed	Active		Frigid	
ZELA	Inceptisols	Aquepts	Humaquepts	Histic Humaquepts	Sandy-skeletal	Mixed			Frigid	
ZIEGENFUSS	Inceptisols	Aquepts	Epiaquepts	Mollic Epiaquepts	Fine	Mixed	Semiactive	Nonacid	Mesic	
ZIMMERMAN	Entisols	Psamments	Udipsamments	Lamellic Udipsamments	Sandy	Mixed			Frigid	
ZURICH	Alfisols	Udalfs	Hapludalfs	Oxyaquic Hapludalfs	Fine-silty	Mixed	Superactive		Mesic	

Soil series (all)	Soil subgroup	Approx. ST	Par materials
Ameliasburg	Orthic Melanic Brunisol	Hapludolls	Till
Atherley	Orthic Humic Gleysol	Endoaquolls	Lacustrine, beach
Athol	Orthic Melanic Brunisol	Hapludolls	Till
Baldwin	Orthic Gray Luvisol	Glossudalfs	Lacustrine, beach
Berrien	Gleyed Brunisolic Gray Brown Luvisol	Endoaqualfs, Epiaqualfs	Glaciolacustrine
Beverly	Gleyed Brunisolic Gray Brown Luvisol	Endoaqualfs, Epiaqualfs	Glaciolacustrine
Bondhead	Brunisolic Gray Brown Luvisol	Hapludalfs	Till
Bookton	Brunisolic Gray Brown Luvisol	Hapludalfs	Glaciolacustrine
Brady	Gleyed Gray Brown Luvisol	Endo-, Epiaqualfs	Glaciolacustrine
Brantford	Brunisolic Gray Brown Luvisol	Hapludalfs	Lacustrine
Breypen	Orthic Melanic Brunisol	Eutrudepts	Till
Brighton	Orthic Melanic Brunisol	Eutrudepts	Outwash
Brisbane	Gleyed Gray Brown Luvisol	Endo-, Epiaqualfs	Glaciofluvial
Brookston	Orthic Humic Gleysol	Endo-, Epiaquepts	Till
Caistor	Gleyed Gray Brown Luvisol	Endo-, Epiaqualfs	Till
Chartrand	Gleyed Gray Luvisol	Endo-, Epiaqualfs	Lacustrine
Chinguacousy	Gleyed Gray Brown Luvisol	Endo-, Epiaqualfs	Till
Colborne	Brunisolic Gray Brown Luvisol	Hapludalfs	Fluvial
Dundonald	Brunisolic Gray Brown Luvisol	Hapludalfs	Till
Dunedin	Eluviated Melanic Brunisol	Eutrudepts	till
Eastport	Orthic Regosol	Udipsamments	Eolian
Edenvale	Gleyed Gray Brown Luvisol	Endo-, Epiaqualfs	Till
Elderslie	Gleyed Eluviated Melanic Brunisol	Humaquepts	Lacustrine
Farmington	Orthic Melanic Brunisol	Hapludolls	Till
Ferndale	Orthic Humic Gleysol	Endo-, Epiaquolls	Glaciofluvial
Fox	Brunisolic Gray Brown Luvisol	Hapludalfs	Lacustrine
Gerow	Regosolic Humic Gleysol	Humaquepts	Till
Gilford	Orthic Humic Gleysol	Endo-, Epiaquepts	Glaciofluvial
Gobles	Gleyed Brunisolic Gray Brown Luvisol	Endoaqualfs, Epiaqualfs	Till
Granby	Orthic Humic Gleysol	Humaquepts	Lacustrine

(continued)

Soil series (all)	Soil subgroup	Approx. ST	Par materials
Grimsby	Brunisolic Gray Brown Luvisol	Hapludalfs	Lacustrine
Guerin	Gleyed Brunisolic Gray Brown Luvisol	Endo-, Epiaqualfs	Till
Gwillimbury	Gleyed Melanic Brunisol	Endo-, Epiaquepts	Glaciofluvial
Haldimand	Gleyed Brunisolic Gray Brown Luvisol	Endo-, Epiaqualfs	Glaciolacustrine
Harkaway	Orthic Melanic Brunisol	Eutrudepts	Till
Harrow	Brunisolic Gray Brown Luvisol	Hapludalfs	Glaciofluvial
Hillier	Orthic Melanic Brunisol	Eutrudepts	Till
Kelvin	Orthic Humic Gleysol	Endoaquepts, Epiaquepts, Endoaquolls, Epiaquolls	Till
Kemble	Gleyed Eluviated Melanic Brunisol	Endo-, Epiaquolls	Till
Kenabeek	Orthic Gleysol	Endo-, Epiaquents	Glaciofluvial
Lansdowne	Gleyed Gray Brown Luvisol	Endo-, Epiaqualfs	Lacustrine
Lily	Orthic Humic Gleysol	Endo-, Epiaquepts	Till
Lincoln	Orthic Humic Gleysol	Endo-, Epiaquepts	Glaciolacustrine
Lindsay	Orthic Humic Gleysol	Endo-, Epiaquolls	Glaciofluvial
Listowel	Gleyed Gray Brown Luvisol	Endo-, Epiaqualfs	Till
Lowbanks	Regosolic Humic Gleysol	Endo-, Epiaquolls	Glaciolacustrine
Mallard	Gleyed Humo-Ferric Podzol	Epi-, Endoquods	Glaciofluvial
Monteagle	Orthic Humo-Ferric Podzol	Haplorthods	Till
Muriel	Brunisolic Gray Brown Luvisol	Hapludalfs	Till
Napanee	Orthic Humic Gleysol	Endo-, Epiaquepts	Lacustrine
Newcastle	Brunisolic Gray Brown Luvisol	Hapludalfs	Glaciofluvial
Niagara	Gleyed Gray Brown Luvisol	Endo-, Epiaqualfs	Glaciolacustrine
Normandale	Gleyed Brunisolic Gray Brown Luvisol	Endoaqualfs, Epiaqualfs	Glaciolacustrine
Ontario	Brunisolic Gray Brown Luvisol	Hapludalfs	Glaciolacustrine
Osprey	Eluviated Melanic Brunisol	Eutrudepts	Till
Otonabee	Orthic Melanic Brunisol	Eutrudepts	Till
Parkhill	Orthic Humic Gleysol	Endo-, Epiaquolls	Till
Perth	Gleyed Brunisolic Gray Brown Luvisol	Endoaqualfs, Epiaqualfs	Till
Plainfield	Brunisolic Gray Brown Luvisol	Hapludalfs	Eolian
Sargent	Orthic Melanic Brunisol	Hapludolls	Till
Smithfield	Gleyed Gray Brown Luvisol	Endo-, Epiaqualfs	Glaciolacustrine
Smithville	Brunisolic Gray Brown Luvisol	Hapludalfs	Glaciolacustrine
Solmesville	Gleyed Gray Brown Luvisol	Endo-, Epiaqualfs	Lacustrine, till
St. Williams	Orthic Humic Gleysol	Endo-, Epiaquolls	Glaciolacustrine
Sullivan	Orthic Melanic Brunisol	Eutrudepts, Hapludolls	**Outwash**
Tavistock	Gleyed Brunisolic Gray Brown Luvisol	Endoaqualfs, Epiaqualfs	Glaciolacustrine
Tecumseth	Gleyed Brunisolic Gray Brown Luvisol	Endoaqualfs, Epiaqualfs	Glaciolacustrine
Tioga	Brunisolic Gray Brown Luvisol	Hapludalfs	Fluvial
Toledo	Orthic Humic Gleysol	Endo-, Epiaquolls	Glaciolacustrine
Vasey	Brunisolic Gray Brown Luvisol	Hapludalfs	Till
Vincent	Brunisolic Gray Brown Luvisol	Hapludalfs	Till
Vineland	Gleyed Brunisolic Gray Brown Luvisol	Endoaqualfs, Epiaqualfs	Glaciolacustrine
Waterloo	Brunisolic Gray Brown Luvisol	Hapludalfs	**Outwash**
Watiford	Brunisolic Gray Brown Luvisol	Hapludalfs	Glaciolacustrine
Wauseon	Orthic Humic Gleysol	Endo-, Epiaquolls	Glaciolacustrine
Welland	Orthic Humic Gleysol	Endo-, Epiaquepts	Glaciolacustrine
Wendigo	Orthic Humo-Ferric Podzol	Haplorthods	Fluvial
Wiarton	Orthic Melanic Brunisol	Endo-, Epiaquolls	Till
Wolsey	Orthic Humic Gleysol	Endo-, Epiaquolls	lacustrine
Wyevale	Orthic Sombric Brunisol	Dystrudepts	Fluvial

Appendix B
Soil-forming factors and lake occurrence for soil series of the Great Lakes Coastal Zone.

Series name	Native vegetation[a]	Parent materials	Parent material particle-size class	Landform	Age (ky)	Lake stage[b]	Lake 1	Lake 2	Lake 3	Lake 4
ABBAYE	Tm	Till	Coarse-loamy	Till plains	11.2	D	SU			
ABSCOTA	Ld	Alluvium	Sandy	Floodplains	13.8	G	MI			
ADAMS	Tm	Glaciolacustrine sandy	Sandy	Glacial lake plains	13	I	ON			
ADRIAN	W	Organic	Sandy or sandy-skeletal	Lake terraces	12.4	A	MI			
AHMEEK	Tm	Till	Coarse-loamy	Till plains	11.2	D	SU			
ALCONA	Tm	Glaciolacustrine fine	Coarse-loamy	Glacial lake plains	11.2	D	SU	MI	HU	
ALDENLAKE	Tm	Loamy mantle/ outwash	**Coarse-loamy over sandy or sandy-skeletal**	Outwash plains	11.2	D	SU			
ALGANSEE	Ld	Alluvium	Sandy	Lake terraces	13.8	G	MI			
ALGONQUIN	Lm	Lacustrine fine	Fine	Lake plains	12.4	A	HU	MI		
ALLENDALE	Tm	Lacustrine fine/ sandy	Sandy over clayey	Lake plains	11.2	A	SU	HU	MI	
ALLIS	Lm	Till	Fine	Lake plains	13.8	W	ER			
ALLOUEZ	Td	Lacustrine sandy	Sandy-skeletal	Beach ridges	11.2	D	SU			
ALPENA	Lm	Glaciofluvial	Sandy-skeletal	Glacial lake plains	11.2	A	MI	HU		
ALSTAD	Td	Till	Fine-loamy	Moraine	12.4	A	HU			
ALTMAR	Td	Outwash	Sandy	Lake terraces	11.2	I	ON			
ALTON	Td	Outwash	Loamy-skeletal	Beach ridges	12.4	I	ON	ER		
AMADON	Tm	Till	Loamy	Bedrock benches	11.2	A	SU	HU		
AMASA	Tm	Outwash	Sandy or sandy-skeletal	Outwash plains	11.2	D	SU			
AMBOY	Td	Glaciolacustrine/ till	**Coarse-silty over sandy**	Till plains	12.4	I	ON			
AMENIA	Td	Till	Coarse-loamy	Till plains	12.4	I	ON			
AMNICON	Tm	Till, clayey	Very-fine	Till plains	11.2	D	SU			
ANGELICA	Lm	Till	Fine-loamy	Till plains	11.2	A	MI	HU		
ANGOLA	Tm	Till	Fine-loamy	Till plains	12.4	WA	ER			
ANNALAKE	Tm	Glaciofluvial	Coarse-loamy	Glacial lake plains	11.2	D	SU			
ANNANIAS	Tm	Glaciolacustrine fine	Coarse-silty	glacial lake plains	11.2	A, D	MI	SU		
APPLETON	Tm	Till	Fine-loamy	Till plains	13	I	ON			
APTAKISIC	Td	Outwash	Fine-silty	Outwash plains	13.8	G	MI			

(continued)

J. G. Bockheim, *Soils of the Laurentian Great Lakes, USA and Canada*,
https://doi.org/10.1007/978-3-030-52425-8

Series name	Native vegetation[a]	Parent materials	Parent material particle-size class	Landform	Age (ky)	Lake stage[b]	Lake 1	Lake 2	Lake 3	Lake 4
ARCADIAN	Tm	Glaciolacustrine fine	Loamy-skeletal	Glacial lake plains	5.3	N	SU			
ARKONA	Td	Lacustrine fine/ sandy	Sandy over clayey	Lake plains	13.8	G	MI			
ARKPORT	Td	Lacustrine sandy	Coarse-loamy	Beach ridges	13	I	ON	MI		
ARNHEIM	Lm	Alluvium	Coarse-loamy	Alluvial fans	5.3	N	SU	MI		
ASHKUM	W	Colluvium/till	Fine	Till plains	13.8	G	MI			
ASHWABAY	Tm	Lacustrine sandy	Sandy	Beach ridges	11.2	D	SU			
ASSININS	Td	Till	Coarse-loamy	Till plains	11.2	D	SU			
AU GRES	Tm	Lacustrine sandy	Sandy	Lake plains	11.2	D	SU	HU	MI	
AU TRAIN	Tm	Glaciofluvial/ bedrock ss	Sandy	Glacial lake benches	11.2	D	SU			
AUBARQUE	Tm	Till	Coarse-loamy	Till plains	5.3	N	HU			
AUGER	Tm	Glaciolacustrine fine	Coarse-silty	Glacial lake plains	11.2	D	SU			
AUGUSTANA	Tm	Eolian/lacustrine/ till	Fine-loamy	Till plains	11.2	D	SU			
AVOCA	Tm	Glaciolacustrine/ till	Sandy over loamy	Till plains	5.3	N	HU			
AZTALAN	P	Glaciolacustrine fine	Fine-loamy	Glacial lake plains	13.8	G	MI			
BACH	Ld	Lacustrine fine	Coarse-silty	Lake plains	5.3	N	HU	MI		
BADAXE	Tm	Till	Coarse-loamy	Till plains	12.4	A	HU			
BADRIVER	Tm	Till, clayey	Fine	Till plains	11.2	D	SU			
BAMFIELD	Td	Till	Fine-loamy	Till plains	12.4	A	HU			
BANAT	Lm	Outwash	Loamy-skeletal	Outwash plains	12.4	A	MI			
BARCELONA	Td	Lacustrine fine	Fine-silty	Lake plains	13	I	ON			
BARRE	Ld	Lacustrine fine	Fine	Lake plains	13	I	ON			
BARRINGTON	P	Lacustrine fine	Fine-silty	Lake plains	13.8	G	MI			
BARTO	Tm	Till	Loamy	Till plains	11.2	D	SU			
BATTLEFIELD	Tm	Lacustrine sandy	Sandy	Beach ridges	11.2	A	HU	MI		
BATTYDOE	Tm	Till	Coarse-loamy	Till plains	11.2	A	HU	SU		
BEAVERTAIL	Lm	Till	Loamy-skeletal	Glacial lake benches	5.3	N	HU	SU		
BEECHER	Sd	Till	Fine	Till plains	13.8	G	MI			
BELDING	Tm	Till	Coarse-loamy	Till plains	11.2	D, A	SU	HU	MI	
BELLEVILLE	Ld	Glaciolacustrine/ till	Sandy over loamy	Till plains	13.8	WA	MI	HU	ER	
BENNINGTON	Td	Till	Fine	Till plains	13.8	WA	ER			
BENONA	Tm	Lacustrine sandy	Sandy	Lake plains	13.8	C, W	MI			
BENSON	Tm	Till	Loamy-skeletal	Till upland	13	I	ON			
BENZONIA	**Tm**	Lacustrine sandy	Sandy	Lake plains	12.4	A	MI			
BERGLAND	Tm	Lacustrine fine	Very-fine	Lake plains	11.2	D	SU	MI		
BESEMAN	Lm	Organic/till	Loamy	Moraine	11.2	D	SU			
BETE GRISE	Lm	Outwash/till/ss bedrock	sandy-skeletal	Glacial lake benches	11.2	D	SU			
BETSY BAY	Lm	Residuum	Sandy	Glacial lake benches	5.3	N	SU			
BIG IRON	Tm	Till	Fine-loamy	Till plains	11.2	D	SU	HU		
BISCUIT	Tm	Lacustrine fine	Coarse-silty over clayey	Lake plains	11.2	A	SU	HU		
BIXLER	Td	Lacustrine fine	Loamy	Lake plains	13.8	WA	HU			
BLAKESLEE	Td	Glaciolacustrine fine	Fine-loamy	Glacial lake plains	13.8	WA	ER			
BLASDELL	Tm	Alluvium	Loamy-skeletal	Alluvial fans	13.8	WA	ER			

(continued)

Series name	Native vegetation[a]	Parent materials	Parent material particle-size class	Landform	Age (ky)	Lake stage[b]	Lake 1	Lake 2	Lake 3	Lake 4
BLOUNT	Td	Till	Fine	Till plains	13.8	G, M	MI	HU		
BLUE LAKE	Tm	Outwash	Sandy	Outwash plains	11.2	A	MI	SU		
BOGART	Td	Outwash	fine-loamy	Lake terraces	13.8	WA	ER			
BOHEMIAN	Tm	Glaciofluvial	Fine-loamy	Glacial lake plains	11.2	A	MI	HU		
BOMBAY	TD	Till	Coarse-loamy	Till plains	13	I	ON			
BONDUEL	Tm	Till	Fine-loamy	Till plains	11.2	A	MI	HU		
BONO	Ld	Lacustrine fine	Fine	Lake plains	13.8	G, W	ER	MI		
BORGSTROM	Tm	Outwash	Sandy	Lake plains	11.2	D	SU	MI		
BOURBON	Td	glaciofluvial	Coarse-loamy	Glacial lake plains	13.8	G	MI			
BOWERS	Td-Tm	Lacustrine fine	Fine	Lake plains	11.2	A	HU	MI		
BOYER	Td	Till/outwash	Coarse-loamy	Till plains	13.8	G,W	MI	HU		
BRADY	Td	Outwash	Coarse-loamy	Lake terraces	13.8	G,W	MI			
BRECKENRIDGE	Lm	Glaciofluvial	Coarse-loamy	Glacial lake plains	13.8	WA	MI	HU		
BRECKSVILLE	Td	Residuum	Fine-loamy	Lake plains	13.8	W	ER			
BREMS	Td	outwash	Sandy	Outwash plains	13.8	G	MI	HU		
BRETHREN	**Tm**	Outwash	Sandy	Lake plains	12.4	A	MI	SU		
BREVORT	Lm	Lacustrine sandy/ fine	Sandy over loamy	Lake plains	12.4	A	MI			
BRIGGSVILLE	Td	Glaciolacustrine fine	Fine	Glacial lake plains	13.8	G	MI			
BRIMLEY	Tm	Glaciofluvial	Fine-loamy	Glacial lake plains	11.2	A, D	MI	HU		
BROCKPORT	Td	Till	Fine	Till upland	13	I	ON	ER		
BROOKSTON	Ld	Silty/till	Fine-loamy	Till plains	13.8	M	MI	HU	ER	
BROWNSTONE	Tm	Residuum, ss	Sandy-skeletal	Bedrock benches	11.2	D	SU			
BRUCE	Tm	Lacustrine fine	Fine-loamy	Lake plains	12.4	A	HU	MI		
BRYCE	W	Lacustrine/Till	Fine	Lake plains	13.8	G	MI			
BUCKROE	Td	Lacustrine sandy	Sandy-skeletal	Bedrock benches	5.3	N	SU			
BURLEIGH	Ld	Lacustrine	Sandy over loamy	Lake plains	12.4	A	MI	HU	SU	
BURSAW	Tm	Till dense	Sandy-skeletal	beach ridges	5.3	N	MI			
BURT	Lm	Residuum	Sandy	bedrock benches	5.3	N	SU			
BUSTI	Tm	Till	Coarse-loamy	Till plains	13.8	WA	ER			
CAFFEY	Tm	glaciofluvial	Sandy over loamy	Glacial lake plains	5.3	N	MI	HU		
CANADICE	Td	Glaciolacustrine fine	fine	Glacial lake plains	13.8	W	ER			
CANANDAIGUA	Lm	Glaciolacustrine fine	Fine-silty	Glacial lake plains	13	W, I	ON	ER		
CANOSIA	Lm	Till	Coarse-loamy	Till plains	11.2	D	SU			
CAPAC	Td	Till	Fine-loamy	Till plains	13.8	G	MI	HU		
CARBONDALE	Lc	Organic		Lake plains	11.2	A, D, W, I	SU	MI	HU	ON
CARLISLE	Lm	Organic		Lake plains	13	W, I	MI	ON	ER	HU
CARLSHEND	Tm	Till/ss bedrock	Loamy	Glacial lake benches	11.2	D	SU			
CASCO	Td	Alluvium/ outwash	Fine-loamy over sandy or sandy-skeletal	Outwash plains	13.8	G	MI			
CASTALIA	Sd	Residuum	Loamy-skeletal	Glacial lake benches	13.8	WA	ER			

(continued)

Series name	Native vegetation[a]	Parent materials	Parent material particle-size class	Landform	Age (ky)	Lake stage[b]	Lake 1	Lake 2	Lake 3	Lake 4
CATHRO	Lc	Organic	Loamy	Lake plains	11.2	D, A	SU	MI	HU	
CAYUGA	Td	Lacustrine/Till	**Fine over coarse-loamy**	Till plains	13	W, I	ON	ER		
CAZENOVIA	Td	Till	Fine-loamy	Till plains	13	I	ON	ER		
CERESCO	Ld	Alluvium	Coarse-loamy	Floodplains	13.8	G, W	MI	ER		
CHABENEAU	Tm	Eolian/outwash/ss bedrock	Coarse-loamy over sandy or sandy-skeletal	Outwash plains	11.2	D	SU			
CHADAKOIN	Td	Till	Coarse-loamy	Till plains	13.8	WA	ER			
CHANNING	Lm	Loamy/outwash	Coarse-loamy over sandy or sandy-skeletal	Outwash plains	11.2	D	SU			
CHARITY	Lc	Lacustrine fine	Fine-silty	Lake plains	12.4	A	HU			
CHARLEVOIX	Tm	Till	Coarse-loamy	Till plains	12.4	A	MI	SU		
CHAUMONT	Lm	Lacustrine fine	Very-fine	Lake plains	10.2	Cs	ON			
CHEBOYGAN	Tm	Till	Coarse-loamy	Lake terraces	12.4	A	HU			
CHEEKTOWAGA	Ld	Lacustrine sandy/fine	Sandy over clayey	Lake plains	12.4	A, I	ON	ER	MI	
CHELSEA	Td	Eolian	Sandy	Dunes	13	I	MI	HU		
CHENANGO	Tm	Outwash	Loamy-skeletal	Outwash plains	13.8	WA	ER			
CHILI	Td	Outwash	Fine-loamy	Beach ridges	13.8	W	ER			
CHINWHISKER	Tm	Glaciofluvial	Sandy	Glacial lake plains	11.2	A	HU	SU		
CHIPPENY	Lm	Organic		Bedrock benches	11.2	A	MI	HU	SU	ON
CHIPPEWA HARBOR	Tm	Till	Coarse-loamy	Bedrock benches	11.2	A	SU			
CHOCOLAY	Tm	Till	Loamy-skeletal	Bedrock benches	5.3	N	SU			
CHURCHVILLE	Tm	Lacustrine/till	**Fine over loamy**	Till plains	13	W, I	ON	ER		
CLAVERACK	Td	Glaciolacustrine sandy/fine	Sandy over clayey	Glacial lake plains	13	I	ON	ER		
COHOCTAH	Ld	alluvium	Coarse-loamy	Floodplains	13.8	G, WA	MI	HU		
COLLAMER	Td	Glaciolacustrine fine	Fine-silty	Glacial lake plains	13	I	ON	ER		
COLOMA	Sm	Outwash	Sandy	Outwash plains	13.8	G	MI			
COLONIE	Td	Glaciolacustrine sandy	Sandy	Glacial lake plains	13	WA, I	ON	ER		
COLWOOD	Ld	Glaciolacustrine fine	Fine-loamy	Glacial lake plains	13.8	G, W	MI	ER	HU	
CONDIT	Td	Till	Fine	Till plains	13.8	W	HU	ER		
CONNEAUT	Td	Glaciolacustrine fine	Fine-silty	Glacial lake plains	13.8	WA	ER			
CONOTTON	Td	Outwash	Loamy-skeletal	Outwash plains	13.8	WA	ER			
COOKSON	Td	Glaciofluvial/till/ ls bedrock	Coarse-loamy	Till plains	11.2	A	SU			
COPEMISH	Td	Eolian	sandy	Beach ridges	11.2	A	MI			
COPPER HARBOR	Tm	Glaciolacustrine sandy	Sandy-skeletal	Glacial lake benches	5.3	N	SU			
CORUNNA	Td	Till/lacustrine fine	Coarse-loamy	Lake plains	12.4	A	HU	ER		
COSAD	Ld	Lacustrine sandy/fine	Sandy over clayey	Lake plains	13	G, I	ON	ER	MI	
COVERT	Tm	Lacustrine sandy	Sandy	Lake plains	13.8	G, WA	MI	HU		
COVINGTON	Tm	Glaciolacustrine fine	Very-fine	Glacial lake plains	10.2	Cs	ON			

(continued)

Series name	Native vegetation[a]	Parent materials	Parent material particle-size class	Landform	Age (ky)	Lake stage[b]	Lake 1	Lake 2	Lake 3	Lake 4
COZY	Tm	Till	Loamy	Till plains	11.2	A	MI			
CROSWELL	Tm	Glaciofluvial	sandy	Glacial lake plains	11.2	A	SU	MI	HU	
CROWELL	Tm	Eolian	Sandy	Beach ridges	11.2	A	HU			
CUBLAKE	Tm	Glaciofluvial	Sandy	Glacial lake plains	11.2	D, A	SU	MI		
CUNARD	Td	Till	Coarse-loamy	Glacial lake benches	11.2	A	HU	MI		
CURTISVILLE	Td	Till	Fine	Till plains	13.8	W	HU			
CUSINO	Td	Glaciofluvial	Sandy	Outwash plains	11.2	A	SU			
CUTTRE	Tm	Till, clayey	very-fine	Till plains	11.2	D	SU			
DAIR	Lm	Glaciofluvial	Sandy	Glacial lake plains	12.4	A	MI			
DANLEY	Td	Till	Fine-loamy	Till plains	13.8	WA	ER	ON		
DARIEN	Tm	Till	Fine-loamy	Till plains	13.8	W	ER	ON		
DAVIES	Lm	Loamy/outwash	**Coarse-loamy over sandy-skeletal**	Glacial drainage channels	11.2	A	SU			
DAWSON	Lc	Organic	Sandy or sandy-skeletal	Lake plains	11.2	D, A	SU	HU	MI	
DEER PARK	Tc	Lacustrine sandy	Sandy	Beach ridges	5.3	N	SU	MI	HU	
DEERFIELD	Tm	Glaciofluvial	Sandy	Outwash plains	13	I	ON			
DEERTON	Td	Residuum, ss	sandy	Bedrock benches	11.2	A	SU			
DEFORD	Lm	Glaciofluvial	Sandy	Glacial lake plains	11.2	D, A	SU	HU	MI	
DEL REY	Td	Lacustrine fine	Fine	Lake plains	13.8	G, WA	ER	MI		
DENOMIE	Tm	Till	Fine-silty	Till plains	11.2	D	SU			
DERB	Tm	Till	Fine-silty	Till plains	13.8	WA	ER			
DETOUR	Tm	Till	Fine-loamy	Glacial lake benches	11.2	A	HU	MI		
DILLINGHAM	Td	Glaciofluvial	Sandy	Till plains	11.2	D	SU			
DISHNO	Tm	Eolian over till	Coarse-loamy over sandy or sandy-skeletal	Moraine	11.2	D, A	SU			
DIXBORO	Td	Glaciofluvial	Coarse-loamy	Glacial lake plains	13.8	G, WA	ER	MI		
DORA	Lm	Organic	Clayey	Lake plains	11.2	A	SU			
DORVAL	Lm	Organic/ lacustrine	Clayey	Lake plains	12.4	A	HU	ON		
DUANE	Tm	Outwash	Sandy-skeletal	Outwash plains	13	I	ON			
DUEL	Td	Till	Sandy	Till plains	11.2	A	MI	HU		
DUNBRIDGE	Td	Till	Fine-loamy	Till plains	13.8	WA	ER			
DUNKIRK	Td	Glaciolacustrine fine	Fine-silty	Glacial lake plains	13	I	ON			
EAST LAKE	Tm	Outwash	Sandy	Lake terraces	11.2	A	MI	HU		
EASTPORT	Tm	Eolian	Sandy	Beach ridges	11.2	A	HU	MI		
EDMORE	Ld	Till	Sandy	Till plains	13.8	W	MI	HU		
EDWARDS	Lm	Organic		Lake plains	13	G, I	MI	HU	ON	
ELCAJON	Tm	Till	Fine-loamy	Glacial lake benches	12.4	A	HU			
ELDES	Ld	Eolian/ glaciolacustrine/ till	Fine-loamy	Till plains	11.2	D	SU			
ELMRIDGE	Tm	Glaciolacustrine fine	Coarse-loamy over clayey	Glacial lake plains	10.2	Cs	ON			

(continued)

Series name	Native vegetation[a]	Parent materials	Parent material particle-size class	Landform	Age (ky)	Lake stage[b]	Lake 1	Lake 2	Lake 3	Lake 4
ELNORA	Tm	Lacustrine sandy		Lake plains	13	WA, I	ER	ON		
EMMET	Tm	Till	Coarse-loamy	Till plains	11.2	A	MI	HU		
ENGADINE	Tm	Lacustrine fine	Coarse-loamy over clayey	Lake plains	11.2	A	MI	HU		
ENSIGN	Tm	Till	Loamy	Glacial lake benches	5.3	N	MI	HU	SU	
ENSLEY	Lm	Till	Coarse-loamy	Lake terraces, wave-cut	11.2	A	MI	HU	ON	SU
EPOUFETTE	Lm	Lacustrine sandy	Coarse-loamy	Lake plains	11.2	A	MI	HU		
EPWORTH	Td	Lacustrine sandy	Sandy	Beach ridges	<13		MI			
ERMATINGER	Lm	Alluvium	Coarse-silty	Floodplains	5.3	N	HU	MI		
ESAU	Lm	Lacustrine sandy	Sandy-skeletal	Beach ridges	5.3	N	HU	MI		
ESCANABA	Tm	Till	Sandy over loamy	Till plains	11.2	A	SU			
ESSEXVILLE	Ld	Lacustrine sandy	Sandy over loamy	Lake plains	11.2	A	HU			
EVART	Lm	Alluvium	Sandy	Floodplains	11.2	A	SU	HU		
FABIUS	Ld	Outwash	Fine-loamy over sandy or sandy-skeletal	Lake plains	11.2	A	MI	HU		
FARMINGTON	Td	Till	Loamy	Till plains	10.2	Cs	ON			
FARNHAM	Tm	Alluvium	Loamy-skeletal	Alluvial fans	13.8	WA	ER			
FARQUAR	Tm	Outwash	Sandy-skeletal	Outwash plains	11.2	D	SU			
FELDTMANN	Tm	Beach deposits	Sandy	Lake plains	5.3	N	SU			
FENCE	Tm	Glaciolacustrine fine	Coarse-silty	Glacial lake plains	11.2	D, A	SU	MI		
FERN	Td	Sandy till/loamy till	Loamy	Till plains	13.8	G	MI			
FIBRE	Lc	Lacustrine sandy	Sandy over clayey	Lake plains	11.2	A	HU			
FILER	Td	Till	Fine-loamy	Till plains	13.8	G	MI			
FILION	Lm	Till	Fine-loamy	Glacial lake plains	5.3	N	HU			
FINCH	Lm	Lacustrine sandy	Sandy	Lake plains	11.2	D, A	SU	MI	HU	
FITCHVILLE	Td	Glaciolacustrine fine	Fine-silty	Glacial lake plains	13.8	WA	ER			
FLINK	Lm	Outwash	Sandy	Lake terraces	11.2	D	SU			
FLINTSTEEL	Tm	Till	Fine-loamy	Till plains	11.2	D	SU			
FOGG	Td	Outwash	Sandy	Outwash plains	13.8	G	MI			
FONDA	Ld	Lacustrine fine	Fine	Lake plains	13	I	ON			
FORBAY	Tm	Eolian/ glaciolacustrine/ till	Fine-loamy	Till plains	11.2	D	SU			
FORTRESS (also LIVONIA)	W	Dredge spoils	Dredgic	Dredge spoils	<0.2		ER			
FOX	Td	Alluvium/ outwash	Fine-loamy over sandy or sandy-skeletal	Outwash plains	13.8	M	MI	HU		
FOXPAW	Lm	Till	Coarse-loamy	Moraine	11.2	D	SU			
FREDA	Td	till	Loamy	Glacial lake benches	11.2	D	MI			
FREDON	Ld	Glaciofluvial/ outwash	Coarse-loamy over sandy or sandy-skeletal	Outwash plains	12.4	WA, I	ER	ON		
FREESOIL	Tm	Glaciolacustrine fine	Coarse-loamy	Glacial lake plains	13	C	MI			
FRIES	Ld	Till	Fine	Lake plains	13.8	WA	ER			
FROHLING	Tm	Till	Coarse-loamy	Till-floored lake plains	11.2	D	SU			

(continued)

Series name	Native vegetation[a]	Parent materials	Parent material particle-size class	Landform	Age (ky)	Lake stage[b]	Lake 1	Lake 2	Lake 3	Lake 4
FULTON	Td	Glaciolacustrine fine	Fine	Glacial lake plains	13.8	WA	ER	HU		
FURLONG	Td	Outwash	Sandy	Outwash plains	11.2	A	SU	MI	HU	
GAASTRA	Tm	Glaciofluvial	Coarse-loamy	Glacial lake plains	11.2	A	SU	HU	MI	
GAGETOWN	**Ld**	Glaciolacustrine fine	Coarse-silty	Glacial lake plains	11.2	A	HU			
GAGEVILLE	Td	Till	Fine-loamy	Till plains	13.8	W	ER			
GALEN	Td	Lacustrine fine	Coarse-loamy	Lake plains	13	I	ON	HU		
GALOO	Tm	Till over ls bedrock	Loamy	Bedrock benches	13	I	ON			
GALWAY	Tm	Till/ls bedrock	Coarse-loamy	Bedrock benches	10.2	Cs	ON			
GARLIC	Tm	Glaciofluvial	Sandy	Glacial lake plains	11.2	D, A	SU	MI		
GAY	Lm	Till	coarse-loamy	Till-floored lake plains	11.2	D	SU	HU		
GETZVILLE	Ld	Lacustrine fine	Fine-silty over sandy or sandy-skeletal	Lake plains	13.8	WA	ER			
GICHIGAMI	Tm	Till	Fine-silty	Till plains	11.2	D	SU			
GIESE	Ld	Till	Coarse-loamy	Till plains	11.2	D	SU			
GILCHRIST	Tm	Outwash/till	**Sandy over coarse-loamy**	Till plains	11.2	A	MI			
GILFORD	W	Outwash	Coarse-loamy	Lake terraces	13.8	G	MI	HU	ER	
GLADWIN	Tm	Glaciofluvial	Sandy	Glacial lake plains	11.2	A	MI	HU		
GLAWE	W	Glaciolacustrine fine	Coarse-silty	Glacial lake plains	11.2	A	MI	HU		
GLENDORA	Ld	Alluvium	Sandy	Floodplains	13.8	G	MI			
GLENFORD	Td	Glaciolacustrine fine	Fine-silty	Glacial lake plains	13.8	WA	ER			
GLYNWOOD	Td	Loess/till	Fine	Till plains	13.8	M	ER	MI	HU	
GOGEBIC	Tm	Eolian/till	**Coarse-loamy over sandy**	Moraine	11.2	D	SU			
GOGOMAIN	Lm	Glaciofluvial/ glaciolacustrine	Coarse-loamy over clayey	Glacial lake plains	11.2	A	MI	HU	SU	
GONGEAU	Tm	Glaciofluvial	Sandy	Glacial lake benches	5.3	N	SU			
GOODHARBOR	Tm	Eolian	Sandy	Beach ridges	5.3	N	MI			
GORVAN	Ld	Alluvium	Fine-loamy over sandy or sandy-skeletal	Floodplains	13.8	G	MI			
GOWDY	Tm	Glaciolacustrine/ till	**Sandy over clayey**	Till plains	12.4	A	MI			
GRACE	**Tm**	Glaciofluvial	Coarse-silty	Glacial lake plains	11.2	A	HU			
GRANBY	Ld	Glaciolacustrine sandy	Sandy	Glacial lake plains	13.8	G, WA	MI	HU	ER	ON
GRAND SABLE	Td	Outwash	Sandy	Outwash plains	5.3	N	SU			
GRATIOT	Tm	Till	Loamy-skeletal	Moraine	11.2	D	SU			
GRATTAN	Tm	**Outwash**	Sandy	Lake plains	11.2	A	MI	HU		
GRAVERAET	Td	Till	Coarse-loamy	Till plains	11.2	D, A	SU	MI		
GRAYCALM	Tm	Glaciofluvial	Sandy	Outwash plains	12.4	A	HU			
GRAYLING	Tm	Glaciofluvial	Sandy	Outwash plains	11.2	A	HU	SU	MI	
GRAYS	Sd	Glaciolacustrine fine	Fine-silty	Glacial lake plains	13.8	G	MI			

(continued)

Series name	Native vegetation[a]	Parent materials	Parent material particle-size class	Landform	Age (ky)	Lake stage[b]	Lake 1	Lake 2	Lake 3	Lake 4
GREENSTONE	**Tm**	Till	Loamy	Lake terraces	11.2	D	SU			
GREENWOOD	Lc	Organic		Lake plains	11.2	D, A	SU	HU	MI	
GREYLOCK	Td	Till	Coarse-loamy	Till plains	11.2	D, A	SU	HU	MI	
GREYSOLON	Tm	Till over gr bedrock	Coarse-loamy	Bedrock benches	11.2	D	SU			
GRINDSTONE	Tm	Till	Fine-loamy	Till plains	11.2	A	HU			
GROTON	Tm	Outwash	Sandy-skeletal	Outwash plains	13	I	ON			
GUARDLAKE	Tm	Outwash	Sandy-skeletal	Outwash plains	11.2	A	MI	HU		
GUELPH	Td	Till	Fine-loamy	Till plains	13.8	W	HU	MI		
GUFFIN	Lm	Lacustrine fine	Very-fine	Lake plains	10.2	Cs	ON			
GULL POINT	Tm	Alluvium/till	Fine-loamy	Till plains	11.2	D	SU			
GUTPORT	Lm	Organic over glaciolacustrine	Fine	Lake plains	5.3	N	HU			
HAGENSVILLE	Lm	Till	Coarse-loamy	Till plains	12.4	A	HU			
HALFADAY	Td	Glaciofluvial	Sandy	Glacial lake plains	11.2	D, A	SU	MI	HU	
HALSEY	Ld	Glaciofluvial/ outwash	Coarse-loamy over sandy or sandy-skeletal	Glacial lake plains	13	I, W	ON	ER		
HARBOR	Td	Glaciolacustrine fine	Coarse-loamy	Glacial lake plains	13.8	WA	ER			
HASKINS	Td	Glaciolacustrine/ till	**Fine-loamy over clayey**	Till plains	13.8	M	ER			
HEALYLAKE	**Tm**	Outwash	Sandy	Lake plains	12.4	A	MI			
HEBRON	Td	Outwash	Fine-loamy	Lake plains	13.8	G	MI			
HEGBERG	Tm	Eolian/lacustrine/ till	Fine-loamy	Till plains	11.2	D	SU			
HEINZ	Tm	Outwash	Sandy	Outwash plains	11.2	A	MI			
HENDRIE	Lm	Glaciolacustrine fine	Coarse-silty	Glacial lake plains	11.2	D	MI	SU		
HERBSTER	Tm	Lacustrine/till	**Fine over coarse-loamy**	Till plains	11.2	D	SU			
HERMANTOWN	Tm	Loamy mantle/till	Coarse-loamy	Till plains	11.2	D	SU			
HESSEL	Lm	Till	Coarse-loamy	Glacial lake benches	11.2	A	HU			
HETTINGER	Lm	Glaciolacustrine fine	Fine-loamy	Glacial lake plains	11.2	A	HU	MI		
HILTON	Td	Till	Fine-loamy	Till plains	13	I	ON			
HINCKLEY	Sm	Outwash	Sandy-skeletal	Outwash plains	13	I	ON			
HOCHHEIM	Td	Loess/till	**Silty over fine-loamy**	Till plains	13.8	G	MI			
HOIST	Td	Till	Coarse-loamy	Till plains	12.4	A	HU			
HORNELL	Tm	Till	Fine	Bedrock benches	13.8	WA	ER			
HORTONVILLE	Td	Till	Fine-loamy	Till plains	13.8	G	MI			
HOUGHTON	W	Organic		Lake plains	13.8	G	MI	HU		
HOWARD	Tm	Outwash	Loamy-skeletal	Outwash plains	13	I	ON			
HOYTVILLE	Ld	Till	Fine	Till plains	13.8	M	ER	HU		
HUDSON	Tm	Lacustrine fine	Fine	Lake plains	13	I	ON	ER		
HULLIGAN	Lm-Tm	Loamy/outwash	**Coarse-loamy over sandy or sandy-skeletal**	Outwash plains	11.2	D	SU			
IARGO	Td	Lacustrine fine	Fine	Lake plains	12.4	A	HU			
INGALLS	Tm	Outwash	Sandy over loamy	Lake plains	11.2	D, A	HU	MI	SU	
IOSCO	Tm	Lacustrine sandy/ fine	Sandy over loamy	Lake plains	11.2	D, A	HU	MI	SU	

(continued)

Series name	Native vegetation[a]	Parent materials	Parent material particle-size class	Landform	Age (ky)	Lake stage[b]	Lake 1	Lake 2	Lake 3	Lake 4
IRA	Td	Till	Coarse-loamy	Till plains	13	I	ON			
ISABELLA	Td	Till	Fine-loamy	Till plains	13.8	WA	HU			
ISHPEMING	Tm	Glaciofluvial	Sandy	Outwash plains	11.2	D	SU			
ITHACA	Td	Till	Fine	Till plains	13.8	G	MI			
JACOBSVILLE	Tm	Till over bedrock	coarse-loamy	Bedrock benches	5.3	N	SU			
JEBAVY	Tm	Glaciofluvial	Sandy	Glacial lake plains	12.4	A	MI			
JESKE	Tm	Glaciofluvial over ss Residuum	Sandy	Bedrock benches	5.3	N	SU			
JIMTOWN	Td	OUTWASH	Fine-loamy	Outwash plains	13.8	WA	ER			
JOHNSWOOD	Tm	Till	Loamy-skeletal	Glacial lake benches	12.4	A	HU	MI		
JUNIUS	Ld	Lacustrine sandy	Sandy	Lake plains	13	I	ON			
KALEVA	**Tm**	Lacustrine sandy	Sandy	Lake plains	11.2	A	MI			
KALKASKA	Tm	Outwash	Sandy	Outwash plains	11.2	D, A	SU	MI	HU	
KANOTIN	Lm	Glaciofluvial	Sandy	Glacial lake plains	12.4	A	HU			
KARLIN	Tm	Outwash	Sandy	Outwash plains	11.2	D, A	MI	SU		
KAWKAWLIN	Td	Till	Fine	Till plains	12.4	A	MI	HU		
KELLOGG	Tm	Lacustrine sandy/ fine	Sandy over clayey	Lake plains	11.2	D, A	SU	HU		
KENT	Tm	Till	Fine	Till plains	13.8	WA	MI	HU		
KEOWNS	Ld	Glaciolacustrine fine	Coarse-loamy	Glacial lake plains	12.4	A	MI			
KERSTON	Ld	Organic/alluvium		Floodplains	13.8	G, W	MI	HU		
KEWAUNEE	Td	Till, clayey	Fine	Till plains	11.2	G, A	MI			
KEWEENAW	Tm	Till	Sandy	Till plains	11.2	D	SU			
KIBBIE	Td	Glaciolacustrine fine	Fine-loamy	Glacial lake plains	13.8	G, WA	ER	MI		
KILMANAGH	**Tm**	Till	Fine-loamy	Till plains	11.2	A	HU			
KINGSBURY	Tm	Lacustrine fine	Very-fine	Lake plains	10.2	Cs	ON			
KINGSVILLE	Ld	Glaciolacustrine sandy	Sandy	Glacial lake plains	13.8	G, WA	MI	HU	ER	
KINROSS	Lm	Glaciofluvial	Sandy	Glacial lake plains	11.2	D, A	SU	HU	MI	
KIVA	Tm	**Glaciolacustrine sandy**	Sandy	Glacial lake plains	11.2	A	MI	HU		
KLACKING	Tm	Outwash	Loamy	Outwash plains	12.4	A	HU			
KOLBERG	Tm	Till over bedrock	Fine	Bedrock benches	11.2	A	MI			
KRAKOW	Tm	Till	Loamy-skeletal	Glacial lake benches	11.2	A	MI			
LACHINE	Tm	Till	Loamy	Glacial lake benches	11.2	A	MI			
LACOTA	Ld	Outwash	Fine-loamy over sandy or sandy-skeletal	Lake plains	12.4	A	HU	MI		
LAGROSS	Tm	Water-sorted shale	Loamy-skeletal	Outwash plains	10.2	Cs	ON			
LAKEMONT	Ld	Lacustrine fine	Fine	Lake plains	13	I	ON	ER		
LAMSON	Lm	Glaciolacustrine fine	Coarse-loamy	Glacial lake plains	13	I, W, G	ON	ER	HU	MI
LAPOIN	Tm	Till/ss bedrock	Fine	Till plains	11.2	D	SU			
LATTY	Ld	Glaciolacustrine fine	Fine	Glacial lake plains	13.8	M	ER	HU		

(continued)

Series name	Native vegetation[a]	Parent materials	Parent material particle-size class	Landform	Age (ky)	Lake stage[b]	Lake 1	Lake 2	Lake 3	Lake 4
LEAFRIVER	Tm	Glaciolacustrine sandy	Sandy	Glacial lake plains	11.2	A, D	HU	MI	SU	
LEELANAU	Tm	Till	Sandy	Till plains	11.2	A	MI			
LENAWEE	Ld	Lacustrine fine	Fine	Lake plains	13.8	M	HU	ER		
LEVASSEUR	Tm	Glaciofluvial/ss bedrock	Sandy-skeletal	Glacial lake benches	5.3	N	SU			
LIMINGA	Tm	Glaciofluvial	Sandy	Glacial lake plains	11.2	D, A	SU	HU		
LINWOOD	Ld	Organic	Loamy	Lake plains	11.2	A, W	HU	MI	ER	
LIVINGSTON	Lm	Glaciolacustrine fine	Very-fine	Glacial lake plains	10.2	Cs	ON			
LOGGERHEAD	Tm	Till	Coarse-loamy	Till plains	11.2	D	SU			
LONDO	Tm	Till	Fine-loamy	Till plains	11.2	A, WA	HU			
LONGRIE	Tm	Till	Coarse-loamy	Glacial lake benches	11.2	A	MI	HU		
LORAIN	Ld	Glaciolacustrine fine	Fine	Glacial lake plains	13.8	WA	ER			
LOXLEY	W	Organic		Lake plains	11.2	D, A	SU	MI	HU	
LUPTON	Lm	Organic		Lake plains	11.2	D, A	SU	HU	MI	
LURAY	Ld	Lacustrine fine	Fine-silty	Lake plains	13.8	WA	ER			
LYONS	Ld	Till	Fine-loamy	Till plains	13	I	ER	ON		
MACKINAC	Tm	Till	fine-loamy	Till plains	12.4	A	MI			
MADALIN	Tm	Glaciolacustrine fine	Fine	Glacial lake plains	13	I	ON			
MADAUS	W	Lacustrine sandy	Coarse-silty over sandy or sandy-skeletal	Lake plains	11.2	A	MI			
MADRID	Tm	Loamy mantle/till	Coarse-loamy	Till plains	13	I	ON			
MAHONING	Td	Till	Fine	Till plains	13.8	W	ER			
MANARY	**Tm**	Lacustrine fine	Fine	Lake plains	12.4	A	HU			
MANAWA	Tm	Till, clayey	Fine	Till plains	11.2	C, A	MI			
MANCELONA	Tm	Outwash	Sandy	Lake plains	11.2	A	MI	HU		
MANIDO	Td	Outwash	Sandy	Lake plains	11.2	D	SU			
MANISTEE	Td	Lacustrine sandy/fine	Sandy over clayey	Lake plains	11.2	D, A	MI	SU	HU	
MANISTIQUE	Lm	Lacustrine sandy	Sandy	Glacial lake benches	5.3	N	MI			
MANLIUS	Tm	Till	Loamy-skeletal	Till upland	13.8	WA	ER	ON		
MARBLEHEAD	Sd	Glaciolacustrine/ls bedrock	Loamy	Glacial lake plains	13.8	WA	ER			
MARKEY	Lm	Organic	Sandy or sandy-skeletal	Lake plains	11.2	D, A	MI	SU	HU	
MARLETTE	Td	Till	Fine-loamy	Till plains	13.8	G, WA	MI	HU		
MARTINSVILLE	Td	Loess	Fine-loamy	Outwash plains	13.8	G	MI			
MARTINTON	P	Lacustrine fine	Fine	Lake plains	13.8	G	MI			
MARTISCO	Lm	Organic over marl	Fine-silty	Lake plains	13	G, I	MI	ON		
MASSENA	Td	Till	Coarse-loamy	Till plains	13	I	ON			
MATTIX	Tm	Outwash	Sandy	Outwash plains	12.4	A	MI			
MAUMEE	W	Lacustrine sandy	Sandy	Lake plains	13.8	G	MI			
MCGINN	Tm	Till	Coarse-loamy	Till plains	12.4	A	HU			
MCIVOR	Tm	Lacustrine sandy	Sandy	Lake plains	11.2	A	HU			
MCMASTER	Td	Outwash	Sandy-skeletal	Outwash plains	11.2	D	SU			
MCMILLAN	Td	Glaciofluvial	Sandy	Outwash plains	11.2	D, A	SU	MI		

(continued)

Series name	Native vegetation[a]	Parent materials	Parent material particle-size class	Landform	Age (ky)	Lake stage[b]	Lake 1	Lake 2	Lake 3	Lake 4
MEEHAN	Tm	Alluvium	Sandy	Glacial lake plains	<11.2	D, A	MI	HU		
MELITA	Td	Outwash	Sandy	Lake plains	11.2	A	HU			
MENOMINEE	Tm	Glaciofluvial	Sandy over loamy	Glacial lake plains	11.2	A	MI	HU		
MENTOR	Td	Glaciolacustrine fine	Fine-silty	Glacial lake plains	13.8	W	ER			
MEQUON	Td	Till	Fine	Till plains	13.8	G	MI			
MERMILL	Ld	Glaciolacustrine fine/till	Fine-loamy	Till plains	13.8	M	ER			
MERWIN	Lc	Organic	Loamy	Lake plains	11.2	D	SU			
MESABA	Tm	Till	Coarse-loamy	Bedrock benches	11.2	D	SU			
METAMORA	Td	Lacustrine/till	Fine-loamy	Till plains	13.8	G, WA	HU	ER	MI	
METEA	Td	Lacustrine sandy/ till	**Sandy over loamy**	Till plains	13.8	M, WA	MI	ER	HU	
MICHIGAMME	Tm	Till	Coarse-loamy	Till plains	11.2	D	SU			
MIDDLEBURY	Td	Alluvium	Coarse-loamy	Floodplains	13	WA, I	ER	ON		
MILFORD	W	Lacustrine fine	Fine	Lake plains	13.8	G	MI	ER		
MILL	Td	Till	Fine-loamy	Moraine	13.8	WA	ER			
MILLGROVE	Ld	Outwash	Fine-loamy	Outwash plains	13.8	M	ER			
MILLSDALE	Pw	Till/ls bedrock	Fine	Lake terraces	13.8	WA	ER			
MILNICHOL	Tm	Lacustrine sandy	Sandy	Lake plains	11.2	A	MI			
MILTON	Td	Till/ls bedrock	Fine	Till plains	13.8	W	ER			
MINER	Td-Ld	Till	Fine	Lake plains, wave-cut	13.8	WA	ER			
MINOA	Td-Ld	Lacustrine fine	Coarse-loamy	Lake plains	12.4	WA, I	ON	ER	HU	
MINOCQUA	Lm	Outwash	Coarse-loamy over sandy or sandy-skeletal	Outwash plains	11.2	A	SU	MI		
MINONG	Tm	Organic		Glacial lake benches	5.3	N	SU			
MISERY	Tm	Till	Coarse-loamy	Lake plains, relict lake basins	11.2	D	SU			
MISHWABIC	Tm	Till	Coarse-loamy	Till plains	5.3	N	SU			
MISKOAKI	Tm	Till, clayey	Very-fine	Till plains	11.2	D	SU			
MITIWANGA	Td	Till	Fine-loamy	Lake plains	13	A, WA	ER	HU		
MOLLINEAUX	Td	Lacustrine sandy	Sandy over loamy	Lake terraces	11.2	A	MI			
MOLTKE	Tm	Glaciolacustrine fine	Coarse-silty	Glacial lake plains	11.2	A	HU	MI		
MONGO	Td	Lacustrine fine	Fine	Lake plains	11.2	A	HU			
MONITOR	Pw	Outwash	Fine-loamy	Outwash plains	13.8	G	MI			
MONTCALM	Tm	Outwash	Coarse-loamy	Outwash plains	13.8	W	MI	HU		
MONTGOMERY	Ld	Lacustrine fine	Fine	Lake plains	13.8	G	MI			
MONTREAL	Tm	Eolian/till	**Coarse-loamy over loamy or sandy**	Till plains	11.2	D	SU			
MOQUAH	Tm	Alluvium	Coarse-loamy	Floodplains	11.2	D, A	SU	MI		
MORGANLAKE	Tm	Outwash/till	Sandy over loamy	Till plains	11.2	D, A	SU	HU	MI	
MORLEY	Td	Loess/till	Fine	Till plains	13.8	G, M	HU	MI		
MOROCCO	Td	Outwash	Sandy	Lake plains	13.8	G	MI			
MOSEL	Sm	Outwash	Fine-loamy	Lake plains	11.2	A	MI			

(continued)

Series name	Native vegetation[a]	Parent materials	Parent material particle-size class	Landform	Age (ky)	Lake stage[b]	Lake 1	Lake 2	Lake 3	Lake 4
MUNDELEIN	P	**Outwash**	Fine-silty	Lake plains	13.8	G	MI			
MUNISING	Tm	Till	Coarse-loamy	Till plains	11.2	D	SU			
MUNUSCONG	Lm	Glaciofluvial/ lacustrine	Coarse-loamy over clayey	Glacial lake plains	11.2	D, A	SU	MI	HU	
MUSSEY	Td	Outwash	Fine-loamy over sandy or sandy-skeletal	Outwash plains	13.8	G, M	MI	HU		
NADEAU	Tm	Outwash	Coarse-loamy	Outwash plains	11.2	A	MI	HU		
NAHMA	Lm	Till	Coarse-loamy	Bedrock benches	11.2	A	MI	SU		
NAMUR	Sm	Till over bedrock	Loamy	Bedrock benches	11.2	A	MI	HU		
NAPPANEE	Td	Till	Fine	Till plains	13.8	W	ER	MI		
NAUMBURG	Tm	Glaciofluvial	Sandy	Glacial lake plains	13	I	ON			
NAVAN	W	Glaciolacustrine fine	Fine-loamy	Glacial lake plains	13.8	G	MI			
NECONISH	Tm	Outwash	Sandy	Lake plains	11.2	A	MI	SU		
NEGWEGON	Tm	Lacustrine fine	Fine	Lake plains	12.4	A	MI	SU	HU	
NENNO	Td	Loess/till	**Silty over fine-loamy**	Till plains	13.8	G	MI			
NESSEN	**Tm**	Lacustrine sandy	Sandy	Lake plains	12.4	A	MI			
NESTER	Td	Till	Fine	Till plains	12.4	G, A	MI	HU		
NET	Tm	Till	Coarse-loamy	Till plains	11.2	D	SU			
NEVENS	Tm	Till	Coarse-loamy	Moraine	11.2	D	SU			
NEWSTEAD	Lm	Till	Coarse-loamy	Till plains	13	I	ON			
NEWTON	W	Lacustrine sandy	Sandy	Lake plains	13.8	G	MI			
NIAGARA	Td	Glaciolacustrine fine	Fine-silty	Glacial lake plains	12.4	WA, I	ON	ER		
NICHOLS	Td	Glaciolacustrine fine	Coarse-silty	Glacial lake plains	13.8	G	MI			
NIPISSING	Tm	Lacustrine fine	Loamy-skeletal	Bedrock benches	5.3	N	SU			
NONESUCH	Td	Till	Loamy-skeletal	Till plains	11.2	D	SU			
NORDHOUSE	Tm	Eolian	Sandy	Dunes	13.8	G	MI			
NORMANNA	Tm	Lacustrine/till	Coarse-loamy	Till plains	11.2	D	SU			
NOSEUM	Td	Outwash	Sandy	Outwash plains	11.2	D, A	SU	MI		
NUNICA	Tm	Lacustrine fine	Fine-silty	Lake plains	11.2	D, A	SU	MI		
NYKANEN	Td	Residuum	Loamy	Bedrock benches	11.2	D	SU			
OAKVILLE	Tm	Eolian	Sandy	Dunes	12.4	G, WA, I	MI	ER	ON	HU
OCQUEOC	Tm	Lacustrine sandy	Sandy over loamy	Lake plains	11.2	D, A	SU	HU		
ODANAH	Tm	Till	Fine	Till plains	11.2	D	SU			
ODESSA	Tm	Lacustrine fine	Fine	Lake plains	13	I	ON	ER		
OGEMAW	Tm	Outwash	Sandy over loamy	Lake plains	11.2	A	MI	HU		
OGONTZ	Td	Glaciolacustrine fine	Fine-silty	Glacial lake plains	13.8	WA	ER			
OLDMAN	Td	Till	Loamy-skeletal	Till plains	11.2	D, A	SU	HU		
OLMSTED	Td	Lacustrine sandy	Fine-loamy	Lake plains	13.8	WA	ER			
OMENA	Td	Till	Coarse-loamy	Till plains	11.2	A	MI	HU		
OMRO	Td	Till	Clayey over loamy	Till plains	11.2	A	MI			
ONAWAY	Td	Till	Fine-loamy	Till plains	11.2	A	MI	HU		
ONEKAMA	Td	Till	Fine	Till plains	13.8	G	MI			
ONOTA	Tm	Till	Coarse-loamy	Glacial lake benches	11.2	D	SU			

(continued)

Series name	Native vegetation[a]	Parent materials	Parent material particle-size class	Landform	Age (ky)	Lake stage[b]	Lake 1	Lake 2	Lake 3	Lake 4
ONTARIO	Td	Till	Fine-loamy	Till plains	13	I	ON			
ONTONAGON	Tm	Glaciolacustrine fine	Very-fine	Glacial lake plains	11.2	D, A	SU	HU		
ORONTO	Tm	Till	Fine-silty	Till plains	11.2	D	SU			
ORPARK	Tm	Till	Fine-loamy	Till plains	13.8	W	ER			
ORRVILLE	Td	Alluvium	Fine-loamy	Floodplains	<13.8	W	ER			
OSHKOSH	Td	Glaciolacustrine fine	Very-fine	Glacial lake plains	13.8	G	MI			
OSHTEMO	Td	Outwash	Coarse-loamy	Beach ridges	13.8	G, WA	MI	ER		
OSSINEKE	Td	Till	Fine-loamy	Till plains	11.2	D, A	MI	HU		
OTEGO	Ld	Alluvium	Coarse-silty	Floodplains	<13.8	WA	ER			
OTISCO	Tm	Outwash	Sandy	Lake plains	11.2	A	MI	HU		
OTISVILLE	Td	Outwash	Sandy-skeletal	Lake terraces	12.4	WA, I	ER	ON		
OTTOKEE	Td	Glaciolacustrine sandy	Sandy	Glacial lake plains	13.8	G, W	MI	HU	ER	
OVID	Td	Till	Fine-loamy	Till plains	13	I	ON			
OZAUKEE	Td	Till	Fine	Moraine	13.8	G	MI			
PAAVOLA	Td	Till	Sandy-skeletal	Till plains	11.2	D	SU			
PAINESVILLE	Td	Glaciolacustrine fine	Coarse-loamy	Glacial lake plains	13.8	WA	ER			
PALMERS	Lm	Till, clayey	Very-fine	Glacial lake plains	11.2	D	SU			
PALMS	Ld	Organic	Loamy	Lake plains	12.4	G, WA, I	MI	HU	ON	ER
PALMYRA	Td	Outwash	Fine-loamy over sandy or sandy-skeletal	Outwash plains	13	I	ON			
PAQUIN	Tm	Glaciofluvial	Sandy	Glacial lake plains	11.2	D, A	SU	MI	HU	
PARKHILL	Ld-Td	Till	Fine-loamy	Till plains	11.2	A	HU			
PELISSIER	Tm	Outwash	Sandy-skeletal	Outwash plains	11.2	D	SU			
PELLA	W	**Glaciolacustrine fine**	Fine-silty	Glacial lake plains	13.8	G, WA	MI	HU		
PENCE	Tm	Alluvium/ outwash	**Coarse-loamy over sandy or sandy-skeletal**	Glacial lake plains	11.2	D, A	SU	HU		
PEOTONE	W	Colluvium/till	Fine	Till plains	13.8	G	MI			
PEQUAYWAN	Tm	Loamy Mantle/ outwash	**Coarse-loamy over sandy or sandy-skeletal**	Outwash plains	11.2	D	SU			
PERRINTON	Td	Till	Fine	Till plains	13.8	G	MI			
PESHEKEE	Tm	Till/gr-gn bedrock	Loamy	Till upland	11.2	D	SU			
PEWAMO	Td	Till	Fine	Lake plains, relict near-shore zone	13.8	G, M	MI	HU		
PHELPS	Tm	Outwash	Fine-loamy over sandy or sandy-skeletal	Outwash plains	12.4	WA, I	ON	ER		
PICKFORD	Lm	Glaciolacustrine fine	Fine	Glacial lake plains	11.2	D, A	SU	MI	HU	
PIERPONT	Td	Till	Fine-silty	Moraine	13.8	W	ER			
PINCONNING	Lm	Lacustrine sandy/ fine	Sandy over clayey	Lake terraces	11.2	A	MI	HU	SU	

(continued)

Series name	Native vegetation[a]	Parent materials	Parent material particle-size class	Landform	Age (ky)	Lake stage[b]	Lake 1	Lake 2	Lake 3	Lake 4
PIPESTONE	Td	Outwash	Sandy	Till plains	13.8	G, W	MI	HU		
PLAINFIELD	Tm	Outwash	Sandy	Outwash plains	12.4	G, A, I	MI	HU	ON	ER
PLATEA	Td	Till	Fine-silty	Till plains	13.8	W	ER			
PLATTERIVER	Tm	Eolian	Sandy	Dunes	13	C	MI			
POMPTON	Tm	Outwash	Coarse-loamy	Outwash plains	13.8	WA	ER			
PORKIES	Tm	Till	Loamy-skeletal	Till plains	11.2	D	SU			
PORTWING	Tm	Lacustrine/till	**Fine over loamy**	Till plains	11.2	D	SU			
POSEN	**Tm**	Till	Loamy-skeletal	Glacial lake benches	11.2	A	HU	MI		
POSEYVILLE	**Tm**	Lacustrine fine	Coarse-loamy	Lake plains	11.2	A	HU			
POTAGANNISSING	Pm	Till	Loamy	Glacial lake benches	11.2	A	SU			
POY	Ld	Glaciolacustrine fine	Clayey over sandy or sandy-skeletal	Glacial lake plains	13.8	G	MI			
POYGAN	Ld	Till	Fine	Till plains	11.2	C, A	MI			
PROPER	Tm	**Eolian**	Sandy	Beach ridges	11.2	A	MI	HU		
PULLUP	Tm	Eolian	Sandy	Beach ridges	11.2	A	MI	HU		
QUETICO	Tm	Till	Loamy	Bedrock benches	10.2	D, Cs	SU	ON		
RAPSON	Tm	Glaciolacustrine sandy/loamy	Sandy over loamy	Glacial lake plains	11.2	A	HU			
RAWSON	Td	Outwash	Fine-loamy	Lake plains	13.8	G, M	ER	MI		
RAYNHAM	Tm	Glaciolacustrine fine	Coarse-silty	Glacial lake plains	12.4	WA, I	ON	ER		
READE	Tm	Till/dolo ss bedrock	Coarse-loamy	Till plains	11.2	D	SU			
RED HOOK	Tm	Glaciofluvial	Coarse-loamy	Outwash plains	13.8	W	ER			
REDRIM	Tm	Beach deposits/ bedrock, ss	Sandy-skeletal	Bedrock benches	5.3	N	SU			
REMSEN	Td	Till	Fine	Till plains	13.8	WA	ER			
REMUS	**Td**	Till	Fine-loamy	Till plains	13	C	MI			
RHINEBECK	Tm	Glaciolacustrine fine	Fine	Glacial lake plains	12.4	WA, I	ON	ER		
RHODY	Lm	Eolian outwash/ss bedrock	Coarse-silty over sandy or sandy-skeletal	Glacial lake benches	5.3	N	SU			
RICHTER	Ld	Glaciofluvial	Coarse-loamy	Glacial drainage channels	11.2	D, A	MI	HU		
RIFLE	Lm	Organic		Lake plains	11.2	D, A, I	SU	MI	ON	HU
RIGGSVILLE	Tm	Lacustrine/till	**Sandy over coarse-loamy**	Lake terraces	11.2	A	HU			
RIMER	Td	Glaciolacustrine/ till	**Loamy over clayey**	Till plains	13.8	G, W	MI	ER		
ROBAGO	Tm	Lacustrine fine	Coarse-loamy	Lake plains	**11.2**	**D**	SU			
ROCKBOTTOM	Tm	Lacustrine fine	Loamy-skeletal	Lake plains	11.2	A	HU			
ROCKCUT	Tm	Outwash/ alluvium	Sandy-skeletal over loamy	Floodplains	5.3	N	HU			
ROCKLAND	Tm	Colluvium/till	Fine-loamy	Till plains	11.2	D	SU			
ROLLINS	Tm	Loamy mantle/ outwash	**Coarse-loamy over sandy or sandy-skeletal**	Outwash plains	11.2	D	SU			
ROSCOMMON	Lc	Lacustrine sandy		Lake plains	11.2	D, A	MI	HU	SU	
ROUSSEAU	Tm	Eolian	Sandy	Lake plains	11.2	D, A	MI	SU	HU	
RUBICON	Tc	Lacustrine sandy	Sandy	Lake plains	11.2	D, A	SU	MI	HU	

(continued)

Series name	Native vegetation[a]	Parent materials	Parent material particle-size class	Landform	Age (ky)	Lake stage[b]	Lake 1	Lake 2	Lake 3	Lake 4
RUDYARD	Tm	Lacustrine fine	Very-fine	Lake plains	11.2	D, A	SU	HU		
RUSE	Lm	Till over ls bedrock	Loamy	Bedrock benches	11.2	D, A, I	MI	HU	SU	ON
SABATTIS	Lm	Till	Coarse-loamy	Till plains	11.2	D	SU			
SAGANING	**Lm**	Outwash	Sandy	Lake plains	5.3	N	HU			
SANBORG	Tm	Till, clayey	Fine	Till plains	11.2	D	SU			
SANDUSKY	Lm	Tufa over lacustrine	Fine-loamy	Lake plains	13.8	WA	ER			
SANILAC	Td	Lacustrine fine	Coarse-silty	Lake plains	13.8	WA	HU			
SARANAC	Ld	Alluvium	Fine	Floodplains	<13.8	C, W	MI	HU		
SATAGO	Td	Till	Fine-silty	Lake plains	5.3	N	SU	HU		
SAUGATUCK	Lm	Glaciofluvial	Sandy	Glacial lake plains	13.8	G	MI	HU		
SAUXHEAD	Td	Glaciofluvial	Sandy-skeletal	Glacial lake benches	5.3	N	SU			
SAYLESVILLE	Td	Lacustrine fine	Fine	Lake plains	13.8	G, WA	MI			
SAYNER	Tm	Outwash	Sandy	Outwash plains	11.2	A	SU			
SCHOHARIE	Td	Lacustrine fine	Fine	Lake plains	13	I	ER	ON		
SCHWEITZER	Tm	Eolian/till	**Coarse-loamy over loamy or sandy**	Moraine	11.2	D	SU			
SCIO	Tm	Lacustrine fine	Coarse-silty	Lake plains	13.8	WA	ER			
SCRIBA	Td	Till	Coarse-loamy	Till plains	13	I	ON			
SEARCH	Tm	Till	Coarse-silty	Moraine	5.3	N	HU	MI		
SEBEWA	Td-Ld	Outwash	Fine-loamy over sandy or sandy-skeletal	Outwash plains	13.8	G	MI			
SEDGWICK	Tm	Alluvium/till, clayey	Coarse-loamy over clayey	Till plains	11.2	D	SU			
SELFRIDGE	Td	Lacustrine sandy	Loamy	Lake plains	13.8	G, WA	MI	ER	HU	
SELKIRK	Tm	Till	Fine	Till plains	13.8	WA	HU	MI		
SEWARD	Td	Glaciolacustrine/till	**Sandy over loamy**	Till plains	13.8	G, W	MI	HU	ER	
SHAG	Tm	Glaciolacustrine fine	Coarse-silty	Glacial lake plains	11.2	D	SU			
SHAKER	Lm	Glaciolacustrine fine	Coarse-loamy over clayey	Glacial lake plains	10.2	Cs	ON			
SHAVENAUGH	**Tm**	Outwash	Sandy	Lake plains	11.2	A	MI			
SHAWANO	Tm	Eolian over outwash	Sandy	Lake plains	11.2	A	MI			
SHEBEON	**Tm**	Till	Fine-loamy	Till plains	11.2	A	HU			
SHELLDRAKE	Tm	**Eolian**	Sandy	Beach ridges	5.3	N	SU			
SHELTER	Tm	Till	Loamy-skeletal	Glacial lake benches	5.3	N	HU	MI		
SHINGLETON	Td	Glaciofluvial	Sandy	Glacial lake benches	5.3	N	SU	MI		
SHINROCK	Td	Glaciolacustrine fine	Fine	Glacial lake plains	13.8	W	HU			
SHIOCTON	Pw	Lacustrine fine	Coarse-silty	Lake plains	13.8	G	MI			
SHUBERTS	Lm	Lacustrine sandy	Sandy	Beach ridges	5.3	N	SU			
SICKLES	Ld	Glaciofluvial	Sandy over clayey	Glacial lake plains	12.4	A	MI			
SIMS	Tm	Till	fine	Till plains	12.4	A	HU	MI		
SISKIWIT	Tm	Till	Sandy	Till plains	11.2	D	SU			
SISSON	Td	**Lacustrine fine**	Fine-loamy	Lake plains	11.2	G, M, A	MI	ER	HU	

(continued)

Series name	Native vegetation[a]	Parent materials	Parent material particle-size class	Landform	Age (ky)	Lake stage[b]	Lake 1	Lake 2	Lake 3	Lake 4
SKANDIA	Lm	Organic over ss bedrock		Bedrock benches	11.2	D	SU			
SKANEE	Td	Till	Coarse-loamy	Till-floored lake plains	11.2	D	SU			
SKEEL	Td	Outwash	Sandy over loamy	Lake plains	12.4	A	HU			
SLADE	Tm	Till	Fine-loamy	Moraine	12.4	A	HU			
SLOAN	Ld	Alluvium	Fine-loamy	Floodplains	13	G, M, I	MI	HU	ER	
SODUS	Td	Till	Coarse-loamy	Till plains	13	I	ON			
SOLONA	Td	Till	Coarse-loamy	Till plains	11.2	A	MI	HU		
SOO	Tm	Lacustrine fine	Fine-silty	Lake plains	11.2	A	HU			
SOUTHWELLS	Td	Till	Sandy	Till plains	11.2	A	MI			
SPARTA	P	Eolian	Sandy	Dunes	13.8	G	MI			
SPEAR	Tm	Glaciolacustrine fine	Coarse-silty	Glacial lake plains	11.2	D	SU			
SPINKS	Td	Eolian or outwash	Sandy	Lake plains	13.8	G, WA	MI	ER	HU	
SPORLEY	Td	Glaciolacustrine fine	Coarse-silty	Glacial lake plains	11.2	D, A	SU			
SPOT	Lm	Glaciofluvial	Sandy	Glacial lake plains	11.2	A	MI	SU	HU	
SPRINGLAKE	Tm	Outwash	Sandy	Outwash plains	11.2	A	MI	HU		
SPRINGPORT	Tm-Lm	Lacustrine fine	Fine	Lake plains	12.4	A	HU			
ST. CLAIR	Td	Till	Fine	Lake plains	13.8	M	ER			
ST. IGNACE	Tm	Till	Loamy	Glacial lake benches	5.3	N	MI	HU		
STAFFORD	Td	Glaciolacustrine sandy	Sandy	Glacial lake plains	13.8	WA	ER	ON		
STANHOPE	Td	Alluvium	Fine-silty	Floodplains	13.8	WA	ER			
STEUBEN	Tm	**Lacustrine/till**	**Coarse-loamy over sandy**	Moraine	11.2	D	SU			
STURGEON	Tm	Alluvium	Coarse-silty over sandy or sandy-skeletal	Floodplains	11.2	D	SU			
STUTTS	Td	Outwash	Sandy	Lake plains	11.2	D, A	SU	MI		
SUGAR	Tm	Lacustrine fine	Coarse-silty over clayey	Lake plains	11.2	A	SU	HU		
SULTZ	Tm	Outwash	Sandy	Lake terraces	11.2	D	SU			
SUMMERVILLE	Tm	Till	Loamy	Glacial lake benches	11.2	D, A	MI	HU	ON	
SUN	Ld	Till	Coarse-loamy	Till plains	13	I	ON			
SUNDELL	Tm	Till	Coarse-loamy	Glacial lake benches	11.2	A	MI			
SUPERIOR	Td	Lacustrine fine/till	Coarse-loamy over clayey	Lake plains	11.2	D	SU	MI		
SWANTON	Tm	Lacustrine fine	Coarse-loamy over clayey	Lake plains	10.2	Cs	ON			
SWORMVILLE	Td	Glaciolacustrine fine	Fine-silty over sandy or sandy-skeletal	Glacial lake plains	13	I	ON			
SYMCO	Td	Till	Fine-loamy	Till plains	13.8	G	MI			
SYMERTON	P	Glaciolacustrine fine	Fine-loamy	Glacial lake plains	13.8	G	MI			
TACODA	Tm	Outwash over lacustrine	**Sandy over clayey**	Lake plains	11.2	A	HU			
TACOOSH	Lm	Organic	Loamy	Lake plains	11.2	D	SU	MI		
TAPPAN	**Tm**	Till	Fine-loamy	Till plains	5.3	N	HU			

(continued)

Series name	Native vegetation[a]	Parent materials	Parent material particle-size class	Landform	Age (ky)	Lake stage[b]	Lake 1	Lake 2	Lake 3	Lake 4
TATCHES	Tm	Lacustrine sandy	Sandy	Lake plains	11.2	A	MI			
TAWAS	Lm	Organic	Sandy or sandy-skeletal	Lake plains	11.2	D, A	SU	HU	MI	
TEDROW	Td	Glaciolacustrine sandy	Sandy	Glacial lake plains	13.8	G, WA	MI	ER		
TEEL	Tm	Alluvium	Coarse-silty	Alluvial fans	13	I	ON	ER		
TEKENINK	Td	Till	Coarse-loamy	Till plains	13.8	G	MI			
THETFORD	Td	Outwash	Sandy	Lake plains	13.8	G, W	MI	ER	HU	
THOMAS	**Ld-Td**	Organic over lacustrine	Fine-loamy	Lake plains	11.2	A	MI	HU		
THOMPSONVILLE	Tm	Lacustrine sandy	Sandy	Lake plains	11.2	A	MI			
TOBICO	W	Outwash	Sandy	Glacial lake benches	5.3	N	HU			
TOIVOLA	Tm	Glaciofluvial	Sandy	Till plains	11.2	D	SU			
TOKIAHOK	Tm	Till	Sandy	Till-floored lake plains	11.2	D	SU			
TOLEDO	Ld	Glaciolacustrine fine	Fine	Glacial lake plains	13.8	WA	HU	ER		
TONKEY	Ld	Glaciofluvial	Coarse-loamy	Glacial drainage channels	11.2	D, A	HU	MI	SU	
TOWERVILLE	Tm	Till/sed bedrock	Fine-loamy	Till upland	13.8	W	ER			
TOWES	Tm	Eolian/outwash/ss bedrock	Coarse-silty over sandy or sandy-skeletal	Glacial drainage channels	5.3	N	SU			
TRIMOUNTAIN	Td	Eolian/till	**Coarse-loamy over loamy or sandy**	Glacial lake benches	11.2	D	SU			
TROUT BAY	Tm	Organic over ss bedrock		Bedrock benches	5.3	N	SU			
TULA	Tm	Till	Coarse-loamy	Till plains	11.2	D	SU			
TUSCOLA	Td	Lacustrine fine	Fine-loamy	Lake plains	13.8	G, WA	ER	MI		
TUSTIN	Tm	Eolian/till	**Sandy over clayey**	Till plains	13.8	G	MI			
TWIG	Ld	Loamy mantle/till	Coarse-loamy	Till plains	11.2	D	SU			
TWINING	Td	Till	Fine-loamy	Till plains	11.2	A	HU			
TYNER	Td	Lacustrine sandy	Sandy	Lake plains	13.8	G, W	ER	MI		
TYRE	Ld	Till	Sandy	Till plains	11.2	A	HU			
UBLY	Td	Till	Coarse-loamy	Till plains	11.2	D, A	MI	SU	HU	
UNADILLA	Td	Alluvium	Coarse-silty	Lake plains	13.8	WA	ER			
VALOIS	Td	Till	Coarse-loamy	Moraine	13.8	W	ER			
VARYSBURG	Tm	Outwash/ lacustrine	Loamy-skeletal over clayey	Lake plains	12.4	WA	ER			
VELVET	Tm	Till	Sandy-skeletal	Moraine	11.2	A	HU			
VERGENNES	Tm	Glaciolacustrine fine	Very-fine	Glacial lake plains	10.2	Cs	ON			
VILAS	Tm	Outwash	Sandy	Outwash plains	11.2	D	SU			
VOELKER	Tm	Glaciolacustrine	Sandy	Till-floored lake plains	11.2	D	SU			
VOLUSIA	Tm	Till	Fine-loamy	Till plains	13.8	WA	ER			
WABUN	Ld	Outwash over lacustrine	**Sandy over clayey**	Lake plains	12.4	A	HU			
WAHBEGON	Ld	Eolian/ glaciolacustrine/ till	Fine-loamy	Till plains	11.2	D	SU			
WAINOLA	Td	Glaciofluvial	Sandy	Glacial lake plains	11.2	A	MI	HU	SU	
WAISKA	Td	Outwash	Sandy-skeletal	Glacial lake benches	11.2	D	SU			

(continued)

Series name	Native vegetation[a]	Parent materials	Parent material particle-size class	Landform	Age (ky)	Lake stage[b]	Lake 1	Lake 2	Lake 3	Lake 4
WAKELEY	Lm	Lacustrine sandy/ fine	Sandy over clayey	Lake plains	11.2	D, A	SU	HU	MI	
WAKEVILLE	Tm	Alluvium	Coarse-silty	Floodplains	13.8	W	ER			
WALLACE	Tm	Lacustrine sandy	Sandy	Lake plains	11.2	A	MI	HU	SU	
WALLINGTON	Tm	Lacustrine fine	Coarse-silty	Lake plains	13	I	ON			
WALLKILL	Ld	Alluvium	Fine-loamy	Floodplains	<13.8	G, M	MI	HU	ON	
WARNERS	Ld	Alluvium	Fine-silty	Floodplains	<13.8	G	MI	HU	ER	
WASEPI	Td	Glaciofluvial	Coarse-loamy	Glacial lake plains	13.8	G, W	MI	HU		
WATSEKA	P	Outwash	Sandy	Outwash plains	13.8	G	MI			
WATTON	Tm	Till	Fine-loamy	Till plains	11.2	D	SU			
WAUCONDA	Td	**Lacustrine fine**	Fine-silty	Lake plains	13.8	G	MI			
WAUSEON	Ld	Glaciolacustrine fine/till	Coarse-loamy over clayey	Till plains	13.8	G, WA	MI	ER	HU	
WAYLAND	Ld	Alluvium	Fine-silty	Floodplains	13	W, I	ER	ON		
WEGA	Td	Glaciolacustrine fine	Coarse-silty	Glacial lake plains	12.4	A	HU	SU		
WESTBURY	Td	Till	Coarse-loamy	Till plains	13	I	HU	ON		
WHEATLEY	Lc	Glaciofluvial	Sandy	Glacial lake plains	11.2	A	HU	MI		
WHITTEMORE	Tm	Glaciofluvial	Sandy	Glacial lake plains	11.2	A	HU			
WICK	Td	Alluvium	Fine-silty	Floodplains (till)	13.8	WA	ER			
WILLETTE	Ld	Organic	Clayey	Lake plains	13.8	G, M	MI	ER	ON	HU
WILLIAMSON	Td	Lacustrine fine	Coarse-silty	Lake plains	13	I	ON	ER		
WILPOINT	Tm	Lacustrine fine	Very-fine	Lake plains	10.2	Cs	ON			
WINDSOR	Tm	Outwash	Sandy	Outwash plains	13	I	ON			
WINTERFIELD	Ld	Alluvium	Sandy	Floodplains	11.2	D	SU	HU	MI	
WISNER	**Tm**	Till	Fine-loamy	Till plains	11.2	A	HU			
WITBECK	Lm	Till	Coarse-loamy	Till plains	11.2	D	SU			
WIXOM	Ld	Outwash over lacustrine	Sandy over loamy	Lake plains	13.8	G, W	HU	MI		
WORMET	Tm	Eolian/outwash	**Coarse loamy over sandy**	Outwash plains	11.2	D	MI			
WURTSMITH	Tm	Lacustrine sandy	Sandy	Beach ridges	11.2	D, A	MI	SU		
YALMER	Td	Outwash/till	**Sandy over loamy**	Moraine	11.2	D	SU			
YELLOWDOG	Td	Residuum	Sandy-skeletal	Bedrock benches	5.3	N	SU			
ZANDI	Tm	Lacustrine fine	Coarse-loamy	Lake plains	11.2	D	SU			
ZEBA	Tm	Till over ss bedrock	Coarse-loamy	Bedrock benches	11.2	D	SU			
ZELA	Lm	Lacustrine sandy	Sandy-skeletal	Beach ridges	5.3	N	MI	HU		
ZIEGENFUSS	Td	Till	Fine	Till plains	13.8	WA	ER	HU	MI	
ZIMMERMAN	Tm	Outwash	Sandy	Lake terraces	11.2	A	HU			
ZURICH	Td	Outwash	Fine-silty	Outwash plains	13.8	G	MI			

[a]Vegetation types: *L* lowland; *P* prairie; *S* savanna; *T* temperate; *W* nonforested wetland; *c* coniferous forest; *d* broadleaved forest; *m* mixed coniferous and broadleaved forest.
[b]Lake stages: *A* Algonquin; *C* Calumet; *Cs* Champlain Sea; *D* Duluth; *G* Glenwood; *I* Iroquois; *M* Maumee; *N* Nipissing; *W* Whittlesey; *WA* Warren

Appendix C
Soil-forming factors and lake occurrence for soil series in the Canadian portion of the Great Lakes Coastal Zone.

Series name	Native vegetation[a]	Drainage class	Parent materials	Parent materials particle-size class	Landform	Age (ky)	Lake stage[b]	Lake 1	Lake 2	Lake 3	Lake 4
Allendale	**L**	Poorly	Fluvial	Moderately coarse	**Alluvial plain**						
Alliston		Imperfectly	Fluvial	Very coarse	**Outwash plain**			HU	ON		
Ameliasburg	Td	Well	Till–calcareous	Fragmental	Moraine		I	ON			
Atherly	**L**	Poorly	Lacustrine, beach	Moderately fine	Lake plain			HU			
Athol	Td	Imperfectly	Till–calcareous	Fragmental	Till plain		I	ON			
Azilda	**L**	Poorly	Lacustrine–calcareous	Medium	Lake plain			HU			
Baldwin	**Td**	Well	Lacustrine, beach	Medium	Lake plain			HU			
Bamford		Imperfectly	Fluvial	Coarse skeletal	**Outwash plain**			ON			
Bass		Imperfectly	lacustrine–calcareous	Fine	Lake plain						
Battersea		Imperfectly	lacustrine	Moderately fine	Lake plain						
Belmeade		Very poorly	Marine	Fine	Coastal plain			STL	HU		
Bennington		Well	Glaciolacustrine–calcareous	Fine	Glacial lake plain			ER			
Berrien	Lm	Imperfectly	glaciolacustrine/till–calcareous	Moderately fine over fine	Glacial lake plain			ER	ON	HU	
Beverly	**Ld**	Imperfectly	Glaciolacustrine–calcareous	Moderately fine	Glacial lake plain			ER	ON		
Binbrook		Imperfectly	Till–calcareous	Fine	Moraine			ER	ON		
Blackwell		Poorly	Glaciolacustrine–calcareous	Moderately fine	Glacial lake plain			HU			
Bondhead	Tm	Well	Till–calcareous	Medium	Till plain		I	ON			
Bookton	Td	Well	Glaciolacustrine/till–calcareous	Moderately fine over medium	Glacial lake plain		I	ON	ER	HU	
Brady	Lm	Imperfectly	Glaciolacustrine–calcareous	Very coarse	Glacial lake plain			HU	ER	ON	
Brant		Well	Lacustrine–calcareous	Medium	Lake plain			ER	ON	HU	
Brantford	**Td**	Well	Lacustrine	Moderately fine	Lake plain			ER	ON		
Breypen	Td	Poorly	Till–calcareous	Fragmental	Moraine			HU			
Bridgman	Tc	Well	Lacustrine–calcareous	Very coarse	Lake plain, dunes			HU	ON		
Brighton	Tm	Well	Outwash	Very coarse	Outwash plain			HU	ON		

(continued)

J. G. Bockheim, *Soils of the Laurentian Great Lakes, USA and Canada*,
https://doi.org/10.1007/978-3-030-52425-8

Series name	Native vegetation[a]	Drainage class	Parent materials	Parent materials particle-size class	Landform	Age (ky)	Lake stage[b]	Lake 1	Lake 2	Lake 3	Lake 4
Brisbane	Td	Imperfectly	Outwash–calcareous	Very coarse	Outwash plain			HU	ER		
Brockport		Well	Till–calcareous	Fragmental	Moraine			ON			
Brooke		Poorly	Till–calcareous	Moderately coarse	Moraine			HU	ON		
Brookston	Lm	Poorly	Till	Moderately fine	Moraine			HU	ER	ON	
Bucke		Well	Glaciofluvial/ glaciolacustrine	Very coarse/fine	Outwash plain						
Burford	Lm	Rapidly	Outwash	Very coarse	Outwash plain			HU	ER		
Buzwah	Tm	Well	Till	Moderately fine	Moraine			HU			
Caistor	**L**	Imperfectly	Till	Fine	Moraine			ER	HU		
Campbell		Well	Lacustrine	Fine	Lake plain						
Cane		Poorly	Lacustrine–calcareous	Medium	Lake plain						
Carp		Imperfectly	Marine	Moderately fine	Coastal plain			STL	HU		
Casey		Mod. Well	Glaciolacustrine/till–calcareous	Moderately Fine/fine	Moraine						
Cashel		Mod. Well	Glaciolacustrine/till–calcareous	Fine/fine	Moraine			ER	ON		
Chartrand	**Td**	Imperfectly	Lacustrine	Moderately fine	Lake plain			HU	SU		
Chesley		Poorly	Lacustrine–calcareous	Fine	Lake plain			HU			
Chinguacousy	Ld-Td	Imperfectly	Till	Moderately fine	Moraine		I	ON	ER		
Churchville		Poorly	Lacustrine–calcareous	Very coarse	Lake plain			ER	HU		
Clyde		Poorly	Lacustrine–calcareous	Fine	Lake plain			HU			
Codrington		Imperfectly	Lacustrine	Medium	Lake plain			ON			
Colborne	Td	Well	Fluvial (spillway)	Coarse skeletal	Alluvial plain		I	ON			
Colwood		Poorly	Glaciolacustrine	Medium	Glacial lake plain			ER	ON	HU	
Cooksville		Imperfectly	Till	Fragmental	Moraine			ON			
Craigleath		Imperfectly	Till	Fine	Moraine			HU			
Darlington		Imperfectly	Till–calcareous	Fragmental	Moraine			ON			
Delamere		Well	Lacustrine–calcareous	Moderately fine	Lake plain						
Deloro		Well	Till	Medium skeletal	Moraine			ON			
Dokise	**Tc**	Well	Outwash	Coarse	Outwash plain			HU			
Donnybrook		Well	Outwash–calcareous	Coarse skeletal	Outwash plain			HU			
Dorking		Very poorly	Till–calcareous	Moderately fine	Moraine			HU	ER		
Dumfries		Well	Till–calcareous	Moderately coarse	Moraine			HU	ON		
Dummer		Well	Till	Medium skeletal	Moraine			ON			
Dundonald	Td	Well	Outwash/till	very coarse over very coarse	Moraine		I	ON			
Dunedin	Td	Well	Till	Fine	Moraine			HU			
Earlton		Imperfectly	Lacustrine	Medium	Lake plain						
Eastport	Tm	Well	Eolian	Very coarse	Dunes			HU	ON		

(continued)

Series name	Native vegetation[a]	Drainage class	Parent materials	Parent materials particle-size class	Landform	Age (ky)	Lake stage[b]	Lake 1	Lake 2	Lake 3	Lake 4
Edenvale	**Td**	Imperfectly	Fluvial/till–calcareous	Coarse over moderately coarse	Moraine		I	ON			
Ekfrid		Imperfectly	Glaciolacustrine–calcareous	Fine	Glacial lake plain			ER			
Elderslie	**Lm**	Imperfectly	Lacustrine–calcareous	Fine	Lake plain			HU			
Elmbrook		Imperfectly	Lacustrine	Fine	Lake plain			ON			
Emily		Imperfectly	Till–calcareous	Fragmental	Moraine			HU	ON		
Englehart		Poorly	Fluvial/lacustrine–calcareous	Moderately coarse/fine	Lake plain						
Evanturel		Well	Lacustrine–calcareous	Medium	Lake plain						
Farmington	Sc	Well	Till–calcareous	Fragmental	Moraine			HU	ON		
Ferndale	Ld	Poorly	Outwash–calcareous	Fine	Outwash plain			HU			
Flamborough		Poorly	Glaciolacustrine	Moderately coarse	Glacial lake plain			ON			
Fox	Td	Rapidly	Glaciolacustrine–calcareous	Very coarse	Glacial lake plain			HU	ER	ON	
Foxboro		Poorly	Outwash–calcareous	Moderately coarse	Outwash plain			ON			
Franktown		Imperfectly	Till–calcareous	Fragmental	Moraine			HU	ON		
Fynxal	**Td**	Well	Lacustrine–calcareous	Moderately fine	Lake plain			HU			
Gananoque		Well	Lacustrine	Fine	Lake plain			ON			
Gerow	Ld	Poorly	Till–calcareous	Moderately fine	Moraine		I	ON			
Gilford	Lm	Poorly	Outwash–plaincalcareous	Very coarse	Outwash plain		I	HU	ER	ON	
Gobles	**Td**	Imperfectly	Till–calcareous	Fine	Till plain			ER			
Gordon		Imperfectly	Lacustrine	Moderately fine	Lake plain						
Granby	Lm	Poorly	Lacustrine–calcareous	Very coarse	Lake plain			HU	ER	ON	
Grenville		Well	Till	Medium	Moraine			STL			
Grimsby	**Tc**	Well	Lacustrine	Coarse	Lake plain		I	ON			
Guelph		Well	Till–calcareous	Medium	Moraine			ON	HU		
Guerin	Lm	Imperfectly	Till	Medium	Moraine		I	ON			
Gwillimbury	**Td**	Imperfectly	Outwash–calcareous	Coarse skeletal	Outwash plain			HU			
Haldimand	**Lm**	Imperfectly	Glaciolacustrine–calcareous	Fine	Glacial lake plain			ON	ER		
Harkaway	Td	Well	Till–calcareous	Medium	Moraine			HU			
Harney		Poorly	Till	Medium skeletal	Moraine			ON			
Harriston		Well	Till–calcareous	Medium	Moraine			HU			
Harrow	**Td**	Well	Outwash–calcareous	Very coarse	Outwash plain			ER			
Highgate		Imperfectly	Lacustrine–calcareous	Moderately coarse	Lake plain			ER			
Hillier	Td	Well	Till–calcareous	Fragmental	Moraine		I	ON			
Howland		Imperfectly	Till–calcareous	Moderately fine	Moraine			HU			
Huron		Well	Till–calcareous	Moderately fine	Moraine			HU	ER		
Jarvis River	Td	Mod. Well	Lacustrine–calcareous	Moderately fine	Lake plain			SU			
Jeddo		Poorly	Till–calcareous	Moderately fine	Moraine			ER	ON		

(continued)

Series name	Native vegetation[a]	Drainage class	Parent materials	Parent materials particle-size class	Landform	Age (ky)	Lake stage[b]	Lake 1	Lake 2	Lake 3	Lake 4
Kagawong		Well	Till–calcareous	Medium	Moraine						
Kelvin	**Ld**	Poorly	Till–calcareous	Moderately fine	Moraine			ER			
Kemble	Ld	Imperfectly	Till–calcareous	Moderately fine	Moraine			HU			
Kenabeek	Lc	Poorly	Outwash	Very coarse	Outwash plain			HU	SU		
Killean		Imperfectly	Till–calcareous	Moderately coarse	Moraine			HU	ON		
Kintyre		Rapidly	Lacustrine–calcareous	Coarse	Lake plain			ER			
Lambton		Imperfectly	Glaciolacustrine/till	Moderately fine/fine	Moraine			HU			
Landsdowne	**Td**	Imperfectly	Lacustrine	Fine	Lake plain		I	ON			
Leech		Imperfectly	Till–calcareous	Moderately fine	Moraine						
Lily	Lm	Poorly	Till–calcareous	Coarse skeletal	Moraine			HU	ON		
Lincoln	**L**	Poorly	Glaciolacustrine–calcareous	Very fine	Glacial lake plain			ON	ER		
Lindsay	**L**	Poorly	Glaciofluvial/till	Medium over fine	Moraine		I	ON			
Listowel	Td	Imperfectly	Till–calcareous	Medium	Moraine			HU			
Little Current		Well	Residuum	Medium	Bedrock						
Lockport		Well	Till–calcareous	Fragmental	Moraine			HU	ON		
Lovering		Imperfectly	Lacustrine–calcareous	Moderately fine	Lake plain			HU			
Lowbanks	**Ld**	Imperfectly	Glaciolacustrine–calcareous	Moderately fine	Glacial lake plain			ER			
Lyons		Poorly	Till–calcareous	Medium	Moraine			ON	HU	STL	
Mallard	Tc	Imperfectly	Outwash	Coarse	Outwash plain			HU	SU		
Malton		Poorly	Glaciolacustrine/till–calcareous	Moderately fine/fine	Moraine			ER	ON		
Maplewood		Poorly	Glaciolacustrine/glaciolacustrine–calcareous	Medium/fine	Glacial lake plain			ER			
Matilda		Imperfectly	Till–calcareous	Medium	Moraine			STL			
Matson		Imperfectly	Outwash	Medium	Outwash plain			ON			
Medonte		Well	lacustrine–calcareous	Moderately fine	Lake plain			HU			
Middlemarch		Imperfectly	Lacustro-till–calcareous	Very coarse	**Glacial lake plain**			ER			
Mill		Poorly	Fluvial/till–calcareous	Very coarse/medium	Moraine			ON			
Milliken		Imperfectly	Till	Medium	Moraine			ON			
Minesing		Poorly	Lacustrine–calcareous	Moderately fine	Lake plain			HU			
Mississauga		Poorly	Till	Fragmental	Moraine			ON			
Monteagle	Tm	Rapidly	Till	Moderately coarse	Moraine			HU	SU	STL	
Morley		Poorly	Till	Fragmental	Moraine			HU	ON		
Muirkirk		Poorly	Outwash–calcareous	Coarse skeletal	Outwash plain			ER			
Muriel	**Ld**	Mod. Well	Till–calcareous	Moderately fine	Moraine			ER			
Musselle	**Lc**	Poorly	Outwash	Moderately coarse	Outwash plain			HU			
Napanee	**L**	Poorly	Lacustrine	Fine	Lake plain		I	ON			

(continued)

Series name	Native vegetation[a]	Drainage class	Parent materials	Parent materials particle-size class	Landform	Age (ky)	Lake stage[b]	Lake 1	Lake 2	Lake 3	Lake 4
Newcastle	Td	Well	Outwash	Medium	Outwash plain		I	ON			
Niagara	**L**	Imperfectly	Glaciolacustrine–calcareous	Very fine	Glacial lake plain			ER	ON		
Noelville	**Td**	Imperfectly	Lacustrine	Medium	Lake plain			HU			
Nolalu	**Tm**	Well	Till	Moderately coarse	Moraine			SU			
Norham		Well	Lacustrine	Medium	Lake plain			ON			
Normandale	**Td**	Imperfectly	Glaciolacustrine–calcareous	Coarse	Glacial lake plain			ER	HU		
North Gower		Poorly	marine	Fine	Coastal plain			STL	HU		
Oneida		Well	Till–calcareous	Moderately fine	Moraine			ER	ON		
Ontario	**Td**	Mod. Well	Glaciolacustrine–calcareous	Very fine	Glacial lake plain			ER	ON		
Osprey	Td	Well	till–calcareous	Medium	Moraine			HU			
Otonabee	Td	Rapidly	Till–calcareous	Moderately coarse	Moraine		I	ON	HU		
Otterskin		Mod. Well	Fluvial/lacustrine–calcareous	Fine/coarse	Lake plain						
Ouellette	**L**	Poorly	Lacustrine	Moderately fine	Lake plain			HU			
Parkhill	Lm	Poorly	Till–calcareous	Medium	Moraine			HU	ON		
Peel		Imperfectly	Glaciolacustrine/till–calcareous	Fine/moderately fine	Moraine			ER	ON		
Perch		Poorly	Lacustrine–calcareous	Fine	Lake plain						
Percy		Well	Outwash	Coarse	Outwash plain			ON			
Perth	Lm	Imperfectly	Till–calcareous	Moderately fine	Moraine			HU	ER		
Petherwick		Poorly	Outwash	Medium	Outwash plain			ON			
Phelans	**Tm**	Well	Outwash	Very coarse	Outwash plain			HU			
Phipps		Poorly	Till–calcareous	Moderately fine	Moraine						
Pike		Well	Lacustrine–calcareous	Fine	Lake plain						
Plainfield	Tm	Rapidly	Eolian–calcareous	Coarse	Dunes			HU	ER		
Pontypool		Well	Fluvial	Very coarse	Alluvial plain			ON			
Port Colborne		Very poorly	Organic	Fabric	Depression						
Rockcroft		Imperfectly	Till–calcareous	Medium skeletal	Moraine			ON			
Rubicon		Imperfectly	Marine–calcareous	Very coarse	Coastal plain			HU	STL		
Sargent	Td	Well	Till–calcareous	Very coarse	Moraine			HU			
Saugeen		Well	Lacustrine–calcareous	Fine	Lake plain			HU			
Schomberg		Well	Glaciolacustrine–calcareous	Moderately fine	Glacial lake plain			ON	HU		
Seeley's Bay		Imperfectly	Lacustrine	Moderately fine	Lake plain						
Shedden		Well	Outwash–calcareous	Very coarse	Outwash plain			ER			
Sidney		Poorly	Lacustrine–calcareous	Fine	Lake plain			ON			

(continued)

Series name	Native vegetation[a]	Drainage class	Parent materials	Parent materials particle-size class	Landform	Age (ky)	Lake stage[b]	Lake 1	Lake 2	Lake 3	Lake 4
Silver Hill		Poorly	Glaciolacustrine/ glaciolacustrine–calcareous	Medium/ medium	Glacial lake plain			ER			
Simcoe		Poorly	Lacustrine	Moderately fine	Lake plain			ON	HU		
Smithfield	Ld	Imperfectly	Glaciolacustrine–calcareous	Moderately fine	Glacial lake plain		I	ON	HU		
Smithville	**Td**	Mod. Well	Glaciolacustrine–calcareous	Very fine	Glacial lake plain			ER			
Solmesville	Ld	Imperfectly	Lacustrine/till	Medium over moderately fine	Moraine		I	ON			
South Bay		Mod. Well	Lacustrine	Fine	Lake plain			ON			
Springvale		Well	Lacustrine–calcareous	Coarse skeletal	Lake plain			ON			
St. Samuel		Poorly	Marine	Coarse	Coastal plain			HU	STL		
St. Williams	**L**	Poorly	Glaciolacustrine–calcareous	Moderately coarse	Glacial lake plain			ER	HU		
Strathburn		Poorly	Glaciolacustrine–calcareous	Very fine	Glacial lake plain			ER			
Sullivan	Td	Well	Outwash	Very coarse	Outwash plain			HU	ER	ON	
Tavistock	**Ld**	Imperfectly	Glaciolacustrine–calcareous	Moderately fine	Glacial lake plain		I	ER			
Tecumseth	Lm	Imperfectly	glaciolacustrine–calcareous	Very coarse	Glacial lake plain		I	ON	HU		
Tennyson		Well	Till	Moderately coarse	Moraine			STL			
Thwaties		Well	Lacustrine/ lacustrine–calcareous	Medium/fine	Lake plain						
Tioga	Tm	Well	Outwash	Very coarse	Outwash plain			HU	ON		
Toledo	Td	Poorly	Glaciolacustrine–calcareous	fine	Glacial lake plain			ER	ON		
Trafalger		Imperfectly	Till–calcareous	Fragmental	Moraine			HU	ON		
Trent		Imperfectly	Fluvial	Moderately coarse	Alluvial plain			ON			
Tuscola		Imperfectly	Glaciolacustrine–calcareous	Medium	Glacial lake plain			ER	ON	HU	
Tweed		Well	Till	Moderately coarse	Moraine			HU	STL		
Uplands		Rapidly	Marine	Very coarse	Coastal plain			HU	STL		
Vasey	Td	Well	Till	Moderately coarse	Moraine			HU			
Vincent	Td	Well	Till–calcareous	Fine	Moraine			HU			
Vineland	**L**	Imperfectly	Glaciolacustrine–calcareous	Medium	Glacial lake plain		I	ON			
Vittoria		Imperfectly	Glaciolacustrine// glaciolacustrine–calcareous	Medium/coarse	Glacial lake plain			ER			
Walsher		Well	Fluvial–calcareous	Medium	Alluvial plain			ER			
Walsingham		Imperfectly	Eolian–calcareous	Coarse	Dunes			HU	ER		
Waterin		Poorly	Eolian–calcareous	Coarse	Dunes			HU	ER		
Waterloo	Td	Well	Outwash	Very coarse	Outwash plain			HU			

(continued)

Series name	Native vegetation[a]	Drainage class	Parent materials	Parent materials particle-size class	Landform	Age (ky)	Lake stage[b]	Lake 1	Lake 2	Lake 3	Lake 4
Wattford	Td	Well	Glaciolacustrine–calcareous	Coarse	Glacial lake plain			ER	HU		
Wauseon	Lm	Poorly	Glaciolacustrine–calcareous	Fine	Glacial lake plain			ER	ON	HU	
Welland	L	Poorly	Glaciolacustrine–calcareous	Very fine	Glacial lake plain			ER	ON		
Wemyss		Imperfectly	Till	Coarse	Moraine			HU	STL		
Wendigo	Tm	Rapidly	Outwash	Very coarse	Outwash plain			HU	SU		
Whitby		Imperfectly	Till–calcareous	Moderately fine	Moraine			ON			
Wiarton	Ld	Imperfectly	Till–calcareous	Medium	Moraine			HU			
Wilsonville		Well	outwash–calcareous	Coarse	Outwash plain			ON			
Woburn		Well	Glaciolacustrine	Medium	Glacial lake plain			ON			
Wolf	**Td**	Well	Lacustrine	Fine	Lake plain			HU			
Wolsey	Lm	Poorly	Lacustrine–calcareous	Fine	Lake plain			HU			
Wyevale	**Td**	Well	Outwash–calcareous	very coarse	Outwash plain			HU			

[a]Vegetation types: *L* lowland; *P* prairie; *S* savanna; *T* temperate; *W* non-forested wetland; *c* coniferous forest; *d* broadleaved forest; *m* mixed coniferous and broadleaved forest.

[b]Lake stages: *A* Algonquin; *C* Calumet; *Cs* Champlain Sea; *D* Duluth; *G* Glenwood; *I* Iroquois; *M* Maumee; *N* Nipissing; *W* Whittlesey; *WA* Warren. *Bold* estimate from drainage class and parent materials 62% calcareous

Appendix D
Benchmark, endemic, rare, and endangered soil series of the Great Lakes Coastal Zone.

Series name	Endemic	Total area (km^2)	Extent class	Benchmark soil	Rare	Endangered
ABBAYE	N	86	1	N	Y	N
ABSCOTA	N	57	1	N	Y	N
ADAMS	N	2521	1	Y	N	N
ADRIAN	N	1740	2	N	N	N
AHMEEK	N	521	2	Y	N	N
ALCONA	N	261	2	N	N	N
ALDENLAKE	N	299	1	N	N	N
ALGANSEE	N	434	1	N	N	N
ALGONQUIN	Y	190	1	N	N	N
ALLENDALE	N	690	2	N	N	N
ALLIS	Y	117	1	N	N	N
ALLOUEZ	Y	12	1	N	Y	Y
ALPENA	N	208	4	N	N	N
ALSTAD	N	273	1	N	N	N
ALTMAR	N	12	1	N	Y	N
ALTON	N	405	1	N	N	N
AMADON	N	88	2	N	N	N
AMASA	N	172	1	N	N	N
AMBOY	N	30	1	N	Y	N
AMENIA	Y	333	1	N	N	N
AMNICON	N	594	2	N	N	N
ANGELICA	Y	360	3	N	N	N
ANGOLA	N	188	1	Y	N	N
ANNALAKE	N	294	2	N	N	N
ANNANIAS	Y	33	1	N	Y	Y
APPLETON	N	632	2	N	N	N
APTAKISIC	N	7	1	N	Y	N
ARCADIAN	Y	133	2	N	N	N
ARKONA	Y	90	1	N	Y	Y
ARKPORT	N	389	1	N	N	N
ARNHEIM	N	291	2	N	N	N
ASHKUM	N	2272	2	Y	N	N
ASHWABAY	N	94	2	N	Y	N
ASSININS	N	65	2	N	Y	N
AU GRES	N	2009	4	N	N	N
AU TRAIN	Y	6	1	N	Y	Y

(continued)

J. G. Bockheim, *Soils of the Laurentian Great Lakes, USA and Canada*,
https://doi.org/10.1007/978-3-030-52425-8

Series name	Endemic	Total area (km^2)	Extent class	Benchmark soil	Rare	Endangered
AUBARQUE	Y	72	2	N	Y	Y
AUGER	Y	27	1	N	Y	Y
AUGUSTANA	Y	100	2	N	Y	Y
AVOCA	N	341	2	N	N	N
AZTALAN	N	94	1	N	Y	N
BACH	Y	182	2	N	N	N
BADAXE	Y	77	1	N	Y	Y
BADRIVER	N	327	2	N	N	N
BAMFIELD	N	69	1	N	Y	N
BANAT	Y	26	1	N	Y	Y
BARCELONA	N	41	1	N	Y	N
BARRE	N	12	1	N	Y	N
BARRINGTON	N	56	1	N	Y	N
BARTO	N	115	3	N	N	N
BATTLEFIELD	N	29	1	N	Y	N
BATTYDOE	N	73	1	N	Y	N
BEAVERTAIL	Y	45	2	N	Y	Y
BEECHER	N	423	1	N	N	N
BELDING	N	222	2	N	N	N
BELLEVILLE	Y	539	1	N	N	N
BENNINGTON	N	2154	1	N	N	N
BENONA	Y	409	2	N	N	N
BENSON	Y	261	1	Y	N	N
BENZONIA	Y	267	1	N	N	N
BERGLAND	Y	160	2	N	N	N
BESEMAN	N	768	1	N	N	N
BETE GRISE	Y	5	1	N	Y	Y
BETSY BAY	N	3	1	N	Y	N
BIG IRON	N	255	3	Y	N	N
BISCUIT	Y	32	1	N	Y	Y
BIXLER	N	72	1	N	Y	N
BLAKESLEE	N	8	1	N	Y	N
BLASDELL	N	171	1	N	N	N
BLOUNT	N	10,735	4	Y	N	N
BLUE LAKE	N	946	2	N	N	N
BOGART	N	227	1	N	N	N
BOHEMIAN	N	11	1	N	Y	N
BOMBAY	N	120	1	N	N	N
BONDUEL	Y	88	2	N	N	N
BONO	N	145	1	Y	N	N
BORGSTROM	N	11	1	N	Y	N
BOURBON	N	119	1	N	N	N
BOWERS	N	153	1	N	N	N
BOYER	N	2173	3	Y	N	N
BRADY	N	655	1	N	N	N
BRECKENRIDGE	N	56	1	N	Y	N
BRECKSVILLE	N	60	1	N	Y	N
BREMS	N	928	2	N	N	N
BRETHREN	Y	62	1	N	Y	Y
BREVORT	N	399	3	N	N	N
BRIGGSVILLE	N	106	1	N	N	N

(continued)

Series name	Endemic	Total area (km^2)	Extent class	Benchmark soil	Rare	Endangered
BRIMLEY	N	55	2	N	Y	N
BROCKPORT	N	47	1	N	Y	N
BROOKSTON	N	4494	1	Y	N	N
BROWNSTONE	N	6	2	N	Y	N
BRUCE	N	37	1	N	Y	N
BRYCE	Y	1159	1	Y	N	N
BUCKROE	Y	11	1	Y	Y	Y
BURLEIGH	N	100	2	N	Y	N
BURSAW	N	2	1	N	Y	N
BURT	Y	18	1	N	Y	Y
BUSTI	N	437	1	N	N	N
CAFFEY	Y	26	1	N	Y	Y
CANADICE	N	197	2	N	N	N
CANANDAIGUA	Y	735	4	N	N	N
CANOSIA	N	296	2	N	N	N
CAPAC	Y	2404	2	Y	N	N
CARBONDALE	Y	2062	4	Y	N	N
CARLISLE	N	1150	3	N	N	N
CARLSHEND	Y	2	1	N	Y	Y
CASCO	Y	913	2	Y	N	N
CASTALIA	N	58	2	N	Y	N
CATHRO	N	6799	4	N	N	N
CAYUGA	Y	152	2	N	N	N
CAZENOVIA	N	461	1	N	N	N
CERESCO	Y	221	1	N	N	N
CHABENEAU	Y	54	1	N	Y	Y
CHADAKOIN	N	307	1	N	N	N
CHANNING	Y	95	1	N	N	N
CHARITY	Y	33	1	N	Y	Y
CHARLEVOIX	N	419	3	N	N	N
CHAUMONT	N	142	2	Y	N	N
CHEBOYGAN	N	295	2	N	N	N
CHEEKTOWAGA	N	30	1	N	Y	N
CHELSEA	N	1428	2	N	N	N
CHENANGO	Y	1757	3	Y	N	N
CHILI	N	1155	1	Y	N	N
CHINWHISKER	Y	106	1	N	N	N
CHIPPENY	N	86	2	N	Y	N
CHIPPEWA HARBOR	Y	23	1	N	Y	Y
CHOCOLAY	Y	20	1	N	Y	Y
CHURCHVILLE	N	244	2	N	N	N
CLAVERACK	N	118	2	N	N	N
COHOCTAH	Y	704	2	N	N	N
COLLAMER	Y	507	4	Y	N	N
COLOMA	N	2197	2	Y	N	N
COLONIE	N	270	4	Y	N	N
COLWOOD	N	769	2	N	N	N
CONDIT	N	430	1	Y	N	N
CONNEAUT	Y	144	3	Y	N	N
CONOTTON	N	176	1	N	N	N
COOKSON	N	40	1	N	Y	N

(continued)

Series name	Endemic	Total area (km^2)	Extent class	Benchmark soil	Rare	Endangered
COPEMISH	Y	5	1	N	Y	Y
COPPER HARBOR	Y	8	1	N	Y	Y
CORUNNA	Y	453	2	N	N	N
COSAD	N	117	3	N	N	N
COVERT	Y	940	3	N	N	N
COVINGTON	Y	337	2	N	N	N
COZY	Y	13	1	N	Y	Y
CROSWELL	N	2000	4	N	N	N
CROWELL	Y	5	1	N	Y	Y
CUBLAKE	N	194	2	N	N	N
CUNARD	N	96	1	N	Y	N
CURTISVILLE	N	18	1	N	Y	N
CUSINO	N	66	1	N	Y	N
CUTTRE	N	676	3	N	N	N
DAIR	N	83	2	N	Y	N
DANLEY	N	67	1	N	Y	N
DARIEN	N	857	2	Y	N	N
DAVIES	N	7	1	N	Y	N
DAWSON	N	1178	3	N	N	N
DEER PARK	N	368	4	N	N	N
DEERFIELD	N	590	1	N	N	N
DEERTON	N	50	2	N	Y	N
DEFORD	N	1313	4	N	N	N
DEL REY	N	1050	2	Y	N	N
DENOMIE	N	31	1	N	Y	N
DERB	Y	64	1	N	Y	Y
DETOUR	Y	79	2	N	Y	Y
DILLINGHAM	Y	47	1	N	Y	Y
DISHNO	N	191	2	N	N	N
DIXBORO	N	161	1	N	N	N
DORA	Y	571	1	N	N	N
DORVAL	Y	122	1	N	N	N
DUANE	N	66	1	N	Y	N
DUEL	N	20	1	N	Y	N
DUNBRIDGE	N	41	1	N	Y	N
DUNKIRK	Y	162	3	N	N	N
EAST LAKE	N	575	3	N	N	N
EASTPORT	N	155	3	N	N	N
EDMORE	Y	56	1	N	Y	Y
EDWARDS	N	484	2	N	N	N
ELCAJON	Y	4	1	N	Y	Y
ELDES	N	78	3	N	Y	N
ELMRIDGE	Y	101	1	N	N	N
ELNORA	N	263	4	N	N	N
EMMET	Y	1670	3	Y	N	N
ENGADINE	N	9	1	N	Y	N
ENSIGN	N	91	2	N	Y	N
ENSLEY	N	762	3	N	N	N
EPOUFETTE	N	235	2	N	N	N
EPWORTH	Y	77	2	N	Y	Y
ERMATINGER	N	56	2	N	Y	N

(continued)

Series name	Endemic	Total area (km^2)	Extent class	Benchmark soil	Rare	Endangered
ESAU	Y	35	1	N	Y	Y
ESCANABA	Y	83	1	N	Y	Y
ESSEXVILLE	N	74	2	N	Y	N
EVART	Y	135	2	N	N	N
FABIUS	Y	58	1	N	Y	Y
FARMINGTON	N	945	2	N	N	N
FARNHAM	N	30	1	N	Y	N
FARQUAR	Y	9	1	N	Y	Y
FELDTMANN	N	1	1	N	Y	N
FENCE	Y	166	2	N	N	N
FER N	Y	115	2	N	N	N
FIBRE	N	38	1	N	Y	N
FILER	N	504	2	N	N	N
FILION	Y	11	1	N	Y	Y
FINCH	N	440	3	N	N	N
FITCHVILLE	N	803	1	Y	N	N
FLINK	N	53	1	N	Y	N
FLINTSTEEL	N	336	3	Y	N	N
FOGG	Y	22	1	N	Y	Y
FONDA	N	116	1	N	N	N
FORBAY	N	52	2	N	Y	N
FORTRESS (also LIVONIA)	Y	6	1	N	Y	Y
FOX	N	2742	2	Y	N	N
FOXPAW	N	57	1	N	Y	N
FREDA	N	3	1	N	Y	N
FREDON	Y	309	3	N	N	N
FREESOIL	Y	10	1	N	Y	Y
FRIES	N	21	1	N	Y	N
FROHLING	N	96	2	N	Y	N
FULTON	N	478	2	N	N	N
FURLONG	N	12	1	N	Y	N
GAASTRA	N	52	1	N	Y	N
GAGETOWN	Y	16	1	N	Y	Y
GAGEVILLE	N	55	2	N	Y	N
GALEN	N	143	2	N	N	N
GALOO	Y	168	2	Y	N	N
GALWAY	N	378	2	N	N	N
GARLIC	Y	309	2	Y	N	N
GAY	N	356	2	N	N	N
GETZVILLE	Y	42	1	N	Y	Y
GICHIGAMI	N	61	2	N	Y	N
GIESE	N	310	2	N	N	N
GILCHRIST	N	4	1	N	Y	N
GILFORD	N	2030	1	Y	N	N
GLADWIN	N	193	2	N	N	N
GLAWE	Y	24	1	N	Y	Y
GLENDORA	N	313	2	N	N	N
GLENFORD	N	657	1	Y	N	N
GLYNWOOD	N	4592	1	Y	N	N
GOGEBIC	N	2372	2	Y	N	N
GOGOMAIN	Y	133	1	N	N	N

(continued)

Series name	Endemic	Total area (km^2)	Extent class	Benchmark soil	Rare	Endangered
GONGEAU	N	11	1	N	Y	N
GOODHARBOR	N	2	1	N	Y	N
GORVAN	Y	8	1	N	Y	Y
GOWDY	Y	28	1	N	Y	Y
GRACE	Y	10	1	N	Y	Y
GRANBY	N	1574	2	Y	N	N
GRAND SABLE	Y	6	1	N	Y	Y
GRATIOT	Y	55	2	N	Y	Y
GRATTAN	N	615	2	N	N	N
GRAVERAET	N	284	2	N	N	N
GRAYCALM	Y	3709	1	N	N	N
GRAYLING	N	2859	2	Y	N	N
GRAYS	N	108	2	N	N	N
GREENSTONE	Y	29	1	Y	Y	Y
GREENWOOD	N	4645	2	N	N	N
GREYLOCK	N	102	2	Y	N	N
GREYSOLON	N	74	2	N	Y	N
GRINDSTONE	N	72	1	N	Y	N
GROTON	Y	71	2	N	Y	Y
GUARDLAKE	N	85	1	N	Y	N
GUELPH	N	491	2	N	N	N
GUFFIN	N	55	2	N	Y	N
GULL POINT	N	59	2	N	Y	N
GUTPORT	Y	2	1	N	Y	Y
HAGENSVILLE	Y	47	1	N	Y	Y
HALFADAY	N	124	2	N	N	N
HALSEY	Y	284	2	N	N	N
HARBOR	N	21	2	N	Y	N
HASKINS	N	870	2	N	N	N
HEALYLAKE	Y	6	1	N	Y	Y
HEBRON	N	89	2	N	Y	N
HEGBERG	Y	148	3	N	N	N
HEINZ	N	11	1	N	Y	N
HENDRIE	Y	51	1	N	Y	Y
HERBSTER	N	137	2	N	N	N
HERMANTOWN	N	240	1	N	N	N
HESSEL	Y	112	2	N	N	N
HETTINGER	Y	88	2	N	Y	Y
HILTON	N	643	3	Y	N	N
HINCKLEY	N	2544	1	Y	N	N
HOCHHEIM	N	966	1	N	N	N
HOIST	Y	76	1	N	Y	Y
HORNELL	Y	325	2	Y	N	N
HORTONVILLE	N	998	2	Y	N	N
HOUGHTON	N	4251	3	Y	N	N
HOWARD	N	871	1	Y	N	N
HOYTVILLE	N	3357	1	Y	N	N
HUDSON	N	673	2	Y	N	N
HULLIGAN	N	81	1	N	Y	N
IARGO	Y	8	1	N	Y	Y
INGALLS	Y	164	3	N	N	N

(continued)

Series name	Endemic	Total area (km^2)	Extent class	Benchmark soil	Rare	Endangered
IOSCO	Y	733	3	N	N	N
IRA	N	255	4	N	N	N
ISABELLA	N	216	1	N	N	N
ISHPEMING	N	131	1	N	N	N
ITHACA	Y	250	1	Y	N	N
JACOBSVILLE	N	27	1	N	Y	N
JEBAVY	N	73	1	N	Y	N
JESKE	N	13	1	Y	Y	N
JIMTOWN	N	238	1	N	N	N
JOHNSWOOD	Y	33	1	N	Y	Y
JUNIUS	N	28	1	N	Y	N
KALEVA	Y	276	2	N	N	N
KALKASKA	N	4261	4	Y	N	N
KANOTIN	Y	8	1	N	Y	Y
KARLIN	N	1057	1	Y	N	N
KAWKAWLIN	N	558	1	N	N	N
KELLOGG	Y	174	3	N	N	N
KENT	N	55	1	N	Y	N
KEOWNS	Y	248	2	N	N	N
KERSTON	N	86	2	N	Y	N
KEWAUNEE	N	1991	3	Y	N	N
KEWEENAW	N	1704	3	N	N	N
KIBBIE	N	629	3	Y	N	N
KILMANAGH	Y	430	2	N	N	N
KINGSBURY	N	362	2	Y	N	N
KINGSVILLE	N	641	2	N	N	N
KINROSS	N	725	3	N	N	N
KIVA	N	108	2	N	N	N
KLACKING	Y	659	1	N	N	N
KOLBERG	N	115	2	N	N	N
KRAKOW	Y	73	1	N	Y	Y
LACHINE	N	8	1	N	Y	N
LACOTA	Y	16	1	N	Y	Y
LAGROSS	N	15	1	N	Y	N
LAKEMONT	N	206	1	N	N	N
LAMSON	N	396	3	N	N	N
LAPOIN	Y	7	1	N	Y	Y
LATTY	Y	988	1	Y	N	N
LEAFRIVER	N	616	2	N	N	N
LEELANAU	N	570	3	N	N	N
LENAWEE	N	869	3	N	N	N
LEVASSEUR	Y	5	1	N	Y	Y
LIMINGA	N	230	2	N	N	N
LINWOOD	N	281	2	N	N	N
LIVINGSTON	N	178	2	N	N	N
LOGGERHEAD	N	235	2	N	N	N
LONDO	Y	863	3	N	N	N
LONGRIE	N	327	4	N	N	N
LORAIN	N	64	1	N	Y	N
LOXLEY	N	2201	2	Y	N	N
LUPTON	N	4987	4	N	N	N

(continued)

Series name	Endemic	Total area (km^2)	Extent class	Benchmark soil	Rare	Endangered
LURAY	N	308	1	N	N	N
LYONS	N	344	1	N	N	N
MACKINAC	N	6	1	N	Y	N
MADALIN	Y	367	2	N	N	N
MADAUS	Y	23	1	N	Y	Y
MADRID	Y	290	1	N	N	N
MAHONING	N	1696	2	Y	N	N
MANARY	Y	15	1	N	Y	Y
MANAWA	N	971	3	Y	N	N
MANCELONA	N	943	2	N	N	N
MANIDO	Y	83	1	Y	Y	Y
MANISTEE	Y	267	3	N	N	N
MANISTIQUE	N	1	1	N	Y	N
MANLIUS	N	553	1	Y	N	N
MARBLEHEAD	N	8	1	Y	Y	N
MARKEY	N	3711	3	N	N	N
MARLETTE	N	2138	1	Y	N	N
MARTINSVILLE	N	669	1	Y	N	N
MARTINTON	N	308	1	N	N	N
MARTISCO	N	86	2	N	Y	N
MASSENA	N	358	2	N	N	N
MATTIX	N	7	1	N	Y	N
MAUMEE	N	764	1	Y	N	N
MCGINN	Y	163	1	N	N	N
MCIVOR	N	20	1	N	Y	N
MCMASTER	Y	13	1	N	Y	Y
MCMILLAN	N	86	1	N	Y	N
MEEHAN	N	1275	2	Y	N	N
MELITA	N	114	2	N	N	N
MENOMINEE	Y	839	2	N	N	N
MENTOR	N	159	1	N	N	N
MEQUON	Y	92	1	N	Y	Y
MERMILL	Y	489	1	Y	N	N
MERWIN	N	390	1	N	N	N
MESABA	N	161	3	N	N	N
METAMORA	Y	580	1	Y	N	N
METEA	N	902	2	N	N	N
MICHIGAMME	Y	603	3	N	N	N
MIDDLEBURY	N	278	1	N	N	N
MILFORD	N	1968	1	Y	N	N
MILL	N	846	1	Y	N	N
MILLGROVE	N	263	1	Y	N	N
MILLSDALE	N	181	1	N	N	N
MILNICHOL	Y	9	1	N	Y	Y
MILTON	N	242	1	N	N	N
MINER	N	144	2	N	N	N
MINOA	N	312	4	N	N	N
MINOCQUA	Y	733	2	N	N	N
MINONG	Y	13	1	N	Y	Y
MISERY	N	45	1	N	Y	N
MISHWABIC	N	3	2	N	Y	N

(continued)

Series name	Endemic	Total area (km^2)	Extent class	Benchmark soil	Rare	Endangered
MISKOAKI	Y	207	4	N	N	N
MITIWANGA	N	166	1	N	N	N
MOLLINEAUX	Y	6	1	N	Y	Y
MOLTKE	Y	32	1	N	Y	Y
MONGO	N	15	1	N	Y	N
MONITOR	Y	21	1	N	Y	Y
MONTCALM	N	1412	2	Y	N	N
MONTGOMERY	Y	266	1	N	N	N
MONTREAL	Y	241	3	N	N	N
MOQUAH	Y	189	2	N	N	N
MORGANLAKE	Y	257	2	N	N	N
MORLEY	N	1961	1	Y	N	N
MOROCCO	N	670	2	N	N	N
MOSEL	N	72	2	N	Y	N
MUNDELEIN	N	176	2	N	N	N
MUNISING	N	840	3	Y	N	N
MUNUSCONG	Y	66	2	N	Y	Y
MUSSEY	Y	41	1	N	Y	Y
NADEAU	N	286	2	N	N	N
NAHMA	N	154	2	N	N	N
NAMUR	Y	150	3	N	N	N
NAPPANEE	N	1519	2	Y	N	N
NAUMBURG	N	664	1	N	N	N
NAVAN	N	83	1	N	Y	N
NECONISH	N	67	1	N	Y	N
NEGWEGON	N	157	1	N	N	N
NENNO	N	44	1	N	Y	N
NESSEN	N	64	1	N	Y	N
NESTER	N	1239	1	Y	N	N
NET	N	215	1	Y	N	N
NEVENS	N	148	2	N	N	N
NEWSTEAD	N	67	2	N	Y	N
NEWTON	Y	139	2	Y	N	N
NIAGARA	N	598	4	Y	N	N
NICHOLS	Y	78	2	N	Y	Y
NIPISSING	Y	22	1	N	Y	Y
NONESUCH	Y	18	1	N	Y	Y
NORDHOUSE	Y	44	2	N	Y	Y
NORMANNA	N	515	3	N	N	N
NOSEUM	N	46	1	N	Y	N
NUNICA	Y	40	1	N	Y	Y
NYKANEN	N	6	1	N	Y	N
OAKVILLE	N	1442	4	Y	N	N
OCQUEOC	Y	108	2	N	N	N
ODANAH	N	147	2	N	N	N
ODESSA	N	431	1	N	N	N
OGEMAW	Y	8	1	N	Y	Y
OGONTZ	N	8	1	N	Y	N
OLDMAN	Y	91	1	N	Y	Y
OLMSTED	N	30	1	N	Y	N
OMENA	N	207	4	N	N	N

(continued)

Series name	Endemic	Total area (km^2)	Extent class	Benchmark soil	Rare	Endangered
OMRO	Y	44	1	N	Y	Y
ONAWAY	Y	1459	3	Y	N	N
ONEKAMA	Y	613	1	Y	N	N
ONOTA	N	8	1	N	Y	N
ONTARIO	N	920	1	N	N	N
ONTONAGON	Y	268	1	Y	N	N
ORONTO	Y	24	1	N	Y	Y
ORPARK	N	149	2	N	N	N
ORRVILLE	N	798	1	N	N	N
OSHKOSH	N	157	1	N	N	N
OSHTEMO	N	3334	1	N	N	N
OSSINEKE	Y	838	2	N	N	N
OTEGO	Y	58	1	N	Y	Y
OTISCO	N	114	1	N	N	N
OTISVILLE	N	129	1	N	N	N
OTTOKEE	N	318	1	Y	N	N
OVID	N	366	2	N	N	N
OZAUKEE	N	1023	2	N	N	N
PAAVOLA	Y	130	3	N	N	N
PAINESVILLE	N	61	2	N	Y	N
PALMERS	Y	19	1	N	Y	Y
PALMS	N	1590	2	Y	N	N
PALMYRA	Y	538	1	Y	N	N
PAQUIN	N	306	2	N	N	N
PARKHILL	Y	2296	3	Y	N	N
PELISSIER	N	89	1	N	Y	N
PELLA	N	1685	1	Y	N	N
PENCE	N	2834	1	Y	N	N
PEOTONE	N	584	1	N	N	N
PEQUAYWAN	Y	140	1	N	N	N
PERRINTON	N	399	1	N	N	N
PESHEKEE	N	129	2	Y	N	N
PEWAMO	N	7572	2	Y	N	N
PHELPS	Y	230	2	N	N	N
PICKFORD	Y	664	2	Y	N	N
PIERPONT	N	66	1	N	Y	N
PINCONNING	N	137	2	N	N	N
PIPESTONE	Y	1503	4	N	N	N
PLAINFIELD	N	5478	4	Y	N	N
PLATEA	N	543	1	Y	N	N
PLATTERIVER	N	7	1	N	Y	N
POMPTON	N	165	2	N	N	N
PORKIES	Y	24	1	N	Y	Y
PORTWING	N	228	2	N	N	N
POSEN	N	101	2	N	N	N
POSEYVILLE	N	106	1	N	N	N
POTAGANNISSING	N	31	1	N	Y	N
POY	Y	122	1	N	N	N
POYGAN	Y	451	2	Y	N	N
PROPER	Y	71	1	N	Y	Y
PULLUP	N	21	1	N	Y	N

(continued)

Series name	Endemic	Total area (km^2)	Extent class	Benchmark soil	Rare	Endangered
QUETICO	Y	271	4	N	N	N
RAPSON	N	32	1	N	Y	N
RAWSON	N	387	2	N	N	N
RAYNHAM	Y	499	2	N	N	N
READE	N	19	1	N	Y	N
RED HOOK	Y	526	2	N	N	N
REDRIM	Y	2	1	Y	Y	Y
REMSEN	N	295	2	N	N	N
REMUS	N	250	2	N	N	N
RHINEBECK	N	820	4	N	N	N
RHODY	N	2	1	N	Y	N
RICHTER	N	103	2	N	N	N
RIFLE	N	3628	3	N	N	N
RIGGSVILLE	N	66	2	N	Y	N
RIMER	N	281	4	N	N	N
ROBAGO	N	43	1	N	Y	N
ROCKBOTTOM	N	6	1	N	Y	N
ROCKCUT	Y	2	1	N	Y	Y
ROCKLAND	Y	156	3	N	N	N
ROLLINS	Y	316	2	N	N	N
ROSCOMMON	N	1753	4	Y	N	N
ROUSSEAU	N	860	3	N	Y	N
RUBICON	N	5528	4	Y	N	N
RUDYARD	Y	293	2	N	N	N
RUSE	Y	105	2	N	N	N
SABATTIS	N	175	3	N	N	N
SAGANING	N	2	1	N	Y	N
SANBORG	N	335	3	N	N	N
SANDUSKY	N	6	1	N	Y	N
SANILAC	Y	118	3	N	N	N
SARANAC	N	244	1	N	N	N
SATAGO	Y	23	1	N	Y	Y
SAUGATUCK	N	141	3	N	N	N
SAUXHEAD	N	27	1	Y	Y	N
SAYLESVILLE	N	76	1	N	Y	N
SAYNER	N	1150	1	Y	N	N
SCHOHARIE	N	410	1	N	N	N
SCHWEITZER	N	79	1	N	Y	N
SCIO	N	379	2	N	N	N
SCRIBA	N	346	2	Y	N	N
SEARCH	Y	12	1	N	Y	Y
SEBEWA	Y	1287	1	Y	N	N
SEDGWICK	N	97	2	N	Y	N
SELFRIDGE	N	802	1	Y	N	N
SELKIRK	N	81	1	N	Y	N
SEWARD	N	94	1	N	Y	N
SHAG	Y	14	1	N	Y	Y
SHAKER	Y	55	1	N	Y	Y
SHAVENAUGH	N	75	1	N	Y	N
SHAWANO	N	402	2	N	N	N
SHEBEON	N	568	2	N	N	N

(continued)

Series name	Endemic	Total area (km²)	Extent class	Benchmark soil	Rare	Endangered
SHELLDRAKE	N	9	1	N	Y	N
SHELTER	Y	230	3	N	N	N
SHINGLETON	Y	7	1	N	Y	Y
SHINROCK	N	62	1	N	Y	N
SHIOCTON	N	233	2	N	N	N
SHUBERTS	N	3	1	N	Y	N
SICKLES	N	39	1	N	Y	N
SIMS	Y	402	1	Y	N	N
SISKIWIT	N	9	1	N	Y	N
SISSON	N	222	2	N	N	N
SKANDIA	N	16	1	N	Y	N
SKANEE	N	270	3	N	N	N
SKEEL	Y	7	1	N	Y	Y
SLADE	N	78	1	N	Y	N
SLOAN	N	1406	2	Y	N	N
SODUS	N	154	3	N	N	N
SOLONA	N	906	3	N	N	N
SOO	Y	54	1	N	Y	Y
SOUTHWELLS	N	39	1	N	Y	N
SPARTA	N	1399	1	Y	N	N
SPEAR	Y	27	1	N	Y	Y
SPINKS	N	2958	3	Y	N	N
SPORLEY	Y	37	1	N	Y	Y
SPOT	N	284	2	N	N	N
SPRINGLAKE	N	112	2	N	N	N
SPRINGPORT	Y	87	1	N	Y	Y
ST. CLAIR	N	241	1	N	N	N
ST. IGNACE	Y	22	1	N	Y	Y
STAFFORD	N	59	1	N	Y	N
STANHOPE	N	66	2	N	Y	N
STEUBEN	N	13	1	N	Y	N
STURGEON	Y	90	2	N	Y	Y
STUTTS	N	149	1	N	N	N
SUGAR	Y	29	1	N	Y	Y
SULTZ	N	229	2	N	N	N
SUMMERVILLE	N	507	3	N	N	N
SUN	N	388	1	N	N	N
SUNDELL	N	90	2	N	Y	N
SUPERIOR	Y	88	2	N	Y	Y
SWANTON	Y	249	1	N	N	N
SWORMVILLE	Y	51	1	N	Y	Y
SYMCO	N	367	3	N	N	N
SYMERTON	N	222	1	N	N	N
TACODA	N	89	1	N	Y	N
TACOOSH	N	1403	1	N	N	N
TAPPAN	Y	1083	3	N	N	N
TATCHES	Y	11	1	N	Y	Y
TAWAS	N	2770	4	N	N	N
TEDROW	N	599	2	N	N	N
TEEL	Y	282	2	Y	N	N
TEKENINK	N	496	2	N	N	N

(continued)

Series name	Endemic	Total area (km^2)	Extent class	Benchmark soil	Rare	Endangered
THETFORD	Y	553	1	Y	N	N
THOMAS	Y	61	2	N	Y	Y
THOMPSONVILLE	Y	11	1	N	Y	Y
TOBICO	N	76	3	N	Y	N
TOIVOLA	Y	27	1	N	Y	Y
TOKIAHOK	Y	32	1	N	Y	Y
TOLEDO	N	757	3	Y	N	N
TONKEY	N	181	2	N	N	N
TOWERVILLE	N	160	2	N	N	N
TOWES	N	1	1	N	Y	N
TRIMOUNTAIN	Y	57	1	N	Y	Y
TROUT BAY	N	18	1	N	Y	N
TULA	Y	451	1	Y	N	N
TUSCOLA	N	174	2	N	N	N
TUSTIN	Y	98	2	N	Y	Y
TWIG	Y	117	1	N	N	N
TWINING	Y	32	1	N	Y	Y
TYNER	N	308	4	N	N	N
TYRE	N	3	1	N	Y	N
UBLY	N	177	2	Y	N	N
UNADILLA	N	259	1	N	N	N
VALOIS	N	1229	1	N	N	N
VARYSBURG	Y	96	1	N	N	N
VELVET	Y	14	1	Y	Y	Y
VERGENNES	N	929	2	Y	N	N
VILAS	N	1491	1	N	N	N
VOELKER	N	37	1	Y	Y	N
VOLUSIA	N	6192	1	Y	N	N
WABUN	N	8	1	N	Y	N
WAHBEGON	Y	56	2	N	Y	Y
WAINOLA	N	699	3	N	N	N
WAISKA	N	168	3	N	N	N
WAKELEY	Y	263	2	N	N	N
WAKEVILLE	N	126	1	N	N	N
WALLACE	N	400	3	N	N	N
WALLINGTON	N	57	2	N	Y	N
WALLKILL	N	211	1	N	N	N
WARNERS	Y	57	1	Y	Y	Y
WASEPI	Y	529	1	Y	N	N
WATSEKA	N	400	1	Y	N	N
WATTON	Y	92	1	N	Y	Y
WAUCONDA	N	77	1	N	Y	N
WAUSEON	Y	199	2	N	N	N
WAYLAND	N	932	2	N	N	N
WEGA	Y	48	1	N	Y	Y
WESTBURY	Y	394	1	N	N	N
WHEATLEY	N	78	2	N	Y	N
WHITTEMORE	N	14	1	N	Y	N
WICK	N	116	1	N	N	N
WILLETTE	Y	182	1	N	N	N
WILLIAMSON	N	234	3	N	N	N

(continued)

Series name	Endemic	Total area (km^2)	Extent class	Benchmark soil	Rare	Endangered
WILPOINT	Y	74	2	N	Y	Y
WINDSOR	N	1921	1	Y	N	N
WINTERFIELD	N	140	1	N	N	N
WISNER	Y	31	1	N	Y	Y
WITBECK	N	206	1	N	N	N
WIXOM	Y	659	1	N	N	N
WORMET	N	69	1	N	Y	N
WURTSMITH	N	306	3	N	N	N
YALMER	N	250	3	N	N	N
YELLOWDOG	N	17	1	N	Y	N
ZANDI	Y	81	2	N	Y	Y
ZEBA	N	46	2	N	Y	N
ZELA	Y	15	2	N	Y	Y
ZIEGENFUSS	N	546	1	N	N	N
ZIMMERMAN	N	1616	2	Y	N	N
ZURICH	N	152	1	N	N	N

Series name	Endemic	Total area (km²)	Extent class	Benchmark soil	Rare	Endangered
SHELLDRAKE	N	9	1	N	Y	N
SHELTER	Y	230	3	N	N	N
SHINGLETON	Y	7	1	N	Y	Y
SHINROCK	N	62	1	N	Y	N
SHIOCTON	N	233	2	N	N	N
SHUBERTS	N	3	1	N	Y	N
SICKLES	N	39	1	N	Y	N
SIMS	Y	402	1	Y	N	N
SISKIWIT	N	9	1	N	Y	N
SISSON	N	222	2	N	N	N
SKANDIA	N	16	1	N	Y	N
SKANEE	N	270	3	N	N	N
SKEEL	Y	7	1	N	Y	Y
SLADE	N	78	1	N	Y	N
SLOAN	N	1406	2	Y	N	N
SODUS	N	154	3	N	N	N
SOLONA	N	906	3	N	N	N
SOO	Y	54	1	N	Y	Y
SOUTHWELLS	N	39	1	N	Y	N
SPARTA	N	1399	1	Y	N	N
SPEAR	Y	27	1	N	Y	Y
SPINKS	N	2958	3	Y	N	N
SPORLEY	Y	37	1	N	Y	Y
SPOT	N	284	2	N	N	N
SPRINGLAKE	N	112	2	N	N	N
SPRINGPORT	Y	87	1	N	Y	Y
ST. CLAIR	N	241	1	N	N	N
ST. IGNACE	Y	22	1	N	Y	Y
STAFFORD	N	59	1	N	Y	N
STANHOPE	N	66	2	N	Y	N
STEUBEN	N	13	1	N	Y	N
STURGEON	Y	90	2	N	Y	Y
STUTTS	N	149	1	N	N	N
SUGAR	Y	29	1	N	Y	Y
SULTZ	N	229	2	N	N	N
SUMMERVILLE	N	507	3	N	N	N
SUN	N	388	1	N	N	N
SUNDELL	N	90	2	N	Y	N
SUPERIOR	Y	88	2	N	Y	Y
SWANTON	Y	249	1	N	N	N
SWORMVILLE	Y	51	1	N	Y	Y
SYMCO	N	367	3	N	N	N
SYMERTON	N	222	1	N	N	N
TACODA	N	89	1	N	Y	N
TACOOSH	N	1403	1	N	N	N
TAPPAN	Y	1083	3	N	N	N
TATCHES	Y	11	1	N	Y	Y
TAWAS	N	2770	4	N	N	N
TEDROW	N	599	2	N	N	N
TEEL	Y	282	2	Y	N	N
TEKENINK	N	496	2	N	N	N

(continued)

Series name	Endemic	Total area (km^2)	Extent class	Benchmark soil	Rare	Endangered
QUETICO	Y	271	4	N	N	N
RAPSON	N	32	1	N	Y	N
RAWSON	N	387	2	N	N	N
RAYNHAM	Y	499	2	N	N	N
READE	N	19	1	N	Y	N
RED HOOK	Y	526	2	N	N	N
REDRIM	Y	2	1	Y	Y	Y
REMSEN	N	295	2	N	N	N
REMUS	N	250	2	N	N	N
RHINEBECK	N	820	4	N	N	N
RHODY	N	2	1	N	Y	N
RICHTER	N	103	2	N	N	N
RIFLE	N	3628	3	N	N	N
RIGGSVILLE	N	66	2	N	Y	N
RIMER	N	281	4	N	N	N
ROBAGO	N	43	1	N	Y	N
ROCKBOTTOM	N	6	1	N	Y	N
ROCKCUT	Y	2	1	N	Y	Y
ROCKLAND	Y	156	3	N	N	N
ROLLINS	Y	316	2	N	N	N
ROSCOMMON	N	1753	4	Y	N	N
ROUSSEAU	N	860	3	N	Y	N
RUBICON	N	5528	4	Y	N	N
RUDYARD	Y	293	2	N	N	N
RUSE	Y	105	2	N	N	N
SABATTIS	N	175	3	N	N	N
SAGANING	N	2	1	N	Y	N
SANBORG	N	335	3	N	N	N
SANDUSKY	N	6	1	N	Y	N
SANILAC	Y	118	3	N	N	N
SARANAC	N	244	1	N	N	N
SATAGO	Y	23	1	N	Y	Y
SAUGATUCK	N	141	3	N	N	N
SAUXHEAD	N	27	1	Y	Y	N
SAYLESVILLE	N	76	1	N	Y	N
SAYNER	N	1150	1	Y	N	N
SCHOHARIE	N	410	1	N	N	N
SCHWEITZER	N	79	1	N	Y	N
SCIO	N	379	2	N	N	N
SCRIBA	N	346	2	Y	N	N
SEARCH	Y	12	1	N	Y	Y
SEBEWA	Y	1287	1	Y	N	N
SEDGWICK	N	97	2	N	Y	N
SELFRIDGE	N	802	1	Y	N	N
SELKIRK	N	81	1	N	Y	N
SEWARD	N	94	1	N	Y	N
SHAG	Y	14	1	N	Y	Y
SHAKER	Y	55	1	N	Y	Y
SHAVENAUGH	N	75	1	N	Y	N
SHAWANO	N	402	2	N	N	N
SHEBEON	N	568	2	N	N	N

(continued)

Index

A
Albic horizon, 36, 65
Alfisols, 37, 38, 43, 44, 46, 49, 51, 59–61, 64, 81, 95, 138, 142, 143, 161, 163–182
Algoma lake stage, 28, 32, 49, 92, 111, 152
Alluvial fan, plains, 68, 139, 161, 186, 190, 201
Anthrosolization, 138, 161
Argillic horizon, 36–38, 49, 59, 61, 63, 65, 71, 72, 138, 161
Argilluviation, 137–139, 161

B
Bayfield Peninsula, 2, 6, 19, 74
Beach ridge, 2–6, 10, 24, 48, 49, 57, 68, 70, 99–101, 104, 105, 109, 111, 114, 115, 121, 138, 141, 145, 149, 152, 161, 162, 185–191, 197–200, 202
Benchmark soil, 143, 161
Bisequal soil, 63, 94
Bluff retreat, 145
Boreal forest, 17, 20, 23, 30, 89, 161
Boulder field, 23
Bruce Peninsula, 22, 91, 111, 112, 116
Brunisols, 39, 40, 43, 47, 51, 78, 85, 89, 149, 161, 182, 183
Buried soil, 2, 10, 12, 36

C
Calcareous, 45, 85, 129, 131, 133, 149, 164, 166, 168, 169, 171, 178, 180, 181, 203–209
Cambic horizon, 36, 37, 49, 66, 68, 137
Cambisolization, 137, 139, 161
Canadian Shield, 22, 111, 112, 144
Cation-Exchange Capacity (CEC) activity class, 43, 46, 51, 57, 60, 63, 64, 68, 70, 72–74, 78, 83, 85, 87
Chicago, Illinois, 90, 99
Chippewa lake stage, 6, 28, 32, 166, 188
Closed (endorheic) basin, 24
Coastal wetlands, 19, 24, 145

D
Drowned forest, 24
Drumlin, 22, 100, 101, 103
Drummond island, 111, 112
Duluth lake stage, 28, 31, 32, 49, 60, 61, 76, 89, 90, 92, 202, 209

E
Early Lake Erie lake stage, 28, 32
Early Lake Ontario lake stage, 28, 32
Endangered soil, 143, 144, 211
Endemic soil, 142
Endoaqualfs, 39, 40, 43–46, 49, 59, 62, 64, 92, 121, 129, 135, 143, 163–169, 171–177, 179, 181–183
Endoaquepts, 40, 43, 44, 46, 49, 65, 75, 78, 91, 99, 121, 127, 129, 163–182
Endoaquolls, 40, 43–46, 49, 70, 72, 84, 85, 91, 111, 115, 121, 129, 143, 164–170, 172–174, 176–182
Entisols, 37, 38, 43, 44, 49, 59, 68–70, 81, 83, 161, 163–182
Eolian, 12, 68, 70, 83, 99, 100, 104, 107, 121, 129, 130, 161, 182, 183, 186, 188–192, 195, 196, 198–202, 204, 207, 208
Epiaquepts, 40, 43, 44, 49, 65, 66, 76, 78, 91, 111, 112, 115, 121, 127, 129, 164, 166, 167, 169–172, 174–179, 182, 183
Esker, 22
Eutrudepts, 40, 43, 44, 49, 65, 66, 76, 78, 89–91, 99, 111, 112, 115, 121, 129, 149, 163, 164, 167–171, 174–177, 179, 180, 182, 183
Extragrade, 43, 45

F
Fluvial, 112, 182, 183, 203–208
Folistic epipedon, 35, 37
Fossil fuel combustion, 145
Fragipan, 36–39, 49, 62, 63, 65, 66, 68

G
Georgian Bay, 19, 22, 30, 111, 112, 115, 149–151, 154
Glacial drainage channel, 189, 198, 201
Glacial Lake Agassiz, 23, 28
Glacial Lake Algonquin, 3, 4, 6, 7, 28, 32, 48, 49, 92, 111, 114, 141, 149, 150, 152, 154, 155, 157, 163, 185, 202, 209, 211, 212
Glacial Lake Chicago, 2, 28, 104
Glacial Lake Iroquois, 7, 32, 28, 48, 92, 129, 133, 141, 149, 157, 202, 209
Glacial Lake Oshkosh, 5, 24, 176, 197, 220, 221
Glacial lake plain, 46, 48, 185–203, 205–209
Glacio-isostatic adjustment, recovery, rebound, 23, 24, 29–31, 150
Glaciolacustrine, 3, 22, 45, 48, 59, 60, 63, 64, 68, 70, 72, 74, 78, 85, 91, 115, 129, 133, 149–151, 155, 156, 161, 182, 183, 186–209
Gleization, 137–139, 161
Glenwood lake stage, 31, 32, 49, 99, 153, 202, 209

J. G. Bockheim, *Soils of the Laurentian Great Lakes, USA and Canada*,
https://doi.org/10.1007/978-3-030-52425-8

Gleysols, 39, 40, 43, 45, 47, 51, 182, 183
Glossic horizon, 36, 38, 39, 59, 60, 65, 138
Glossification, 137–139, 161
Glossudalfs, 2, 40, 43–46, 49, 59, 61, 64, 90, 111, 138, 143, 163–182
Greatlakean (glacial) advance, 22
Great Lakes Coastal Zone (GLCZ), 2, 3, 7, 9–12, 17, 18, 22, 24, 30, 37, 39, 40, 43, 45–51, 55–63, 65, 66, 68–70, 72, 73, 75, 78, 81, 85, 99, 129, 137–139, 141–146, 149, 151, 161–163, 185, 203, 211
Green Bay lobe, 24

H

Haplorthods, 4, 5, 40, 43–46, 49, 57, 61, 62, 66, 73, 74, 89–91, 94, 98, 99, 101, 102, 111, 112, 115, 129, 138, 141, 143, 163–181, 183
Haplosaprists, 43, 44, 46, 49, 75, 81, 86, 87, 89–91, 94, 99, 100, 102, 104, 111, 115, 121, 129, 132, 141, 163, 165–168, 171–173, 176, 179–181
Hapludalfs, 40, 43–46, 49, 57, 59, 60, 64, 89–92, 94, 99–101, 103, 104, 107, 111, 115, 121, 127, 129, 133, 135, 138, 143, 163–183
Headland, 23, 149, 152
Hinge line, 23, 141
Histic epipedon, 36, 39
Histosols, 37, 38, 43, 44, 49, 51, 59, 75, 81, 87, 138, 161, 163, 165–168, 170–174, 176, 177, 179–181
Humification, 137

I

Ice-wedge polygon, 149, 151
Inceptisols, 37, 38, 43, 44, 49, 59, 65, 68, 78, 81, 137, 161, 163–182
Indiana Dunes National Lakeshore, 10, 13, 20
Inlet lake, 23
Intergrade, 39, 43, 45
Invasive plant, 138, 145
Isle Royale, 6, 66, 69, 75–77, 81, 82, 84, 86

K

Kame, 22
Keweenaw Peninsula, 17, 19, 69, 94, 142, 149, 152

L

Lacustrine, 3, 12, 30, 45, 48, 51, 52, 59, 60, 62, 63, 68, 70–74, 78, 79, 83, 87, 89–92, 94, 99–101, 104, 111, 112, 114, 115, 121, 127, 129, 130, 132, 135, 138, 141, 149, 155, 161, 182, 183, 185, 186–196, 198–209
Lake-effect snow, 18
Lake Erie, 1–3, 7, 19, 22, 23, 29, 31, 32, 48, 49, 52, 59, 61, 65, 66, 68, 70, 75, 78, 111, 121–124, 126, 127, 129, 142, 145, 151, 155, 156, 161
Lake Huron, 2–5, 7, 8, 10, 17, 19, 22, 30, 32, 48, 49, 52, 57, 59–61, 63, 65, 68, 70, 75, 85, 99, 111–118, 121, 137, 141, 142, 145, 147, 149–152, 155, 156, 158, 161, 205
Lake Michigan, 1, 2, 4, 5, 7, 9, 10, 12, 13, 17, 22, 25, 28, 30–32, 36, 48, 49, 52, 57, 59, 61, 63, 65, 66, 68, 70, 75, 77, 90, 99–101, 104–106, 108, 109, 111, 113, 114, 121, 122, 141, 142, 145–147, 150, 161
Lake Nipissing, 6, 7, 28, 31, 32, 49, 81, 92, 111, 141, 149, 151–153, 156, 157, 175, 196, 202, 209, 219, 220
Lake Ontario, 1–3, 5–7, 9, 10, 19, 22, 28, 30–32, 45, 48, 49, 52, 59, 61, 65, 66, 68–70, 75, 91, 92, 121, 129–131, 133, 135, 141–143, 145, 149, 151, 152, 155, 157, 161, 176, 183, 197, 207, 220, 221
Lake Superior, 1, 2, 6–8, 17, 19, 20, 22, 25, 28, 29, 32, 48, 49, 59–61, 63, 65, 66, 68–70, 75, 77, 89, 90, 93, 94, 96, 97, 125, 137, 141–143, 145, 149, 152, 161, 179, 200, 222–224
Lamellae, 39, 66, 69, 101, 107
Laurentian Mixed Forest, 17
Leverett, Frank, 1, 2, 12
Luvisols, 39, 40, 43, 45, 47, 51, 64, 89, 161, 182, 183

M

Manitoulin Island, 10, 22, 91, 111, 112, 115, 149
Marine, 203, 204, 207, 208
Marquette (glacial) advance, 6, 22, 61, 66, 71, 86, 89, 90, 94, 95, 125
Maumee lake stage, 121
Melanization, 137, 139, 161
Midwest Broadleaf (Carolinian) Forest, 17
Milwaukee, Wisconsin, 2, 90, 99
Mineral class, 43, 46, 51, 63, 68, 70
Mollic epipedon, 72
Mollisols, 37, 38, 43, 44, 49, 59, 70, 72, 81, 85, 137, 138, 161, 163–181
Moraine, 1, 2, 5, 22, 46, 63, 68, 89, 111, 121, 134, 161, 185, 186, 189–191, 195–209
Municipal sludge, 145

N

Niagara Escarpment, 22, 27, 30, 99, 111, 112, 133, 134
Niagara Falls, 22, 159
North Bay, 28

O

Ochric epipedon, 36
Organic soils, 37, 39, 43, 161
Ortstein, 36, 37, 39, 49, 63, 65, 73, 165, 167–169, 171, 173, 175–181
Outburst, 23, 29
Outwash, 3, 4, 22, 30, 45, 48, 49, 51, 52, 60–63, 66, 68, 70, 72, 73, 79, 87, 94, 95, 100, 101, 104, 107, 112, 143, 182, 183, 185–209
Outwash fan, 22
Outwash plain, 3, 48, 63, 68, 70, 73, 79, 87, 89, 138, 161, 185–209

P

Paludization, 137–139, 161
Particle-size class, 43, 45, 46, 48, 51, 60, 63, 68, 70, 138, 185, 203
Patterned ground, 149, 157
Pedodiversity, 104, 141, 143, 144, 162
Plant hardiness zone, 17, 18
Podzolization, 137, 138
Podzols, 39, 40, 43, 45, 47, 51, 73, 89, 161, 183
Pollutant, 145
Port Huron (glacial) advance, 7, 22, 91, 111
Pothole, 23, 24
Prairie Parkland, 17

R

Raised beach, 24, 57, 149–152, 157
Rare soil, xii
Regosols, 39, 40, 43, 45, 83, 161, 182
Residuum, 48, 52, 85, 86, 186, 187, 189, 193, 196, 202, 206
Roche moutonnée, 22

S

Saginaw Bay, 2, 161
Sand dune, 9, 12, 52, 68, 115, 121, 129
Sand mining, 147

Sand spit, 24
Sault Ste. Marie, 10, 89–91, 111
Sea cave, 23
Sea stack, 23, 149, 150, 155
Shoreline erosion, 145, 146
Sleeping Bear Dunes, 104
Soil chronosequence, 10, 12, 57, 162
Soil classification, 35–37, 39, 40, 161
Soil, definition, 1
Soil, family, 35, 37, 43, 49, 143
Soil, great group, 35, 37–40, 43–46, 49–51, 59, 61, 63, 65, 68–70, 72, 75, 81, 90, 111, 115, 121, 127, 129, 135, 139, 143, 161, 163
Soil map, 1, 2, 4, 6–9, 12
Soil, order, 35, 37, 39, 43–45, 51, 56, 59, 61, 65, 75, 161, 163
Soil, series, 1, 2, 7, 9, 11, 12, 17, 22, 35–37, 39, 40, 43–51, 55, 56, 59–66, 68–72, 75–78, 80–87, 89, 90, 94–105, 107–109, 111–115, 121, 123–125, 129, 132–134, 141–144, 161–163, 182, 185, 203, 211
Soil, subgroup, 35, 37–40, 43, 45, 47, 49, 51, 81, 161, 163, 182
Soil, suborder, 35, 37, 38, 43, 44, 49, 161, 163
Soil survey, 1–3, 7–12, 35, 40, 60, 62, 63, 66, 69–71, 75–77, 80–82, 84, 86, 93, 94, 96–98, 101–109, 113, 114, 116, 118, 122–126, 130–133, 138
Soil taxonomic system, 39, 137
Solifluction lobe, 149, 150
Spillway, 23, 31, 204
Spodic horizon, 36, 49, 63, 65, 100
Spodosols, 13, 37, 38, 43, 44, 46, 49, 51, 59, 61, 63, 65, 73, 81, 95, 138, 142, 143, 161–181
Stanley lake stage, 2, 28, 32, 92, 121, 151, 158
St. Lawrence River, 6, 29, 64, 73, 142, 144
Storm runoff, 146
Straits of Mackinac, 99
Superior, Wisconsin, 1, 8, 94

T
Talbot Trail, 155, 159
Tension zone, 17, 19, 23, 30, 46, 99, 111, 142–144
Till, 1–4, 6, 7, 22, 30, 45, 48, 49, 51, 52, 59, 60, 62–66, 68, 70–74, 78, 79, 82, 83, 85, 87, 89–92, 94–96, 99–101, 103, 104, 111, 112, 115, 121, 127, 129–131, 135, 138, 141, 143, 149, 182, 183, 185–209
Till-floored glacial lake plain, 4–6, 191, 200, 201
Till plain, 3–6, 46, 60, 63–65, 68, 70, 72–74, 77–79, 85, 87, 89, 99–101, 104, 113, 121, 138, 141, 161, 185–203, 205
Tombolo, 24, 149, 152
Toronto, Ontario, 48, 52, 129
Two Creeks soil, 2, 10, 13, 28

U
Udipsamments, 2, 4, 5, 40, 43–45, 49, 57, 68, 70, 80, 83, 89–91, 94, 99–101, 104, 111, 115, 121, 125, 127, 129, 141, 143, 162, 163, 165–170, 173–178, 180–182
Umbric epipedon, 35–37, 68

V
Varved lake, 24
Vertization, 137–139, 161

W
Warren lake stage, 7, 31, 32, 48, 49, 92, 111, 121, 202, 209
Wave-cut bench, bluff, terrace, 24, 83
Whittlesey lake stage, 28, 31, 32, 49, 92, 121, 202, 209

The manufacturer's authorised representative in the EU is Springer Nature Customer Service Centre GmbH, Europaplatz 3, 69115 Heidelberg, Germany. If you have any concerns regarding our products, please contact ProductSafety@springernature.com

Printed and bound by CPI Group (UK) Ltd, Croydon, CR0 4YY
30/09/2026
02225476-0001